AF631245

ENCYCLOPEDIA OF DRUG AND DRUG DELIVERY

ENCYCLOPEDIA OF DRUG AND DRUG DELIVERY

Vol. 5

DRUG DESIGNING

By

Dr. Kamlesh Dashora
Associate Professor
Institute of Pharmacy
Vikram University
Ujjain (M.P.)
(India)

&

Dr. Darshan Dubey
Assistant Professor
Institute of Pharmacy
Vikram University
Ujjain (M.P.)
(India)

D P H

DISCOVERY PUBLISHING HOUSE PVT. LTD.
NEW DELHI-110 002

Published by:
Tilak Wasan
DISCOVERY PUBLISHING HOUSE PVT. LTD.
4383/4A, Ansari Road, Darya Ganj
New Delhi-110 002 (India)
Phone: +91-11-23279245, 43596064-65
Fax : +91-11-23253475
E-mail: parul.wasan@gmail.com
discoverypublishinghouse@gmail.com
web : www.discoverypublishinggroup.com

***First Edition:* 2012**

ISBN: 978-93-5056-038-9 (Set)
978-93-5056-039-6 (Vol. 1)
978-93-5056-040-2 (Vol. 2)
978-93-5056-041-9 (Vol. 3)
978-93-5056-042-6 (Vol. 4)
978-93-5056-043-3 (Vol. 5)

Encyclopedia of Drug and Drug Delivery
© 2012, Authors

All rights reserved. No part of this publication should be reproduced, stored in a retrieval system, or transmitted in any form or by any means: electronic, mechanical, photocopying, recording or otherwise, without the prior written permission of the author and the publisher.

This book has been published in good faith that the material provided by authors is original. Every effort is made to ensure accuracy of material, but the publisher and printer will not be held responsible for any inadvertent error(s). In case of any dispute, all legal matters are to be settled under Delhi jurisdiction only.

Printed at:
Shree Balaji Art Press
Delhi

Preface

The Present title "**Encyclopedia of Drug and Drug Delivery**" is the amazing advancement of pharmaceutical studies. It provides the various fundamental aspects of drug designing to be understood adequately. It has been compiled for undergraduate, postgraduate students, research scholars, teachers, technicians, industrial microbiologists and scientists involved in pharmaceutical research and development. The basic concepts have been clearly explained and their functions are adequately highlighted. The presentation of text is simple and systematic. The selection of chapters and topics is according to the specified syllabi of several Indian Universities and are so structured as to enable the student to move easily from the fundamental to the complex. It describes the ideas and techniques that are creating or new domain of science and technology. It holds the promise of revolutionizing field ranging from medical science to pharmaceutical industries. It is our earnest hope that this title will be of great value to all our students.

Organizational and economic changes in the pharmaceutical industry have led to a marked reduction in the supply of new chemical entities. This combined with an increase in the expiration of patents and tighter regulatory control has lead many companies to rethink existing product design. The growing trend to improve the efficacy of existing drug products, to both extend parent life and allow competition with opportunistic generic companies, has allowed for a greater understanding of formulation design. This trend is set to continue for some time as economic and political pressure increases.

There have been many improvements in understanding the theoretical and practical aspects of drug discoveries. Numerous therapeutic compounds have been investigated to treat a variety of different illnesses. There are many pharmaceutical methods requiring a variety of formulation options. Advanced formulation design may be thought of as a way to optimize the mode of drug designing to a patient that will ultimately lead to a more effective therapy and approach an optimal therapeutic outcome for a particular drug. This book is designed to take the reader through various cutting-edge areas of advanced formulation design by describing therapeutic categories that may be beneficial. The reader will be introduced to some novel technologies and updated on the state of the art in areas designed to overcome poor efficacy issues.

To make the work more comprehensive and informative, the authors have consulted many authoritative books, research journals, abstracts, monographs etc., so there can be no claim to originality except in the manner of treatment.

The authors express their thanks to their friends and colleagues whose continue inspirations have initiated them to bring out this book.

The authors express their gratitude to Mr. Wasan and staff of M/s Discovery Publishing House Pvt. Ltd., for their whole hearted co-operation in the publications of this book.

The authors will remain sincerely responsible for any shortcomings of the book and be grateful to the readers for their suggestions and constructive criticism for the continuous betterment of the book. They have this opportunity to appeal to the readers to send their suggestions straightaway to this publisher.

–Authors

Contents

1

Introduction

Over the past several years, the fraction of new drug products that are new chemical entities has steadily decreased, reflecting the tremendous cost required to bring new chemical entities to the marketplace. Increased understanding of drug metabolic and toxicologic factors, such as the effect of the patient age on drug distribution, the genetic factors that may result in dramatic intersubject variability in metabolism, short-term versus long-term exposure toxicities, and the potential for teratogenic, mutagenic, and embryo- toxic effects, has increased the scrutiny under which governmental agencies view the new chemical entity. This careful inspection is intended to minimize the possibility of toxic reaction(s) and to demonstrate the safety and efficacy of new drug products. The regulatory process has also resulted in significantly more costly and time-consuming testing prior to commercialization.

BIOCHEMICAL FEATURES

This increased emphasis on safety has placed an additional burden on those who are involved in the development of new drugs, while increasing financial pressures have led to the need for decreased development time. The investigation of approved drugs has resulted in enhanced patient safety and therapeutic efficacy by directing research efforts toward the more efficacious delivery of known pharmacologically active agents to the appropriate physiologic site. This trend has caused pharmaceutical researchers to seek the most suitable methods to deliver both new and existing compounds in the most pharmacologically appropriate manner. The methods may be designed to optimize bioavailability, minimize toxicity and side effects, and improve stability. The objective of this article is to present approaches that have been employed to improve bioavailability and/or minimize the toxicity and side effects of various drugs.

A rational approach to dosage form design requires a complete understanding of the physicochemical and biopharmaceutical properties of the drug substance. For example, the successful design of an efficacious oral dosage form requires an understanding of the pathways of physiologic disposition of the drug. On oral administration of an immediate release dosage form (the most common delivery system), the dosage form must disintegrate, the drug must dissolve in the gastrointestinal (GI) fluids, cross the GI mucosa, enter the mesenteric blood system, and pass through the liver prior to reaching the systemic circulation and the site of action.

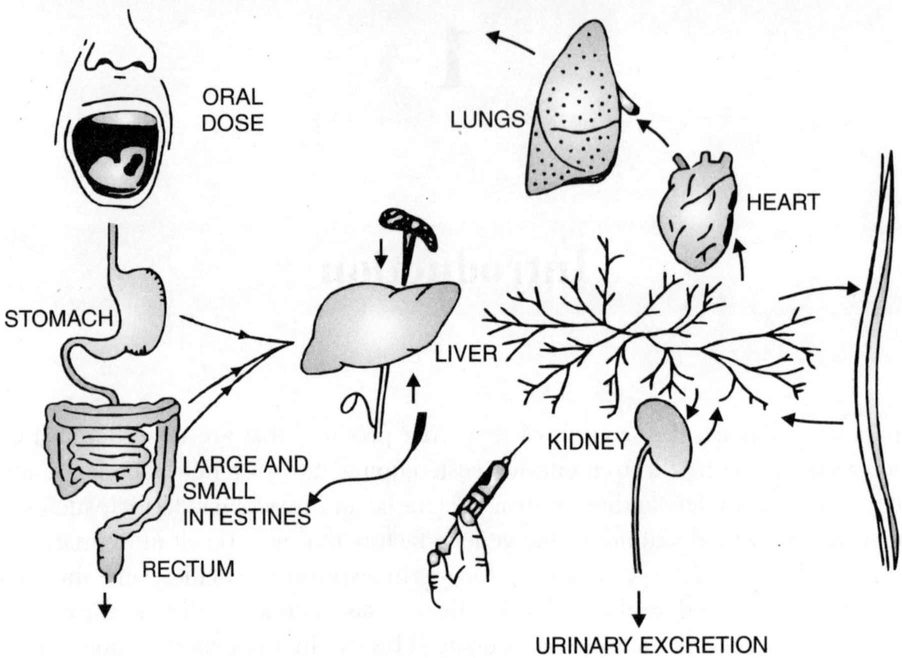

Fig. 1.1. Physiological factors associated with bioavailability.

The drug may be metabolized by GI fluids, by enzymes in the gut wall, or by hepatic metabolism prior to reaching the systemic circulation. The net result is incomplete bioavailability due to first-pass metabolism (inactivation), and/or metabolic formation of a pharmacologically active species. The physicochemical and biopharmaceutical properties of the drug can have a tremendous impact on its bioavailability and, hence, on its efficacy and toxicity profile. Thus, understanding these parameters is often tantamount to the selection and development of the optimum dosage form.

These properties of the drug are its:

1. pH solubility profile and dissolution rate
2. Partition coefficient between lipoidal barriers and aqueous physiologic media
3. Stability and/or degradation rate in the physiologic fluids
4. Susceptibility to metabolic inactivation
5. Mechanism of transport through biologic membranes.

The aqueous solubility of a drug in the 2–8 pH range has a direct influence on its oral and parenteral formulations. A drug with poor solubility (i.e., less than 0.1 mg/ml) in acidic media may show poor and erratic oral bioavailability due to the dependency of absorption processes in GI fluids. Intravenous dosing requires that the drug be administered in a soluble form. The

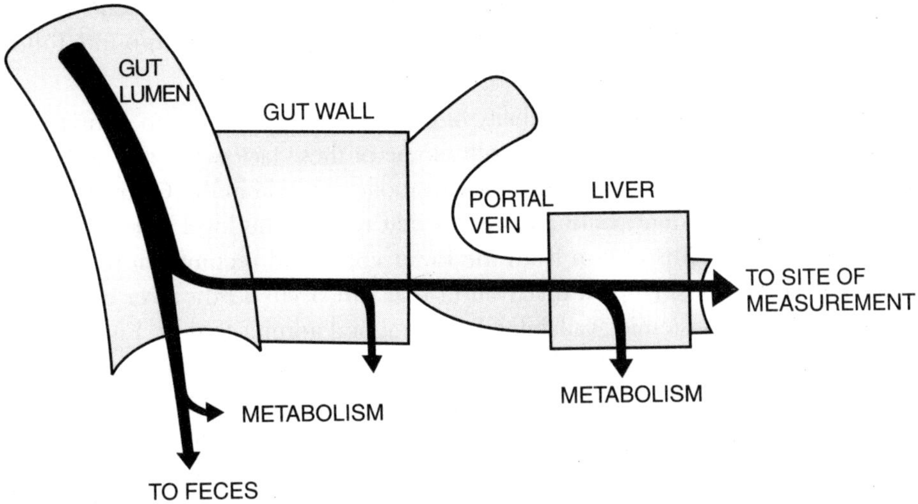

Fig. 1.2. A drug, given as a solid encounters several barriers and sites of loss in its sequential movement during gastrointestinal absorption.

adjustment of pH, the addition of a cosolvent or a ligand for complexation, or the formation of an emulsion may permit solubilization, but each of these techniques has limitations. Rapid intravenous administration of a solubilized drug can result in rapid dilution in an environment in which the drug is insoluble, resulting in incomplete availability and a delayed response due to the formation of particulate matter within the vascular system. Poor aqueous solubility is not always a limitation; in fact, it may be a desirable feature for a sustained effect after oral or parenteral administration. Oral sustained release may be achieved if the drug combines poor aqueous solubility with the ability to be adsorbed throughout the GI tract.

Parenteral sustained release can be achieved after intramuscular administration of a suspension of a drug with low solubility under physiologic conditions or from a drug that precipitates from an aqueous vehicle or that forms a reservoir or depot from an oil-containing dosage form. The lipid-aqueous partition coefficient of a drug molecule affects its absorption by passive diffusion. In general, octanol/pH 7.4 buffer partition coefficients in the 1–2 pH range are sufficient for absorption across lipoidal membranes. However, the absence of a strict relationship between the partition coefficient of a molecule and its ability to be absorbed is due to the complex nature of the absorption process. Absorption across membranes can be affected by several diverse factors that may include the ionic and/or polar characteristics of the drug and/or membrane as well as the site and capacity of carrier-mediated absorption or efflux systems. Compounds that are intended for oral administration and can undergo rapid degradation at low pH may require protection from the acidic environment of the stomach.

Protection can often be afforded by administering the drug in the form of an acid-insoluble chemical species or in a dosage form with an acid- resistant coating. The insoluble chemical species must remain insoluble and unavailable for solution degradation as it passes through the stomach and must dissolve upon reaching the chemically more stable environment of the intestine at a higher pH. To be effective, an acid-resistant coating must remain intact and protect its contents until it reaches the required pH to dissolve the coating and release the contents in the intestine where the drug may be more stable. Metabolic inactivation of a compound following oral administration can occur in the GI lumen, the GI mucosa, or the liver.

The site of metabolism and the susceptibility of the metabolic processes to saturation are factors that may influence oral bioavailability. Occasionally, some of these factors may be altered to optimize oral bioavailability. For example, segment specific metabolic sites within the GI tract may be avoided through the use of pH-dependent coating materials that rely on the local pH environment of the GI tract to release the drug. Absorption from the lower colon and rectum can reduce exposure to the portal circulation and the first-pass inactivation that can occur in the liver and, thus, provide the opportunity to improve systemic availability following oral administration. Enzyme systems may be saturated by the rapid release of the contents of the dosage form at a local site or by coadministration within the dosage form of a competitive inhibitor. Once the physicochemical and biopharmaceutical properties of the drug are determined and the desired plasma concentration profile is defined, the pharmaceutical scientist can select and develop an efficacious dosage form by utilizing a formulation approach, a prodrug approach, a device approach, or an alternative administration route approach.

Bioavailability

Use of formulation techniques can improve the bioavailability and/or minimize the toxicity and side effects of drugs. Factors to consider include those that impact on solubility and dissolution rates, chemical and enzymatic stability, and absorption capability. Several parameters, including particle size, crystalline habit, and salt form, can affect the solubility and dissolution rates of the drug. The effect of particle size on the dissolution rate of relatively insoluble compounds becomes significant when the drug is administered as a suspension or solid dosage form and the material is well dispersed within the GI tract. However, caution must be exercised when compressing the material into a final dosage form because excessive force can result in particle agglomeration and an actual increase in the effective particle size. Polymorphism is the ability of a chemical species to crystallize in more than one distinct crystal habit.

The pharmaceutical applications of polymorphism have been reviewed by several authors. The differences in dissolution rate and solubility that polymorphs can produce may have a dramatic impact on bioavailability when dissolution is the rate-limiting step in the absorption process. Tawashii investigated the GI absorption of two polymorphs of aspirin, the stable and metastable forms, forms I and II, respectively. He found that the metastable form produced a 70% higher total serum salicylate levels than the stable form I. The selection of a salt form directly influences the physicochemical and biopharmaceutical properties of a compound. The impact of salt selection has been reviewed. Nelson examined the dissolution of theophylline salts and commented on their impact on oral administration.

The dissolution rates of the theophylline salts proceeded independently of the pH of the medium but was governed by the diffusion layer pH. The choline and isopropanolamine salts dissolved three to four times faster than the ethylenediamine salt and produced higher and prolonged blood levels. For highly insoluble amine bases, such as ergotamine and certain antimalarials, the drug may precipitate in the small intestine as soon as the pH rises following stomach emptying.

The authors' experience has been that committing the effort and expense required to develop a new, more soluble salt form of this type of drug may be futile, especially if the new salt is only two to three times more soluble than existing salts. Thus, the selection of the most appropriate salt form should be made early in the development process to optimize bioavailability.

The stability of a drug in the gut is influenced by both chemical and enzymatic factors. Protection from chemical degradation may be accomplished via coating techniques, and enzymatic protection may be achieved with enzymatic inhibitors. An enteric coating protects the drug during transit through the acidic medium of the stomach. Upon entering the higher pH environment of the duodenum, the coating is dissolved, and the drug becomes available for absorption. Such a coating also provides protection for the gut mucosa when the drug is capable of producing GI irritation. This method has been employed for a number of drugs, including potassium chloride, ammonium chloride, aspirin, diethylstilbestrol, erythromycin, and divalproex. Nishimura et al. employed enteric coating of levodopa to improve its bioavailability. Levodopa is absorbed from the upper portion of the small intestine but undergoes rapid degradation in the intestinal mucosa by levodopa decarboxylase. The authors developed an enteric-coated dosage form with an effervescent core.

The dosage form remains intact through the stomach, dissolves upon reaching the small intestine, and bursts open due to the effervescence of the incorporated sodium bicarbonate. The rapid release in the duodenum results in concentrations of levodopa sufficient to saturate the enzyme at the absorption site. The limitations of enteric-coated dosage forms include the possibility of duodenal irritation from caustic drugs and an increase in intersubject variability due to the presentation to the small intestine of a dosage form that needs to undergo disintegration and dissolution versus a disintegrated and partially dissolved drug substance. Enzyme inhibitors compete with the active drug for the enzyme and, thereby, reduce the degradation of the drug and deliver it more efficiently to the systemic circulation. An example is the carbidopa–levodopa combination. Carbidopa competes for levodopa decarboxylase, thereby reducing the levodopa degradation and improving the low bioavailability of levodopa. Alternatively, the enzyme in the gut can be utilized to control the release of the active drug in the gut. For example, sulfasalazine, which is employed in the treatment of ulcerative colitis, is a combination of sulfapyridine and 5-aminosalicylate chemically linked via an azo bond.

It remains absorbed and intact throughout the GI tract until it reaches the large intestine, where bacterial azoreductase enzymes degrade the azo bond and release sulfapyridine and 5-aminosalicylate to act locally on the lesions. Altering the availability for absorption permits tailoring of the concentration-time profile for a drug. In the case of synthetic contraceptive steroids and theophylline dosage forms, the approach should be to slow the release rate so that undesirable spikes in the plasma levels are minimized. This effect can be accomplished with a sustained-release formulation. Informulating a drug in a sustained-release dosage form, the following must be taken into consideration: the effective drug plasma level, the rate of absorption of the compound, the rate of elimination of the compound, and the site(s) of absorption. Having established these parameters, calculating the required release rate from the dosage form is a relatively simple matter and formulation techniques are readily available to produce the desired release rate. The mechanism

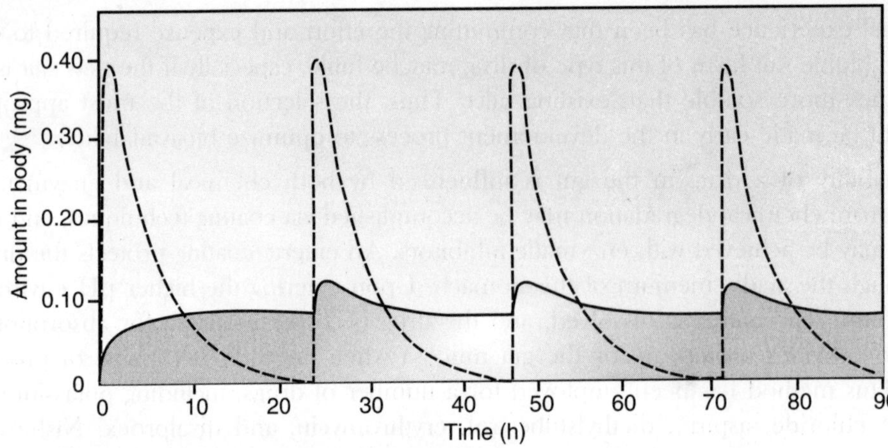

Fig. 1.3. Computer simulation of a plasma level-time profile of norethindrone from fast-releasing tablet (- - -) and sustained release tablet (–), from a dose of 0.5 mg/tablet and an elimination half-life of 3 h.

of absorption must always be evaluated when a sustained-release dosage form is considered. A drug that is passively absorbed throughout the GI tracts is an ideal candidate for sustained release. Drugs such as riboflavin, folic acid, aminopenicillins, amino-β-lactams and nucleoside analogs, which have windows of absorption due to site-specific and/or active transport processes, may have incomplete bioavailability when formulated in oral, sustained-release dosage forms.

Formulation Approach

An alternative to the formulation approach is the prodrug approach. A prodrug is defined as a drug that is prepared by chemically modifying a pharmacologically active species to form a new chemical entity that undergoes transformation to the active species within the body. The modification alters the physicochemical and biopharmaceutical properties of the drug in some beneficial manner. The ideal prodrug should have the following characteristics:

1. Possess no pharmacologic activity
2. Be eliminated more slowly than its rate of cleavage to the parent
3. Be non-toxic
4. Be inexpensive to prepare.

Prodrugs can be used to increase or decrease the aqueous solubility, mask bitterness, increase lipophilicity, improve absorption, decrease local side effects, and alter membrane permeability of the parent molecule. For example, chloramphenicol has an aqueous solubility of 2.5 mg/ml, but chloramphenicol sodium succinate, a prodrug, has an aqueous solubility of 100 mg/ml. Hydantoins also pos-

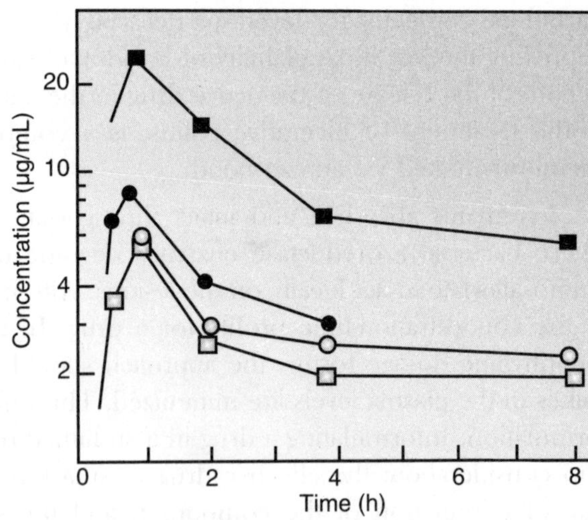

Fig. 1.4. Plasma acetaminophen concentration following the oral administration of acetaminophen and its ethyl vinyl ether prodrug.

sess low aqueous solubilites that result in low and variable availability and precipitation following injection. In an effort to increase the aqueous solubility of phenytoin, Stella et al. prepared the ethyl and triethylamine esters of diphenylhydantoic acid. The esters improved aqueous solubility by adding an amine function to the molecule, and under physiologic conditions, the cyclization to phenytoin is irreversible, rapid, and complete. The perception of taste requires that some minimum aqueous concentration be exceeded so that the taste can be detected (perceived). As a result, bitterness can be masked by reducing solubility. The technique has been applied to acetaminophen by blocking the phenolic substituent with ethyl vinyl ether.

Chemical hydrolysis to acetaminophen is rapid in the acidic conditions of the stomach. Plasma acetaminophen concentrations in dogs following oral administration of acetaminophen and the prodrug were found to be similar. Membrane permeability is governed in part by the lipophilicity of a compound. Highly polar compounds have low lipophilicities and, therefore, low membrane permeability. Epinephrine is a compound of this type. It is very effective in the treatment of glaucoma, but it produces a myriad of side effects. Ocular side effects include hyperemia, mydriasis, corneal edema, and allergic sensitivity. Systemic side effects, such as cardiac arrhythmias, elevated blood pressure, cerebral vascular accidents, dizziness, fear, and restlessness, are observed frequently. In an attempt to enhance the absorption and minimize the side effects of epinephrine, a prodrug, the dipivaloyl ester, was synthesized. It was found to be devoid of cardiac symptoms, with no effect on heart rate or blood pressure and was effective in *lowering intraocular pressure* (IOP).

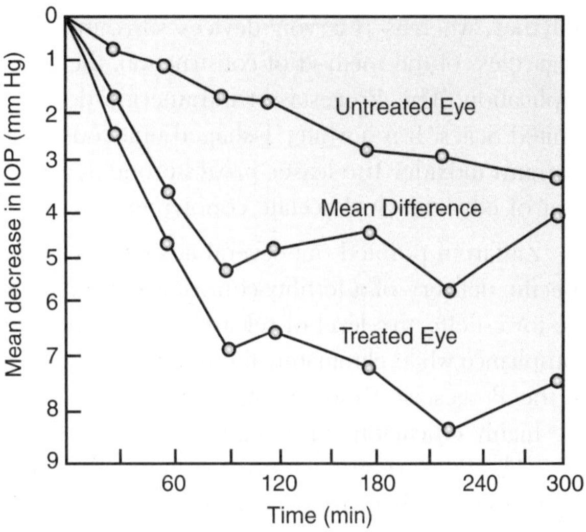

Fig. 1.5. The mean effect of one drop of 0.025% solution of the DPE of epinephrine on the intraocular pressure of nine glaucomatous individuals.

On the other hand, highly lipophilic compounds, such as hormones, can be solubilized via the prodrug approach. For example, the rate of transdermal absorption of the highly lipophilic drug, testosterone, was enhanced over 50-fold by forming water-soluble, yet lipophilic, prodrug ester. The prodrug testosteronyl-4-dimethyl-aminobutyrate was found to penetrate human skin tissue, in vitro, 54 times faster than testosterone itself. Furthermore, the prodrug was found to generate testosterone rapidly in biological fluids by enzymatic hydrolysis. Oral administration of aspirin can result in gastrointestinal bleeding. The bleeding has been attributed to local irritation due to the acidic nature of the carboxylic acid substituent.

In an attempt to reduce the gastric irritation of aspirin, acylal prodrugs were synthesized. In vitro, the prodrugs generated rapidly aspirin and in a pH-independent fashion. The conversion to aspirin was very sensitive to changes in dielectric constant: decreases caused a corresponding decrease in the reaction rate. This sensitivity is strongly indicative of an S_N1 mechanism. The authors postulated that GI irritation would be reduced due to blockage of both the charge and the acidic nature of the aspirin.

Controlled Device

Controlling the release of medication at the site of action is often desirable, especially for compounds that are absorbed rapidly through mucous membranes or are removed rapidly from the site of action. The approach normally reduces the systemic side effects of the agent. The application of this approach can be illustrated with the Progestasert and Pilocarpine Ocusert dosage forms. Diffusion-controlled devices may be designed for continuous release and usually use either a matrix or reservoir construction. In matrix systems, the drug is dispersed randomly throughout a polymer, whereas reservoir devices surround the drug with an intact rate-controlling membrane. Regardless of the method of construction, the system must be safe and biocompatible for biological application. The Progestasert intrauterine device (IUD) is a contraceptive IUD marketed in the United States. It is a white, T-shaped unit constructed of ethylene-vinyl acetate copolymer containing titanium dioxide. It releases progesterone at a rate of 65 μg/day for 1 year, controlled by an outer coat of ethylene-vinyl acetate copolymer.

Zaffaroni pointed out several advantages to the uterine progesterone system. It permits target-specific delivery of a fertility-control agent for up to 1 year by utilizing a single natural hormone at the lowest effective level of release of hormonal activity. The 1-year duration gives improved patient compliance while eliminating the pulsing seen with repetitive oral or injectable regimens. The efficacy of the Progestasert was investigated by Aznar and Giner. Their results indicated that the systems are highly efficacious in avoiding accidental pregnancy and result in decreased menstrual blood loss. The Pilocarpine Ocusert is a contact lens-shaped device that is inserted into the lower cul-de-sac of the eye. It provides continuous release of pilocarpine at a rate of 10 or 20 μg/h, over a 1-week period, for the treatment of open-angle glaucoma.

The system is capable of producing significant lowering of ocular pressure and constriction of the pupillary diameter. Its use greatly enhances patient compliance and the convenience of the therapeutic regimen while reducing local and systemic side effects. Armaly and Rao examined the clinical effects of the Pilocarpine Ocusert systems with different release rates. The systems exhibited a dose-response relationship such that increases in the release rate above 50 μg/h resulted in increased ocular hypotensive effects with no appreciable change in ocular pressure. The reduction in pressure observed at the 50 μg/h rate was comparable with the changes seen after the administration of 4–8% of pilocarpine solutions.

Routes for Drug Administration

The administration of drugs by alternative routes avoids absorption and metabolic barriers that may be present in the GI tract. The routes can also provide systematic availability when oral administration is contraindicated due to a physiologic condition, or the route may provide for a concentration-time profile that approaches intravenous dosing profiles. The ophthalmic, nasal, pulmonary, buccal, transdermal, and rectal routes provide one or more of these advantages. The ophthalmic route has been used traditionally for topical application for local effects. However, with the increasing number of peptide drugs being developed, the ophthalmic route has been considered for systemic drug delivery. After topical administration in the eye, peptides can be absorbed from the mucosa during tear turnover as well as via the blood vessels of the conjunctiva.

The route suffers from the hesitancy of practitioners to place a drug into the eye for any reason other than to produce ophthalmic effects. Another drawback is the sensitivity of the eye to irritation by foreign substances. Nasal administration produces rapid blood levels and rapid responses that approach those obtained from intravenous dosing. In addition, the absorbed drug does not pass through the liver before reaching the systemic circulation, and, thus, first-pass metabolism is avoided. In recent years, the nasal route has received a great deal of attention as a convenient and reliable route for the systemic administration of drugs, especially those that are ineffective orally and must be given by injection. Recently, butorphanol tartrate was introduced commercially in a nasal spray dosage form for the relief of pain, such as migraine headache. It would appear that the nasal route could be considered for drugs that meet the following criteria: are ineffective orally; are used chronically; are used in small doses; and are desirable to have rapid entry to the general circulation. Published work carried out in many laboratories has shown that, with the notable exception of the peptides, drugs with a wide variety of chemical structures are well absorbed through the nasal membranes of animals and man.

It is believed by many authors that in vivo nasal absorption of compounds with molecular weights less than 300 daltons is not significantly influenced by the physicochemical properties of the drug molecule. Factors, such as the size of the molecules and, in the case of peptides, their ability to hydrogen bond with the component(s) of the membrane, are more important than their lipopohilicity and their ionization state. For example, the in vivo rate of absorption of the very lipophilic drug progesterone is similar to that observed for sodium benzoate, and the in vivo rate of absorption of benzoic acid is independent of the pH of the medium. Although many drugs are absorbed rapidly and quantitatively following nasal administration, peptides have generally shown low bioavailabilities. Hussain et al. examined the nasal bioavailability of leucine enkephalin.

The low bioavailability of this pentapeptide was attributed to hydrolysis in the nasal cavity, with dipeptides causing significant inhibition of the hydrolysis. They concluded that polar compounds, such as peptides, can cross the nasal mucosa, and that administration of low concentrations results in extensive hydrolysis in the nasal mucosa and that hydrolysis of leucine enkephalin can be reduced by concomitant administration of peptidase labile peptides. If the presence of a pharmacologic activity for a peptide is the only criterion for its use in therapy, then nasal administration may be employed, even though bioavailability is not 100%. However, when bioavailability, as determined by plasma level profiles, is considered, the nasal route for peptide administration is not optimal unless enhancers are employed. Enhancers may cause irritation and reduced membrane integrity.

Other drawbacks of nasal delivery include the small volume permitted (i.e., 0.2 ml or less), the need for highly potent drugs with low doses, the potential for irritation, and the unknown consequences of long-term nasal administration of drugs, adjuvants, and other formulation components. The buccal and sublingual routes of administration permit rapid delivery to the systematic circulation. Absorption from the buccal and sublingual vasculature and lymphatics bypasses hepatic circulation and, thereby, reduces first-pass metabolism. The driving force of absorption is the high thermodynamic activity of the compounds. Organic nitrates and testosterone have been administered by these routes to produce rapid plasma concentrations and to minimize hepatic metabolism. The sublingual administration of methyltestosterone was examined by Alkalay et al. The sublingual tablet produced a 50% higher relative bioavailability when compared with the oral tablet or oral solution.

The increased bioavailability was attributed to the avoidance of first-pass hepatic metabolism due to absorption from the sublingual vasculature and lymphatics. The limitations of the buccal and sublingual routes include the requirement for successful candidates to possess high thermodynamic activities, the restrictive size of the buccal pouch and sublingual area, and the concern over the palatability and local irritation by the compound. Transdermal administration can avoid first-pass metabolism as well as provide a large surface area for continuous-controlled administration of drugs with short biological half-lives and narrow therapeutical indices. The route has been used for nitroglycerin ointments, and transdermal therapeutical systems (patches) have been developed for scopolamine, nitroglycerin, clonidine, estradiol, and nicotine.

Scopolamine can cause side effects of dry mouth, drowsiness, tachycardia, central nervous system (CNS) disturbances, and amnesia if plasma concentrations exceed the levels required to alleviate motion sickness. The application of scopolamine to the posterior auricular areas (the most permeable anatomical site) via a microporous patch permits constant delivery of 0.5 mg over 3 days to prevent motion sickness and avoid side effects. The transdermal route suffers from the inability of the skin to deliver large doses, the relatively slow plasma increase when compared with other routes, and the potential for irritation. However, several products utilizing this route of administration are currently on the market.

Absorption can be enhanced with molecules that maintain an appropriate hydrophilic-lipophilic balance for efficient passage through the barrier of the stratum corneum. Rectal administration of systemic effect has traditionally been limited to clinical situations where oral intake is restricted due to a physiologic condition (e.g., vomiting) or to compounds that are irritating to the gastric mucosa. The route has been utilized for aspirin, acetaminophen, aminophylline, promethazine, per- chlorpromazine, chlorpromazine, and indomethacin. The rectal administration of drugs is limited by increased interpatient variability and patient acceptability. This section has reviewed some of the dosage form design methods available to the pharmaceutical scientist that have been shown to improve the therapeutic efficacy of certain drugs. The key to optimal dosage form design lies in the prerequisite understanding of the physicochemical and biopharmaceutical properties of the drug and the available routes of administration. Once the compound has been characterized and the problem has been defined, the technology is available to optimize bioavailability.

COMMERCIAL USES

The creation and manufacture of dosage forms has been at the center of pharmacy practice for the past thousand years. For American pharmacists of the nineteenth century, *secundem artem*, or the acronym "S.A." in physicians' prescriptions, instructed them to use their special skills "according to the art" of their profession to compound a medicine; it was out of this art, rather than science, that almost all of today's major dosage forms arose. Tablets, capsules, injectables, and oral solutions were all known to pharmacists and physicians a century ago. In addition, there were scores of specialized dosage forms that attempted to meet the medical needs of patients, even if the drugs administered in these doses were ineffective or designed to treat symptoms rather than the underlying disease.

The origins of most of these dosage forms are lost in history. For this reason, the authors have elected to forego a contrived narrative tying together the few facts at hand with an equally large amount of speculation about the history of dosage forms. Rather, we have assembled a glossary of terms used in orthodox Western medicine to describe both common and unusual modes of drug administration. For most of its history, the field of pharmacy was much more concerned with drug preparations than with the resulting dosage forms.

Up to the sixteenth century, almost all drugs were derived from plants and were made into preparations that served as the ingredients for medicines; these preparations were called *galenicals*, after the great central figure of Western therapeutics, Claudius Galen of Pergamon. Strictly speaking, galenicals are pharmaceutical preparations obtained by macerating or percolating crude drugs with alcohol or some other menstruum to remove only the desired principles and leave the inert constituents undissolved. Examples of galenicals include decoctions, extracts, fluidextracts, fluidglycerates, infused oils, infusions, oleoresins, resins, tinctures, and vinegars.

The term is used very loosely today to designate any type of simple pharmaceutical preparation, whether it is an extract of a crude drug or a solution of chemicals. Because galenicals were often administered without alteration, the pursuit of new extraction and other preparative techniques sometimes led to developments in dosage forms as well. The goal of medicine preparers since Galen's time has been to create dosage forms that are stable, free from inert material, therapeutically efficacious, and concentrated to facilitate handling and administration. Preparations are divided into liquids and solids.

The liquids are further subdivided as: (1) general aqueous solutions and preparations; (2) sweet or viscid aqueous solutions and preparations; (3) general nonaqueous solutions and preparations; (4) alcoholic or ethereal solutions and preparations; (5) oleaginous solutions and preparations; and (6) parenteral solutions and preparations. The solids are subdivided as: (i) medicated solids; (ii) medicated particulate solids; (iii) medicated solid applications; (iv) oral individual dosage forms; and (v) nonoral individual dosage forms. The dosage forms are arranged alphabetically and include Latin titles and synonyms. The historical development of individual dosage forms is traced to about 1950; it was about then that modern pharmaceutical science was applied in depth to the problems of dosage forms, and the term itself gained general currency in the literature of pharmacy.

Methods of Preparations

Abstracts: Abstracts are powdered extracts of crude drugs prepared by percolating the drug with an appropriate menstruum, reserving a certain portion, evaporating the weak percolate to a thin extract, blending this extract with the reserve portion and lactose, and evaporating the mixture until dry. The mass is weighed and enough lactose added to make the finished product exactly half the weight of the drug from which it is derived. The resulting product represent twice the strength of the original drug. Although they were convenient for compounding prescriptions, abstracts gained little popularity.

Acids, diluted: Diluted acids are acid preparations, usually 5-10% strength, used for both internal and external medicines. Acids became official with the first United States Pharmacopoeia (1820). The term comes from the Latin acidus meaning "sharp" or "sour."

Aerosols: Aerosols are a system of finely divided liquid or solid particles dispersed in and surrounded by a gas. The roots of the modern aerosol go back to 1862 and J.D. Lynde, who received a patent for a valve, complete with dip tube, designed to dispense an aerated liquid from a bottle. In 1899, Helbing and Pertsch added liquefied gases. Patents in the early twentieth century usually were related to spraying perfumes. Aerosol fire extinguishers came into use in the 1930s, and insecticide sprays (bug bombs) appeared during World War II. Introduced in 1947, medical aerosol use increased greatly during the 1950s. In 1952, fewer than half a million such aerosols were produced; by 1963, this figure had risen to almost 40 million.

Ampuls: Ampuls are small, flask-shaped, hermetically sealed glass containers containing a sterile medicinal liquid intended for hypodermic injection, either subcutaneously, intramuscularly, or intravenously. Also, ampul is the class name adopted by the National Formulary V (N. F.) (1926) for the solutions in these containers. The ampul was invented in 1886 by the French pharmacist Stanislas Limousin (1831–1887) in response to a need by physicians to conserve their stocks of injectable solutions, which were difficult to transport and deteriorated rapidly due to the development of mold. In his classical essay, "Ampoules hypodermiques; nouveau mode de preparation des solutions hyperdermiques," published in Archives of Pharmacy (1886), Limousin outlined the essential directions for their manufacture: These ampoules have the form of a small ovoid balloon. They are terminated by a tapered glass tube, and their capacity is a little greater than one cc.

Although great advances have been made in the techniques and mechanics of ampul production, Limousin's simple rules capture the basic underlying principles. In the United States, ampuls, or "hermetically sealed containers which are filled with a medicinal liquid in a sterile condition, intended for parenteral use," became official in the National Formulary V (1926); that same year, the United States Pharmacopeia X included a chapter on sterilization but no monograph for individual ampuls. Iodine Ampuls, N.F., containing Iodine Tincture, United States Pharmacopeia, in sealed containers, intended to be broken and the liquid applied topically for the emergency disinfection of cuts or wounds, remained official through the National Formulary XIII (1970). The French term ampoule came from the Latin ampulla, which originally designated an earthen jar container for perfume.

Auristilla: Auristilla is a preparation used for medication of the ear canal. The National Formulary V (1926) Compound Oil of Hyoscyamus is an illustration of the type, closely resembling the Baume Tranquille of the French Codex (1908).

Bacillules: Bacillules are rod-shaped lozenges, prepared by massing the lozenge material, rolling it into cylinders or pill pipes, and cutting the cylinders into sections approximately twice the length of the diameter. Licorice lozenges are frequently prepared in this form, a popular example of which was formerly known as Wister's lozenge.

Balneum: Balneum is a bath for general application.

Balsams: Balsams are natural solutions of resin in an essential oil, which may also contain benzoic or cinnamic acids. The term has ancient roots in words referring both to spices and to embalming. In premodern times, balsams (baume in French) were any resinous vegetable juices or gums, acquiring their modern meaning in the 1800s. Several official preparations, such as Balsam of Copaiba, do not meet this definition but still carry the name because of their outward similarities with true balsams.

Bandages: Bandages are strips or ribbons of muslin gauze or other material employed in surgery for the retention of dressings and for the compression, protection, or support of diseased or injured parts. Bandages may be classified as inelastic, semielastic, elastic, or splint bandages. Inelastic bandages include muslin ribbon or roller bandages, gauze bandages, and water dressing bandages; semielastic bandages include flannel bandages; elastic bandages include rubber, rubberized, or crepe bandages; splint bandages include plaster-of-Paris bandages and crinoline bandages (gauze stiffened with dextrin) or starch bandages (gauze stiffened with starch paste), used to form the base upon which plaster-of-Paris is applied.

Baths: The word baths refers to the external application of water to the body, one of the oldest therapeutic techniques. Although some drugs were added, the simple water bath, or balneum, was most common. In the nineteenth century, when the therapeutic application of water was at its peak, baths were divided into hot or vapor (above 36.1°C), warm (29.4–36.1°C), tepid (18.3–29.4°C), and cold (0–18.3°C). The hot bath was a stimulant; the warm bath was soothing; the tepid bath was for treating skin problems; and the cold bath could be used as a stimulant, tonic, or sedative, depending on the administration technique.

Boluses: Boluses are large pills, over 325 mg (5 grains) in weight. The term comes from the Greek bolos meaning "*lump*".

Bougies: Bougies are instruments or shaped, solid medications for insertion into the urethra or other body cavities. The term comes from the French bougie, signifying a thin wax candle named for the Algerian city, Bougie.

Cachets: Cachets are lenticular or spoon-shaped rimmed disks pressed from rice-flour wafer sheets, used to administer bitter or nauseating drug powders. The powder is deposited in dry concave cachets, and the rims are moistened with water; empty convex cachets are placed on top, sealing the cachets and enveloping the powder. The term "*cachet*" is from the French cacher, "to hide." The cachet de pain was invented by the French pharmacist Stanislas Limousin (1831–1887) in 1873 as an improvement over wafers and wafer envelopes. Limousin also developed a perforated board, accommodating three sizes of cachets, a powder measurer, powder funnels, and wooden "*wetter and pressers*" to speed extemporaneous cachet production by the pharmacist. A popular brand of cachets and cachet apparatus was manufactured around 1885 under the trade name Konseals by the J.M. Grosvenor Company of Boston around 1885.

The Konseal apparatus consists of three perforated, nickel-plated metal plates, hinged together to form a cover plate, a base plate, and a shield plate. Saucer- shaped rice-flour Konseals are pressed into the perforations on the cover and base plates, while the shield plate is folded back over to protect the sealing edges of the Konseals in the base plate. The Konseals in the base plate are filled with powdered drug with the help of special funnels and are tamped down with thimble compressors. The shield plate is lifted, and a moistened roller is passed over the edges of the empty Konseals in the cover plate, which is closed over the base plate, sealing the Konseals. The Konseals are made of thinner material than ordinary cachets, and the finished product is less bulky and neater in appearance. Johann Schmidt later introduced rimless Dry Seal Cachets resembling flattened capsules, which were sealed by pressure without moistening.

Capsules: Capsules are telescoping, interconnecting shells of hard gelatin and sugar used for the administration of solids, masses, and liquids. The term comes from the Latin *capsula,* which is the diminutive of capsa, meaning "box." Successor to the soft capsules invented by the French pharmacist François Mothes in 1833, the telescoping hard gelatin capsule was invented and patented in 1847 by James Murdoch of London. In the United States, the New York firm of H. Planten manufactured two-part capsules sometime after 1836, but their usefulness was impaired by their poor fit. In 1863, the firm developed jujube paste capsules intended for dispensing powders alone, offering them to the trade before 1870. Another manufacturer, Dundas Dick, experimented in the same direction and secured a patent on cone-shaped capsules in 1865. Twelve years later, inspired by reports from Italy, the Detroit pharmacist F.A. Hubel made molds of iron wire mounted in blocks of wood, which could be dipped in a gelatin solution.

Using pins of different diameters for the body and cap allowed Hubel to produce capsule sections which would telescope into each other, producing a small cylinder closed at both ends—the prototype of the modern hard gelatin capsule. Hubel sold his entire output to Parke, Davis & Co. in 1875, securing the first in a series of patents in 1877. Despite this protection, several small competing firms soon emerged that made capsules for other companies to sell or for their own sales organizations, most of which were consolidated by James Wilkie into the U.S. Capsule Company. Around 1901, this firm and its subsidiary, the M.L. Capsule Company, were purchased by Parke, Davis & Co., which expanded and improved its manufacturing processes under Wilkie's supervision. Wilkie is also credited with introducing phosphorbronze wire for capsule molds, a material superior to iron and one which remained in use until replaced by stainless steel in the 1930s. Capsule manufacturing received its greatest impetus in the 1920s when automatic filling devices were perfected. Pharmacists extemporaneously fill capsules at the prescription counter by "*punching*" capsule bodies into a smooth layer of medicated powder and filler at a uniform level of compression, replacing the caps, and checking the weight of the filled capsule.

Capsules, soft: Soft capsules are elastic globular shells of gelatin, containing sufficient glycerin to retain permanent flexibility, and intended primarily for the administration of irritating or nauseating oily liquids. They were invented in 1833 by French pharmacist François Mothes; an improved capsule was patented by Mothes and Joseph Dublanc the following year and perfected in 1840 by the court apothecary Adolph Steeger of Bucharest. Soft capsules are easier to swallow than hard capsules and are desirable for administering large volumes of liquids. The original Mothes capsules were hollow globes of soft gelatin with an opening at the top for filling. Once filled, the capsules are sealed with a drop of hot gelatin. In 1846, Giraud introduced olive-shaped elastic capsules with an elongated neck, made by dipping molds into a warm glycerogelatin solution. After drying, the molds were stripped and the capsule necks sealed for future use. Pharmacists cut off the necks of the capsules as needed, filled them with a dropper or syringe, and sealed them with a drop of warm glycerogelatin or the blade of a hot spatula. The first official formula for the manufacture of soft gelatin capsules appeared in the French Codex Medicamentarius of 1866.

Cataplasms: Cataplasms are moist substances intended for external application and are of a consistency as to accommodate themselves accurately to the surface to which they are applied, without being so liquid as to spread over the neighboring parts, or so tenacious as to adhere firmly to the skin. Cornelius Celsus (50–25 B.C.) described early Roman softening ointments or malagma (from the Latin ma lasso, meaning "soften") made by boiling flour in water to make a stiff paste that was admixed with melted gums or wax. The word "*cataplasm*" derives from two Greek words, kata, meaning "down," and plasso, meaning "to mold." Modern cataplasms are made by rubbing together glycerin and a dry powder, such as kaolin, to make a very firm mixture. Since cataplasms (or poultices) were commonly used in the United States in domestic practice and made in the home, they were rarely prepared by pharmacists. A kaolin cataplasm became official in the United States Pharmacopeia VIII (1905) but was transferred to the National Formulary IV in 1916.

Caustics: Caustics are local remedies that destroy life on the part of the body to which they are applied. The term comes from the Greek kaustikos, meaning "to burn." The strongest common caustic (causticum commune acerrimum) was potassium hydroxide, or caustic potash, used to form issues on the body or to open abscesses. The United States Dispensatory (1836) advised that the most convenient mode of employing a caustic to form an issue was: To apply to the skin a piece of linen spread with adhesive plaster, having a circular opening in its centre corresponding to the intended size of the issue, and then to rub upon the skin within the opening a piece of the caustic previously moistened at one end. The application is to be continued until the life of the part is destroyed, when the caustic should be carefully washed off by a wet sponge or wet tow, or neutralized by vinegar.

Caustic potash was sometimes used to remove strictures in the urethra or, in solution, as an application to the spine in treating tetanus. A contemporary caustic, silver nitrate, was described in the seventh century A.D. by Geber; it remains official in the United States Pharmacopeia as Fused Silver Nitrate. Christopher Glaser, apothecary to Louis XIV, first prepared it in sticks for use as a caustic.

Cements: The 7th edition of the National Formulary (1942) listed a formula for a Cement of Zinc Compounds and Eugenol, which was widely used by dentists as a temporary filling. The cement was supposed to exert a sterilizing effect and protect the dentine from further destruction. The formula passed out of the National Formulary with its 9th edition (1950).

Cerates: Cerates are unctuous substances for external application as dressings for inflamed surfaces. Derived from the ancient Greek keroma (from keros, "*wax*"), cerates are generally made with oil, lard, or petrolatum as a basis, with sufficient wax, paraffin, spermaceti, or resin added to raise the melting point of the oils and fats employed. Cerates should be of such consistency that they may be easily spread at ordinary temperatures upon muslin or a similar material with a spatula and, yet, not so soft as to liquefy and run when applied to the skin. The most widely used cerate was Ceratum Cantharidis, a blistering plaster made of Spanish flies that was official in the United States Pharmacopeia and later in the National Formulary until 1950.

Cerecloths: Cerecloths are an early form of dipped plasters, described by William Salmon in his Pharmacopoeia Londonensis (1691), used chiefly to lay upon issues–small ulcers produced by caustics or cutting, the discharge from which was encouraged to fulfill certain therapeutic indications. Also known as spasma-draps or sparadraps.

Coatings, pill: Rhazes (850–923) used a mucilage of psyllium seed to coat offending pills; a century later, Avicenna (980-1037) introduced silver and gold coatings for pills not merely to mask bad taste but to enhance the supposed medicinal effect. Later, the influential seventeenth-century Parisian physician and pharmacist Jean de Renou recommended that pills with a bitter taste should be gilded and mixed among some powdered spices. Coating pills with gold and silver leaf was commonly practiced in France, other parts of Europe, and the United States until well into the nineteenth century but then fell into disuse. Mohr, Redwood, and Procter's classic Practical Pharmacy (1849) reported that foil-coated pills were "still occasionally administered, but much less frequently than formerly."

Sugar-coated and gelatin-coated pills had their first acceptance as an indirect consequence of the invention of the gelatin capsule. In 1838, M. Garot, a Parisian pharmacist, coated offending pills with gelatin. A year earlier, the French pharmacist Labelonie had recommended that pills of cubeb and copaiba be covered with sugar, a process that was patented by Adolphe Fortin of Paris in 1837. Over the next several years, other French pharmacists secured patents for coating pills with various combinations of sugar, honey, and acacia. By 1862, England's Bernard Proctor could list 45 distinct processes for coating pills with a variety of substances. In the United States, the first manufacture of sugar-coated pills was associated with the patent-medicine industry, probably as early as 1845. New Yorkers Cornelius V. Clickener and Zadoc Porter and a Philadelphia physician named Swayne all claimed to be the "*inventor*" of the sugar-coated pill but doubtlessly adapted French technology.

By 1857, five different sugar-coated patent- medicine pills were available to American pharmacists, but no process was available for pharmacists who wished to sugar-coat pills extemporaneously. About that same time, Henry A. Tilden and William R. Warner independently developed processes for manufacturing sugar-coated pills and began selling them in bulk; in 1866, Warner began manufacture under his own firm name, William R. Warner & Company. That same year, Henry Wathew of Philadelphia developed and patented a prototype of the angled coating pans still employed today. By the mid-1870s, a variety of small mechanical coating pans were available to American pharmacists, although the quality of the extemporaneous coatings that could be achieved was generally inferior to that of the manufactured variety, which caused pharmacists to relinquish the practice.

Coatings, tablet: After the introduction and acceptance of the compressed tablet as a dosage form superior to pills, the same techniques that had been employed for sugar-coating pills were adapted to tablets. Moreover, compression coating techniques were also introduced by Charles Carter of Philadelphia as early as 1858. In 1896, P.J. Noyes invented an apparatus incorporating a movable die cavity into which the coating material was fed; the powder was compressed to form a coating around the pill or tablet by striking the upper punch with an automatic hammer. The following year, Noyes patented an improved machine that not only applied the coating but also performed the preliminary compression of the tablet. F.J. Stokes received a patent for a similar machine in 1917, although the compression coating of tablets did not become widely adopted until the 1950s.

Collodions: Collodions are liquid, external preparations with a base of pyroxylin dissolved in a mixture of alcohol and ethyl oxide or similar solvent. They were used medicinally soon after the discovery of gun cotton by Schoebein in 1846, first as a surgical dressing by Maynard in 1847.

Condita: Condita are comprised of candied or preserved roots or fruits. They are made by boiling the plant parts until tender, soaking them in hot syrup, and then pouring off the syrup; the procedure was repeated several times, reusing the syrup after boiling it down to a thick consistency. Condita can also be a compound of wine, honey, and spices (often pepper). Condita are also known as confitures or sweet-meats.

Cones, medicated: Cones are light, porous hemispherical masses of sucrose and egg albumin, used as a vehicle for homeopathic medications. The cones, also called disks, are designated (in millimeters) according to size by the diameter of the base. The common size should absorb about 2 drops of dispensing alcohol. Cones are medicated by adding a sufficient quantity of the dilution to saturate them and pouring off the excess liquid.

Confections: Confections are saccharine, soft solids, in which one or more medicinal substances are incorporated to provide an agreeable form of administration and a convenient method for preservation. In the thirteenth century, some apothecaries were called *confectionarii* from confectio meaning "a composition." Confections are made by adding medicinal ingredients in either the form of a smooth paste, a fine powder, or a liquid to a basis of finely powdered sugar. Confection of Rose and Confection of Senna were official in the National Formulary through the 5th edition (1926).

Conserves: Conserves are confections prepared from fresh medicinal agents and refined sugar, beaten into a uniform mass. The United States Dispensatory (1836) noted that "as active substances even thus treated undergo some change, and those which lose their virtues by desiccation cannot be long preserved, the few conserves now retained are intended rather as convenient vehicles of other substances, than for separate exhibition." Conserves are also known as preserves.

Cordials: Cordials are sweetened alcoholic preparations of high alcoholic content. They are also a tonic medicine formulated to stimulate the heart.

Creams: Creams are semisolid emulsions, usually medicated, intended for external application. The term "*cream*" has been used to refer to a wide variety of opaque, soft semisolids or thick liquids intended for external use. The British Pharmacopoeia classifies creams as medicated liquid emulsions consisting of a mixture of anhydrous lanolin, olive oil (or other fixed oil), and lime water; milk of magnesia is often referred to as a cream, and many cosmetic preparations are called creams. Claudius Galen prepared the first *unguentum refrigerans* or "*cold cream*," an ointment containing olive oil, rose oil, white wax, and a small quantity of water. This was a prototype for other cosmetic ointments introduced by Johann Mesue, Jr., in the thirteenth century. A modern version, consisting of almond oil, spermaceti, white wax, and rose water, passed into the United States Pharmacopeia as "Rose Water Ointment" or "Galen's Cerate." Pharmaceutical creams became official with the introduction of the formula for a "*Sun Cream*" in the National Formulary VIII (1946), a product designed to prevent sunburn but permit tanning. Cosmetic creams, which are usually not medicated, include preparations classified as all-purpose creams, baby creams, barrier creams, bleaching creams, cleansing (or rolling) creams, cold (or fatty) creams, foundation (or vanishing) creams, hair creams, and hand creams.

Cucufa: Cucufa is a cap, dusted on the inside with medicinal powder, which is applied to the head to strengthen the brain.

Decoctions: Decoctions are a solution of vegetable drugs, obtained by boiling the substances in water; they are also known as apozemes.

Dentifrices: Dentifrices are powders, pastes, washes, or medicated soaps used for cleaning the teeth. Dentifrices are usually flavored with aromatic oils, frequently contain soap, almost always some form of chalk, and are applied with a toothbrush. The National Formulary carried official formulas for a liquid dentifrice through its 5th edition (1926) and a powder dentifrice through its 11th edition (1960).

Douches: Douches are a column of fluid, of a certain nature and temperature, allowed to fall on a part of the body. Air can also be a used as a douche.

Dragees: Dragees are candied or preserved roots and fruits described by the Arab Najm ad-dyn Mahmoud in the eighth century and reintroduced in the eighteenth century by the famous French pharmacist Moyse Charas. By the middle of the nineteenth century, the term was extended to include a type of sugar-coated pill formed by repeatedly shaking slightly moistened, tiny 6-mg (1/10 grain) sugar granules (or nonpareils) in a basin of finely powdered drug mixed with sugar. After a sufficient number of layers had been built up, the dragées would receive a final coating of sugar or copal and tolu balsam, a painstaking process described by Ernest Agnew in the American Journal of Pharmacy (1870).

Draughts: Draughts are liquid medicines usually prepared to be taken in a single dose or "draught." Draughts are also known as potions.

Dressings, medicated: Medicated dressings are external applications resembling ointments in consistency but remaining semisolid at body temperature. Paraffin Dressing, formerly official in the National Formulary VI(1936), was employed as an air-excluding, soft, pliable, analgesic, splint-like covering for surfaces denuded by burns. A wide range of materials have been used as coverings or protectives to apply heat or medicaments to a diseased or injured part, to prevent wound infection, and to absorb and prevent decomposition of wound discharges.

Antiseptic or Medicated Cottons include borated cotton, iodoform cotton, iodized cotton, and styptic cotton; Antiseptic or Medicated Gauzes include borated gauze, corrosive sublimate gauze, carbolated gauze, iodoform gauze, and picric acid gauze; Antiseptic or Medicated Lints include borated lint; Cellulose Waddings or wood wools include moss, peat, sawdust, jute, and oakum or marine lint.

Dressings, protective: Protective dressings are comprised of a wide range of materials employed as wound coverings, either to shield the parts from external infection or to prevent the escape of fluids contained in the dressing. Protective dressings are also used as a covering for poultices and for the retention of heat. Common protective dressings include oiled silk, oiled muslin or cambric, waxed paper, gutta percha tissue, rubber dams, mackintosh or jaconet, and rubber sheeting.

Drops: Drops are pharmaceutical mixtures meant to be given in small amounts. Before the twentieth century, the term applied to solutions used in small quantities expressed in "drops." These were commonly strong medicines "dropped" into water, such as Vinegar of Opium or "*black drop*." In modern pharmacy, the term became more associated with the need to get a medicine into an appropriately small amount of vehicle for application to the eye (ophthalmic), ear (otic), or passages of the nose (nasal). As a dosage unit, the drop is troublesome because it can vary greatly in size, depending on the size of the dropper orifice and the surface tension of the liquid. The United States Pharmacopeia IX (1910) set the official dropper at 20 drops per gram of water at 15°C ± 10%.

Drops, toothache: Drops for toothache are comprised of a solution of phenol in oil of cinnamon and methyl salicylate. The solution entered the National Formulary V (1926). A mixture of phenol, creosote, or volatile oils dissolved in paraffin, with a few filaments of cotton added, and molded into sticks constitutes Dental Wax.

Electuaries: Electuaries are confections prepared from dried medicinal agents, especially powders, combined with syrup or honey in order to render them pleasant to the taste and convenient for internal use. The United States Dispensatory (1836) noted that electuaries "should not be so soft... as to allow the ingredients to separate, nor so firm... as to prevent them from being swallowed without mastication." French writers recommend using brown sugar syrup to prepare electuaries, because it is less apt to crystallize than that made from refined sugar. The term comes from the Greek words ek, meaning "out," and leichein, "to lick".

Elixirs: Sweetened, hydroalcoholic, flavored liquid medicines, which became popular in mid-nineteenth-century America. The word is derived apparently from the Arabic al-iksir, which is an Arabic form of the Greek, xirion. Originally the term meant "*dry powder*." Elixirs came into medicine through their connection with alchemy. Elixir Rubrum, one of the most renowned alchemical compounds, could supposedly turn mercury to gold or prolong life. The term was picked up by followers of Paracelsus and became applied to liquid preparations. European elixirs were generally bitter. One of the first American elixirs was Cordial Elixir of Quinine, made by John T. Heinitsh of Lancaster, Pennsylvania. After the Civil War an "*elixir craze*" began, which led to scores of companies competing for business. As much as any other development, the "craze" led to the publication of the first National Formulary in 1888.

Emulsions: Emulsions are a preparation consisting of two immiscible liquids, usually water and oil, one of which is dispersed as small globules in the other. Before the late seventeenth century, the term only applied to natural emulsions, such as ground almonds and water, which resembled milk. In 1674, a physician named Grew reported the preparation of oils and egg yolk to the Royal Society of Great Britain. In the 1700s, other emulsions were made with acacia, honey, tragacanth, and other natural emulsifying agents. In the 1800s, the wet-gum (ca. 1850) and dry-gum (ca. 1870) methods were established as standard preparation techniques. Interest in medicinal emulsions peaked in the early to mid-twentieth century with the development of several new emulsifying agents.

Enemas: Enemas are injections of liquid, either plain or containing drugs, into the rectum and colon to empty the lower intestine or to introduce food or medicine for therapeutic purposes.

Enemas are one of the most ancient and widely used methods of introducing therapeutic substances into the body. The origins of use are lost in prehistory, but written records in Egypt before 1000 B.C. describe enemas being used to both cleanse the bowel and administer medicines. These early enemas consisted of three parts: a vehicle (usually water, beer, or milk), an emollient (usually oil or honey), and a medicinal substance.

The Greek historian Herodotus (fifth century B.C.) attributed the general good health of the Egyptians to their use of enemas and claimed they had achieved their expertise through their experiences with injecting embalming fluids via the anus. Pliny (first century A.D.), however, argued that the Egyptians had learned to administer enemas by watching the ibis use its curved bill to inject itself with Nile River water as a purge. Hippocrates recommended enemas to treat fevers and constipation. Other authors of antiquity, such as Galen and Oribasius, wrote at length about what substances could be introduced via enemas. Enema apparatus was first described in detail by Arabian physicians of the eleventh and twelfth centuries. Albulcasis described a device made of an anal tube or funnel attached to a bag made from an animal bladder or sheep skin. The metal piston syringe came into use in the 1400s and supplanted these clumsy bags.

The enema syringe soon became the object of much medical ingenuity, particularly in France. In 1480, Louis XI suffered a severe stroke and recovered, giving credit to the enemas he had received, beginning a 400-year period of fascination with the enema among the French, who refined the syringe apparatus. Molie're referred to enemas repeatedly in his works, and Ambrose Pare devised a syringe instrument about 1580 that allowed enema self-administration. During the seventeenth century, other special apparatus were designed to allow administrations of tobacco smoke via enemas. The Dutch anatomist Regner de Graaf completed the first book- length study of enemas in 1668.

In the mid-eighteenth century, gum rubber began to be used in enema apparatus, replacing skin bags. In 1820, John Read developed a two-way syringe with ball valves (the first modern stomach pump), which was also used for enemas. Spring-loaded syringes and other advances followed, but by the late nineteenth century the dangers of improper enema use became apparent, especially the consequences of high-pressure administration. By the mid-twentieth century, enemas were again administered with simple funnels and gravity, a reflection of their prehistoric origin. Enemas are also known as clysters or glysters.

Enteric-coated doses: Enteric-coated doses are dosage forms that have been coated or chemically treated to prevent disintegration in the stomach but which disintegrate in the intestinal tract. Enteric coating is employed when the medicinal substances would be decomposed or rendered inactive by gastric enzymes or when they would irritate the gastric mucosa. The Pacific Medical and Surgical Journal (1867) noted that collodion protects pills from dissolving in the stomach, but gastric insolubility as a basis for medication is generally credited to the German dermatologist Paul Unna, who introduced keratin-coated pills in 1884; Ceppi introduced salol as an enteric coating in 1891. Glutoid capsules are a special form of enteric coating prepared by subjecting soft or hard gelatin capsules to the action of formaldehyde until they become insoluble in the stomach but not in the intestine. The hardening process was developed in Switzerland by Weyland in 1895 and patented by Hausmann in Germany that same year. Pharmacists have extemporaneously coated capsules, pills, and tablets with salol, keratin, casein, and shellac, or with a mixture of n-butyl stearate, carnauba wax, and stearic acid to create enteric coatings.

Early patented enteric coatings included keratin, fat-covered capsules enclosed in membranous sacs, and benzoin-coated capsules. Later patented coatings utilized cellulose nitrate and acetate, cellulose esters and ethers with saponifiable organic compounds, ammoniacal bleached shellac, stearic acid, carnauba wax, petrolatum, elm bark, and agar, and abietic, oleic, and benzoic acids with methyl abietate. Since 1940, research on enteric coatings has focused on the synthesis of resinous polymers, which are insoluble in acids, such as cellulose acetate phthalate and a glycerol-stearic acid-phthalic anhydride ester.

Epispastics: Epispastics are local remedies, the application of which produce a serous discharge beneath the cuticle, forming a blister. Epispastics are also known as vesicatories.

Epithema: Epithema consists or topical applications other than ointments or plasters. Liquid epithema include fomentations, soft epithema include cataplasms, and dry epithema include bags filled with dried drugs.

Essences: The term essences sometimes refers to a volatile oil or to a simple tincture; most often, the term is used interchangeably with spirits.

Extracts: Extracts are either pasty or semisolid masses or dry, solid, or powdered products prepared by exhausting drugs with appropriate solvents, carefully evaporating the products to fixed standards. An extract is intended to preserve the useful constituents of a drug in a concentrated, relatively uniform, permanent condition, and in a form suitable for medication. The Edinburgh Pharmacopoeia (1817) and Dublin Pharmacopoeia (1826) distinguish between extracts prepared from infusions, decoctions, or tinctures, and those prepared from the expressed juices of plants, calling the latter succi spissati or inspissated juices. Three forms of extracts are recognized: semiliquid or those of syrupy consistency; plastic masses known as pilular or solid extracts; and dry powders known as powdered extracts.

Fluidextracts: Fluidextracts are concentrated liquid preparations representing the therapeutically active principles of vegetable drugs. They are formulated in such a way that the activity of one gram of the drug is contained in one milliliter of the fluid-extract. They are generally prepared by some form of percolation, using alcohol in the menstruum. Fluid-extracts first became official in 1850 when five were entered in the United States Pharmacopeia by their heyday in the late nineteenth century, almost 100 were official. Joseph Remington called fluidextracts "American preparations," because they were developed by such native pharmaceutical scientists as William Procter, Jr., and Edward Squibb. Fluidextracts were perhaps the ultimate galenical class because of their permanence, concentrated form, and uniform relationship between the strength of the extract and the drug it contained. With the dominance of synthetic chemical drugs in the twentieth century, fluidextracts all but disappeared, except for those used for flavoring purposes.

Fluidglycerates: Fluidglycerates are a class of fluidextracts in which a mixture of glycerin and water is used as the primary menstruum during percolation instead of alcohol and water. The preparation of these extracts was suggested by Beringer in 1908. They were briefly official from the 5th-7th editions of the National Formulary (1926-1942).

Fomentations: Fomentations consists of an external application of cloths dampened with hot water or a medicinal decoction. Narcotic drugs were sometimes used. Dry fomentations were heated bricks wrapped in cloth and applied externally.

Frontalia: Medicines applied to the forehead.

Gargles: A gargle is a liquid medicine intended to be retained in the mouth and placed in contact with the back of the throat by throwing back the head and agitated by air released from the larynx.

Gels: Gels are semisolid organic or inorganic colloids rich in liquid, consisting of hydrated threads or granules of the dispersed phase intimately associated with the dispersion medium. Although Francesco Selmi studied inorganic colloids in the 1 840s, modern colloid science began in 1861 with the work of Thomas Graham, who investigated diffusion and dialysis and introduced such terms as colloid, glue, sol, gel, peptization, and syneresis. In the early 1900s, Freundlick introduced the terms *lyophilic* and *lyophobic* to describe colloids in which the dispersed phase has a high or a low affinity, respectively, for the dispersion medium. In 1950, Weiser divided gels into inorganic gels, which include gelatinous precipitates (such as Milk of Magnesia) and inorganic jellies (such as Bentonite Magma), organic gels or jellies (such as Pectin Paste), and crystalline or amorphous jellylike networks in which both solid and liquid phases are continuous.

Glycerites: Glycerites are solutions of medicinal substances in glycerin introduced in the 5th revision of the United States Pharmacopoeia (1873). Glycerites afford a rapid and simple method of making aqueous solutions of phenol, tannic acid, tar, and other substances that are not otherwise easily soluble. This class of preparations is called glycerins in Great Britain.

Glycerogelatins: Glycerogelatins are soft, medicated masses, usually molded into the form of blocks, which melt at body temperature and have as a base a mixture of gelatin, glycerin, and water; Glycerated Gelatin United States Pharmacopoeia is generally used as a base. At the time of application, the blocks are melted, and the liquid applied to the skin with a soft brush. Four glycerogelatins were introduced into the 3rd edition of the National Formulary (1906), and a general formula remained official through the 8th edition (1946). Under the title Gelatinum, the British Pharmaceutical Codex (1922) included preparations that were either similar to glycerogelatins, that more closely resembled dermatologic pastes, or that were intended for internal administration and called *jellies*.

Granules: Granules are small spheres of sugar pellets that are saturated with liquid medication before being swallowed. They can also be very small pills of 0.06 g or less, sometimes called parvules, made by slightly moistening sugar granules (or nonpareils) with a syrup in which an active ingredient has been dissolved. Granules containing 1 mg of powerful medications, such as arsenious acid or aconite, found favor among physicians in Europe, particularly in Italy, because they provided powerful drugs in small, precise doses.

Gums: Gums are mucilaginous, amorphous, transparent, or translucent glucosidal principles of plants used internally (as demulcents or expectorants), externally (as emollients or protectives), or for their emulsifying action. Three distinctive types of gums are recognized: arabin, which is completely soluble in water; bassorin and cerasin, which are partially soluble or swell in contact with water; and mucilages and pectins, which swell to form jellies. Many gums are complex mixtures of several of these types.

Infusions: Infusions are aqueous solutions obtained bysoaking vegetable drugs in cold or hot (not boiling) water.

Inhalants: Inhalants are products consisting of finely powdered or liquid drugs that are carried into the respiratory passages by the use of powder blowers or low-pressure aerosol containers holding a suspension of the drug in a liquefied propellent. A dry inhalation is a product consisting of finely powdered drugs that are carried into the respiratory passages with the help of special devices. Inhalants can also refer to drugs or a combination of drugs which, by virtue of their high vapor pressure, can be carried by an air current into the nasal passages where they exert their effect. In the latter form of inhalant, the drug is absorbed on fibrous material and enclosed in an inhaler—a plastic or metal tube fitted with a cap to prevent loss of medicament when not in use. The patient removes the cap, inserts the nasal tip into a nostril, and breathes the air drawn through the inhaler to obtain the drug.

Inhalations: Inhalations are medicinal agents administered by breathing in gases and vapors. A wide variety of techniques and purposes fall under this category. The isolation of pure gases by Priestley, Scheele, and others in the late eighteenth century motivated Thomas Beddoes to found the Pneumatic Institute (1798) in England to treat lung diseases. The gases used for inhalation therapy included oxygen, nitrous oxide, and ether. Gases were applied as surgical anesthesia in the 1 840s largely through the efforts of the Americans Crawford Long, Horace Wells, and W.T.G. Morton; their innovations were the first great contributions to medical science by citizens of the United States. The therapeutic use of gases continued throughout the nineteenth century, especially for the treatment of tuberculosis.

During this period, devices were designed to vaporize liquid drugs with steam for the treatment of lung disorders. The first apparatus developed to atomize medicinals, made by Berson in 1860, was a combination inhaler and atomizer operated by steam that was generated by heating water in a closed vessel. These devices were quite popular for home use, and their modern counterparts remain a part of home health care. The term inhalation has also been applied to preparations more properly called inhalants, such as Compound Eucalyptus Inhalation, or sprays, such as Epinephrine Inhalation.

Injections: Injections are sterile solutions or suspensions used for administering pharmaceutical preparations by intravenous, subcutaneous, intramuscular, and intraspinal injection. The term injections is also the class name adopted by the 5th edition of the National Formulary (1926) for solutions in ampuls intended for hypodermic injection. Although ancient humans may have invented the concept of introducing drugs through punctures in the skin by attempting to recreate the effects of venomous snake and insect bites through the use of poisoned arrows, perhaps the first introduction of medication through the skin for medicinal purposes was inoculation for smallpox. Human inoculation with the smallpox virus by pricking the body with needles dipped in pus from an active case of the disease was practiced for centuries among people of the Orient but was only introduced into Western medicine about 1717. In 1796, Edward Jenner (1749–1823) performed his first vaccination with material from a cowpox sore. In 1657, Sir Christopher Wren was the first to inject a drug intravenously, a practice successfully adopted by the English practitioner Johan D. Major in 1662 under the title "chirurgica infusoria." Physicians experimented with the injection of water, opium, arsenic, cinnamon, oil of sulfur, and other substances with limited and often fatal results.

Injections of purging medicines, such as jalap resins, were particularly popular in the treatment of syphilis. Nevertheless, the successful utilization of intravenous injection awaited the proof of the germ theory of disease and the discovery of sterile methods by Pasteur, Koch, Lister, and others in the latter half of the nineteenth century; the introduction of the hypodermic syringe, which was suggested by Charles G. Pravez of Lyons in 1853, popularized by Alexander Wood of Edinburgh and Charles Hunter of London in 1855–1858, and improved by Luer in 1894; and the invention of the ampul by Stanislas Limousin in 1886. A committee of the Royal Medical and Chirurgical Society of London gave approval to hypodermic injections in 1867, the same year the British Pharmacopoeia published a monograph for the first official injection, Injectio Morphinae Hypodermica. No attempt was made to sterilize these solutions, but E.R. Squibb (1873) and others recommended that parenteral solutions could be preserved by the addition of small amounts of carbolic acid, salicylic acid, chloroform, or camphor. At about the same time (1875), John Tindall developed the process of sterilization by discontinuous heating, which bears his name. Nevertheless, hypodermic routes of administration were slow to gain widespread recognition, for physicians became increasingly aware of fevers and other toxic symptoms following the use of crudely prepared injections. Ehrlich's introduction of hypodermic injections of salvarsan for syphilis (1910) provided the strongest impetus for the development of parenteral administration, stimulating a series of rapid advancements in technique.

In 1911, Martindale and Wynn discussed the pharmaceutical manipulation of salvarsan, emphasizing sterilization and aseptic techniques. That same year, Hort and Penfold applied the term pyrogens to describe substances that cause a febrile reaction upon injection. They found that distilled water sealed in sterile containers gave rise to toxic symptoms, whereas freshly distilled water caused no reaction; later that same year, Wechselmann showed that febrile reactions could be eliminated if solutions were made from sterile distilled water. Seibert confirmed these findings in 1923, noting that poorly constructed stills could produce pyrogenic distilled water. In 1930, Rademaker formulated a rigid set of aseptic precautions and rules, governing the preparation of parenteral fluids, which are still valid today, and setting the stage for modern intravenous medicines. Injections are also known as *parenterals*.

Insessia: Insessia refers to a vapor bath, usually administered by having the patient sit on a perforated chair, beneath which is placed a large container filled with hot water or a hot decoction of a plant drug.

Insufflations: Insufflations are powders used for blowing into the nose, preferably by means of one of the various kinds of powder blowers or insufflators made for the purpose. They may also be applied directly in the way in which snuff is usually taken. Also, a snuff. The term comes from the Latin insufflare meaning "to breath into" or "to blow into."

Inunctions: Inunctions are ointments applied with friction, intended for local application and quick absorption. The term was formerly applied to preparations consisting of wool fat in which mercury or other medicinal agents were incorporated. A Compound Menthol Inunction remained official in the National Formulary until 1960.

Juices: Juices are liquids obtained by expression from the fresh parts of plants. The British Pharmaceutical Codex (1949) contained monographs for the juices of garlic, lemon, and taraxacum. In the United States, Cherry Juice remained official through the 16th edition of the National Formulary (1985) as a flavoring agent.

Juleps: A julep is a sweet drink, usually a demulcent, acidulous, or mucilaginous mixture. Much more popular in Europe than in the United States, a typical julep of the late nineteenth century (*mistura gummosa*) contained 10 parts acacia triturated with 30 parts syrup of acacia, 100 parts water, and 10 parts orange flower water.

Konseals: Konseals is a trade name for the brand of cachets and cachet apparatus manufactured by the J.M. Grosvenor Company of Boston about 1885, originally introduced as "Morstadt's cachets" by Karl Morstadt of Prague.

Lamels: Lamels are small disks, about 3 mm in diameter, cut or stamped from thin films of glycerinated gelatin, containing definite quantities of various medicaments used in ophthalmology. Lamels, or eye disks, are applied with a camel's-hair brush to the inner surface of the lower eyelid, where they are immediately dissolved in the lachrymal fluid. Four lamels were official in the British Pharmacopoeia until 1953.

Linctus: A linctus is a thick viscid liquid that must be licked from a spoon or licorice stick, from the Latin lingere, "to lick".

Liniments: Liniments are external preparations of a consistency thicker than water, but thinner than ointments, usually applied to the skin with a gentle rubbing of the hands. Liniments are among the oldest of dosage forms, along with related forms, such as plasters and ointments. The term came to its present use about 1600. Drying liniments are preparations which dry when smeared on the skin, forming a medicated film removable by water.

Liniments, dental: Dental liniments were introduced in the National Formulary V (1926) through cooperation between dental and pharmaceutical authorities; these concentrated, often poisonous liniments, were designed to be rubbed into the gums.

Lohochs: Lohochs are thick syrupy medicines, usually used to fight a cough, to be sipped slowly or licked, sometimes called a looch. If in the latter dosage form, it is usually called a *linctus*.

Lotions: Lotions are fluid preparations, usually containing suspended insoluble material and applied externally. They are different from liniments by being aqueous, rather than oleaginous or alcoholic in nature. Lotions were official in the first National Formulary (1888). In the mid-nineteenth century, lotions were often applied by wetting linen and placing on the affected area.

Magisteries: Certain precipitates from saline solutions bore this title, but it was usually applied to secret remedies.

Magmas: In modern pharmacy, the term magma means an aqueous preparation containing precipitated inorganic material in a fine state of subdivision. Magma Bismuthi and Magma Magnesiae (milk of magnesia) were the first official magmas in the United States Pharmacopeia IX (1916). In the nineteenth century and previous eras, this term referred to the residue obtained after expressing organic substances to extract their fluid parts, usually referred to as a marc.

Masses: Masses are plastic, semisolid pharmaceutical preparations composed of active medicinal substances combined with a diluent or filler and an excipient, capable of being shaped into pills with little or no further treatment. The three essential requirements of a pill mass are: adhesiveness (the mass must be sufficiently adhesive to retain its shape and yet be soft enough to be worked by the fingers or suitable apparatus into the desired form); firmness (the mass must possess sufficient firmness to permit the pills to retain their shape); and plasticity (a natural result of the proper degree of adhesiveness and firmness). Two masses appear to have survived the changes in modern medicine: Mass of Mercury (or Blue Mass), a cholagogic preparation last official in the National Formulary IX (1950), and Ferrous Carbonate Mass (or Vallet's Mass), a hematinic last official in the National Formulary X (1955).

Milks: Historically, the term milk has been applied generally to any liquid that possesses the outward appearance of milk, such as milk of magnesia. Legend holds that the class of beauty preparations called toilet milks may have arisen from Cleopatra and her milk baths. The modern milks are oil-in-water emulsions, named for their appearance and use as additives to baths. Moreover, actual milk was used and modified, especially with the addition of malt, as a medicinal beverage. Fermented Milk, or kumyss, is fresh cow's milk, to which sugar and yeast are added for fermentation. Fermented milk was official in the 3rd through the 5th edition of the National Formulary (1906–1926). Humanized Milk was a combination of cow's milk and fresh cream, plus Humanizing Milk Powder, which contained a small amount of Compound Pancreatic Powder and lactose, made official in National Formulary III (1906).

Mixtures: Mixtures are aqueous preparations for internal use containing insoluble, nonfatty substances. They differ from emulsions in containing no fat and from liniments in being used internally. Both mixtures and emulsions were originally grouped under Mistura in the United States Pharmacopoeia; in the 7th revision (1890), emulsions were given separate recognition. Examples include Compound Mixture of Opium and Glycyrrhiza (Brown Mixture), Carminative Mixture (Dalby's Carminitive), Mixture of Copaiba (Lafayette Mixture), Mixture of Copaiba and Opium (Chapman's Mixture), Mixture of Magnesia, Asafetida, and Opium (Dewees' Carminative), Oleobalsamic Mixture (Hoffman's Balsam), Compound Mixture of Opium and Chloroform (Squibb's Diarrhoea Mixture), and Expectorant Mixture.

Moxa: Moxa are cones of combustible matter used for cauterization by burning. Moxibustion, the burning of moxa, was an ancient method of counterirritation or cautery arising out of China. Small cones of combustible organic material (originally Artimesia moxa or common mugwort) were placed on certain areas of the skin, ignited, and allowed to burn down, leaving a blister. Moxa entered Western medicine in the seventeenth century as a treatment for gout but fell into disuse a century later along with other forms of cautery.

Mucilages: Mucilages are viscid preparations made by dissolving or suspending gummy substances in water. The term comes from the Latin *mucus*. The gummy substances, if natural, are carbohydrates obtained from the exudates of trees or shrubs. Gums have been used since ancient Egypt. Hippocratic works (ca. 400 B.C.) mention acacia. Mucilage of Acacia was official in the first United States Pharmacopoeia (1820), with Mucilage of Tragacanth becoming official in the 1st revision (1830). Both have been used as thickening agents or to prevent the creaming of emulsions.

Mulls: Mulls are ointments of high fusion points, containing the desired medicinal agent, and spread on soft muslin or mull in a manner similar to that of ordinary spread plasters. The most suitable base for preparing mulls is a mixture of suet and lard, with the occasional addition of wax or lead oleate plaster. Mulls are prepared extemporaneously by tacking unsized mull over a sheet of moistened parchment paper and spreading a melted, partially cooled ointment over the mull with a broad, flat bristle brush and smoothing the surface with two warmed elastic spatulas. When cooled, the mull is covered with waxed paper and rolled into a cylinder for dispensing. Salicylic Acid Mull, Salicylated Creosote Mull, Corrosive Mercuric Chloride Mull, and Zinc Mull were last official in the National Formulary V (1926).

Oil sugars: Mixtures of sugar with fixed and volatile oils, rendering the oils miscible to water to an certain extent, offered as a convenient mode of administering medicines to children. The National Formulary offered a general formula for 2% oil sugars through its 7th edition (1946).

Oils: Any liquid that greases; i.e., leaves, when dropped on a cloth, a stain which water does not wash out; this stain makes paper translucent. If a solid substance exhibits similar properties, it is called a fat. Oils are called volatile or fixed, according to whether this stain disappears on warming or is permanent because of the nonvolatility of the oil. A few medicated oils, such as Phenolated Oil N.F. and Phosphorated Oil N.F., were briefly official.

Oils, infused: Oleaginous preparations for external use made by macerating a drug with alcohol and ammonia water, and digesting the mixture with sesame oil at 60–70°C until the alcohol and ammonia water have evaporated. The most common, Infused Oil of Hyoscyamus, was used to make Compound Oil of Hyoscyamus. This preparation was popular in France for the treatment of earache under the name baumetranquille.

Ointments: Semisolid preparations intended for application to the skin with or without inunction. In addition to serving as vehicles for the topical application of medicinal substances, ointments may also serve as emollients for the skin and as protectives to prevent contact of the skin surface with aqueous solutions and skin irritants. Ointments are made by fusion, incorporation, or chemical reaction. Ancient humans attributed special powers to the fats of animals and humans, and their mixtures with resins, waxes, powdered herbs, and minerals represent one of the earliest dosage forms employed. A greaseless ointment, consisting of hartshorn beaten up with incense and flour and mixed with sweet ale appears in the Ebers papyrus (1500 B.C.).

The Greeks did not distinguish between liquid or semisolid preparations used for an ointment, but rather according to use or ingredients. For example, malagma were softening ointments, whereas keroma were wax ointments, the predecessor of the later cerates. Plant mucilages, balsams, and oils mixed with wax were also classified as ointments. Galen's rose water ointment (or cold cream) was an early departure (second century A.D.) from the entirely fatty type of preparation. By the thirteenth century, pharmacists distinguished between olea or oils (liquid oily ointments), emplastra or plasters (masses sticking firmly to the skin), and unguenta or ointments (semisolid smears), a concept that remained unchanged for nearly 500 years. The first United States Pharmacopoeia (1820) recognized lard as the chief ingredient of the first official ointments, which were rendered to the consistency of butter by the addition of suet, wax, or spermaceti. By the middle of the nineteenth century, however, natural ointment bases began to be replaced by artificial bases, introduced with special regard to the purposes they were to serve.

Schacht introduced Glycerite of Starch in 1858, a translucent jelly prepared by heating glycerin and starch in certain proportions to a certain temperature, and W.A. Miller introduced Petrolatum in 1873 as "Cosmolin and Paraffin Ointment," both of which preparations were adopted by the United States Pharmacopeia in 1880. In 1885, the pharmacologist Oscar Liebreich rediscovered the therapeutic value of wool fat, the oesypus of the ancient Greeks, which he called lanolin; it was recognized by the Pharmacopeia in 1893. Later, the Russian chemist Lifschuetz discovered that the emulsification power of lanolin depended upon the free alcohols he had isolated as a group (1895–1898).

In 1907, the dermatologist Paul G. Unna introduced eucerin, a new ointment base consisting of 1 part of Lifschuetz's alcohols, 20 parts of paraffin ointment, and 20 parts of water, the forerunner of the American "Aquaphor." Between 1920 and 1944, hydrogenated oils, sulfated and sulfonated hydrogenated oils, as well as stearic acid, sodium stearate, self-emulsifying glyceryl stearate mixtures, polymers of glycols (such as polyethylene glycol 4000), and esters of these glycols (such as polyethylene glycol monostearate) became important, followed by such modern bases as Plastibase, attapulgite, Veegum, guar gum, Carbopol, and the silicones, which were introduced between 1945 and 1959.

Ointments, ophthalmic: Sterile ointments designed for application to the eyelids. Petrolatum, petrolatum- mineral oil, and petrolatum-anhydrous lanolin bases are often used in ophthalmic ointments because of their low irritating potential. Finely powdered, sterile active ingredients are aseptically incorporated into a sterile base, using sterile utensils, and dispensed in sterile ophthalmic-tipped tubes to reduce the possibility of contamination.

Oleates: Usually liquid preparations made by dissolving alkaloids in oleic acid. Oleate of Mercury, however, is an ointment-like product of mercuric oxide in oleic acid. Oleic acid was named by the pharmacist Chevreul after olives.

Oleoresins: Extracts of plant drugs prepared by percolation using a selective solvent (usually ether or acetone), followed by complete removal of the solvent by evaporation. The term first arose in the 1820s (Buchner and Peschier), and the oleoresins first appeared in the United States Pharmacopeia of 1860 through the efforts of William Procter, Jr.

Oleovitamins: Preparations using fish liver oil, fish liver oil diluted with an edible vegetable oil, or a solution of vitamin concentrate in fish liver oil or in an edible vegetable oil. Oleovitamins were created during World War II to fill a therapeutic gap created by the interruption in cod liver oil supplies. The class became official in the second supplement to the United States Pharmacopeia XI (1942) as a source of vitamins A and D.

Paints, medicinal: Liquid medicinal preparations possessing antiseptic, caustic, soothing, or stimulating properties, usually applied by means of a brush. Paints intended to remain in contact with a specified surface are usually prepared with collodion, glycerin, glycerin and water, egg albumin in alcohol, or gutta percha. Paints intended to be absorbed are prepared with oleic acid or fatty oils. Caustic substances are usually applied dissolved in distilled water, alcohol, or ethereal vehicles, whereas resinous substances, such as benzoin, storax, tolu balsam, or sandarac dissolved in ether, are employed as bases for medicated varnishes, and used for application to the skin and raw mucous surfaces.

Papers, medicated: Preparations intended primarily for external application, either by saturating paper with medicinal substances or by applying the latter to the surface of the paper by the addition of some adhesive liquid. Potassium Nitrate Paper remained official in the National Formulary through its 5th edition (1926), but the most widely used paper is Mustard Paper, commonly called mustard plaster, a mixture of powdered black mustard and a solution of rubber spread on paper, cotton cloth, or other fabric, which remained official in the National Formulary through its eleventh edition (1960). Mustard poultices or cataplasms were formerly called *sinapisms*, after the botanical name for black mustard, Sinapis nigra, and were described in the United States Dispensatory (1836) as Powerfully rubefacient ... usually becoming insupportably painful in less than an hour.... As a general rule, the poultice should be removed when the patient complains much of the pain; and in cases of insensibility should not, unless greatly diluted, be allowed to remain longer than one, or at most two hours, as violent inflammation, followed by obstinate ulceration, is apt to occur. Home-made mustard plasters (equal parts of mustard and flour, moistened with tepid water to form a paste and applied to the skin in a muslin bag) still play a role in folk medicine.

Papers, waxed: Parchment-like paper treated with melted wax or paraffin used largely as an economical substitute for more expensive protective dressings.

Parvules: Small sugar-coated pills, of 0.06 g or less, sometimes incorrectly called *granules*.

Pastes: Ointment-like mixtures of starch, dextrin, zinc oxide, sulfur, calcium carbonate, or other medicinal substances made into a smooth paste with glycerin, soft soap, petrolatum, lard, or other fats, and medicated with antiseptic or astringent agents, designed for external use. Early pastes, such as Pasta Glycyrrhizae and Pasta Althaeae, were internal preparations, most of which were of gum-like consistency. The modern pastes were introduced by the noted dermatologists Paul G. Unna and Oskar Lassar around 1900. Dermatologic Pastes normally contain a higher proportion of powdered material than that included in ointments and are less greasy but more absorptive than other preparations for external application. The British Pharmaceutical Codex groups Witch Hazel Cream, Vanishing Cream, Tannic Acid Jelly, Catheter Lubricant, and a wide assortment of "medicated preparations for external application, employed principally as antiseptic, caustic, cooling, protective or soothing dressings" under the title "Pastes." Pastes entered the National Formulary III in 1906.

Pastilles: A form of lozenge, particularly those which are chocolate flavored; also, combustible cones of aromatic drugs used for fumigation. The term came into English usage around 1650 from the French pastille, which was derived from the Latin pastillus, meaning "little loaf."

Pearls: Round or oval capsules made by enclosing liquids, solids, or tablets in a shell of glycerogelatin material. Pearls are less elastic than soft capsules and contain no air space, the glycerogelatin shell being completely filled with the medicinal substance. Pharmacists made pearls extemporaneously by laying a softened sheet of glycerogelatin over a warmed molding plate containing a specified number of semicircular (or other shaped) depressions. The sheet was covered with a measured quantity of medicinal liquid, and the liquid with a second sheet of glycerogelatin to exclude air.

The whole assembly was covered with a matching molding plate, and compressed with a mechanical press to form the pearls, an exacting and time-consuming process requiring special apparatus. This process has been superseded in industry by a continuous automated process in which a liquid is injected between two ribbons of gelatin while passing between revolving dies.

Pellets: Small spheres of sucrose saturated with an alcoholic tincture, primarily used in homeopathic medicine. Pellets are made in different sizes, designated according to the diameter of ten pellets measured in millimeters. Remington's Practice of Pharmacy (1926) states that pellets "should be made of the purest materials, should be perfectly white and odorless and able to withstand all the tests prescribed for sucrose or cane sugar".

Pencils, medicated: Cylinders used in dermatologic practice to apply medicinal agents directly to the skin. The medicinal agent is incorporated into a paste consisting of starch, dextrin, tragacanth, and sucrose with sufficient water to form a plastic mass, which is rolled into cylinders about 5 mm in diameter, cut into sections about 5 cm long, dried on parchment paper at room temperature, and wrapped in tinfoil. Medicated pencils intended as a caustic application, such as sticks of silver nitrate, are sometimes referred to as escharotica. Salicylic Acid Pencils, the last official medicated pencils, appeared in the National Formulary V (1926); also known as Antiseptic, Astringent, Caustic, Salve, or Styptic Pencils.

Pessaries: Medicated vaginal suppositories, globular or oviform in shape, weighing between 4 g (if made from oil of theobroma or cocoa butter) and 10 g (if made from glycerated gelatin). The term derives from a Greek word describing the small stones used for playing the game of draughts.

Petroxolins: Fluid preparations for external use with a base of liquid petrolatum and ammonium oleate. The preparations became official in the 6th edition of the National Formulary (1936).

Pills: Small, solid masses of a globular, ovoid, or lenticular shape intended for oral administration. Pills are prepared by incorporating medicinal agents with other materials to form a cohesive, plastic mass, which is divided into the requisite number of portions, each of which is formed into the desired shape. Pills usually range in weight from 0.10 to 0.30 g. Exceptionally large pills of 0.60 g or more are referred to as boluses; very small sugar-coated pills of 0.06 g or less are known as parvules or granules; pellets, globules, or orbicules are small spheres of sugar saturated with an alcoholic tincture, largely used in homeopathic medicine. When the pill came into use in ancient Mesopotamia and Egypt, it offered for the first time a definite dose corresponding with a desired therapeutic action.

The Greeks named the little balls of medicine *katapotia* ("something to be swallowed"), later Latinized to *catapotium*; by the first century A.D., the term *pilula* came into use In Rome. Pills persisted as a major dosage form for a remarkably long period of time. For example, over half of the prescriptions dispensed at a Charlestown, Massachusetts, pharmacy during the years 1872–

1875 were for pills. Moreover, certain combinations of drugs persisted for thousands of years in pill form. The Pills of Rufus, for example, were originally a hiera (bitter powder) made into pill form by the Arabs and popularized by Avicenna (980–1033); a modern version, Pills of Aloe and Mastic, were last official in the 8th revision of the United States Pharmacopeia (1905).

Plasters: Substances intended for external application, made of such materials and of such consistency as to adhere to the skin. Adhesive plasters afford protection and mechanical support, whereas medicated plasters furnish an occlusive and macerating action, bringing the medication into close contact with the skin; when spread on perforated cloth, the product is called a porous plaster. Plasters are among the most ancient of all pharmaceuticals. Primitive humans may have used plasters of mud and leaves to help heal wounds or relieve pain. The Ebers papyrus (1500 B.C.) describes several plasters and poultices for treating burns, and the Greeks assigned a special place within their temples where plasters were prepared. Indeed, the word plaster is derived from the Greek emplastron meaning "to smear on" or "to mold on."

The famous diachylon plaster, made from oil, litharge, and certain plant juices was compiled by Menecrates, physician of the Emperor Tiberius about 39 A.D. and passed on in verse form to Claudius Galen (131–201), who developed a number of practical formulas for plasters and procedures for their preparation which endured for centuries. A modern version of diachylon plaster survived as Lead Oleate Plaster (or Lead Plaster) in the United States Pharmacopeia and later in the National Formulary through its 8th edition (1946). Originally, plasters were spread directly on the affected part of the body; by the sixteenth century, plaster material was being spread on linen or leather or dipped. In 1514, Giovanni da Vigo popularized spread plasters called sparadraps or spasmadraps; in 1691, William Salmon described dipped cerecloths in his Pharmacopoeia Londonensis.

Plasters were spread with the aid of an offset spatula or plaster iron heated by means of a spirit lamp, an arduous and time- consuming process which exhausted the patience of pharmacists who struggled trying to keep the refractory plaster masses warm and malleable. Although Elisha Perkins of Baltimore is credited with obtaining the first American patent for a manufactured plaster (1830), John C. De La Cour of Camden, New Jersey, is generally credited as among the earliest producers of machine-spread adhesive plasters (1836), one of the first commercially prepared dosage forms manufactured in the United States. In 1852, an English apothecary named Mather patented a method of spreading plasters on leather by the use of heated rollers, producing a thinner and more uniform product, leading Edward Parrish to observe in his classic text, American Pharmacy (1856), that "the spreading of plasters which was formerly an important part of the business of the apothecary has now... been monopolized by manufacturers who bring machinery to their aid." The only plaster which still finds significant use today is Salicylic Acid Plaster, United States Pharmacopeia–the common corn plaster (also see Plasters, adhesive, and Plasters, porous).

Plasters, adhesive: A mixture of rubber, resins, and waxes, with a filler of absorbent powder, such as zinc oxide, orris root, or starch, mechanically mixed and spread on cotton cloth. Early adhesive plasters were composed of resin and litharge or diachylon spread on calico, muslin, or linen. In 1843, B.C. Rowland of Liverpool, England, reported on an "India Rubber Court Plaster" made by dissolving India rubber in naphtha or turpentine and spreading the liquefied rubber on silk or satin. Two years later, Horace Day and William Shecut of New York City patented a combination of India rubber and gums as a plaster mass. In 1852, Benjamin Nickels of Surrey, England, patented an "*elastic plaster*" combining adhesive material on an elastic fiber.

In 1863, Joshua Melvin of Lowell, Massachusetts, patented the manufacture of adhesive plaster in roll or cylindrical form. By the early twentieth century, two types of rubber-based adhesive plasters emerged: surgeons' adhesive plaster, a plain yellow-colored mass, and zinc oxide adhesive plaster, a white mass containing zinc oxide. Modern adhesive plasters, consisting of vinyl resin, plasticizers, and other chemical additives have an excellent ability to remain adhered under severe conditions of moisture and heat, and rarely cause skin irritation. Adhesive plaster remained official through the 19th revision of the United States Pharmacopeia (1975).

Plasters, Blister: Plasters designed to produce inflammation, blisters, or issues (running sores). Many blistering agents were used to prepare blister plasters, but the potent and powerful cantharides (or Spanish flies), was the most widely applied. It was used internally by the ancients, who were well aware that it produced hematuria even when applied externally. Aretaeus introduced an external preparation of cantharides as a blistering agent in the second century to the shaved head to relieve headache. This treatment was also used for epilepsy and vertigo, and persisted through the centuries, but only after the patient had been given milk to drink for 3 days to protect the bladder from injury. "In many constitutions the strangury will ensue, especially where the discharge of serous juice is too great," an English textbook on materia medica advised in 1730. "But, however it be, such applications are necessary when a patient proves delirious, as frequently happens in high fevers."

A century later, the United States Dispensatory (1836), in discussing the use of Cerate of Cantharides, "the common blistering plaster of the shops," remarked that "When the full operation of the flies is desirable, and the object is to produce a permanent effect, the application should be continued for twelve hours.... It should then be removed, and followed by a bread and milk poultice, or some other emollient dressing, under which the cuticle rises, and a full blister is usually produced. By this management the patient will escape strangury, and the blister will very quickly heal after the discharge of the serum." Elisha Perkins of Baltimore obtained a patent in 1830 for an apparatus to prepare blister plasters. George W. Carpenter of Philadelphia sold and recommended Perkin' s blister cloth in 1831 as "a very convenient article for the country physician, being ready spread for immediate use." Ceratum Cant ha rides was official in the United States Pharmacopeia and, later, in the National Formulary until 1950.

Plasters, porous: Adhesive plasters spread on perforated cloth. Sir William Butts, royal physician to Henry VIII, devised a "spasmadrap or dypped plaster" which was to be poked "full of smalle hoolys."In 1845, Horace Day and William Shecut of New York City patented a rubber-based "porous plaster" rendered full of minute holes to "allow the free escape of the perspiration." In 1854, Somerville Scott Alison of London patented a porous or "perforated Lambskin... prepared according to the process called chamois curing." The porosity is considered a mechanical advantage in that it prevents the plaster from slipping from the point of application, each opening serving as a stop. Porous plasters are also far more comfortable than the nonporous variety, which they have superseded (also see Plasters and Plasters, adhesive).

Politzer plugs: Greased pellets of cotton about the size of a coriander seed with a thread attached, for insertion into the ear as a protective. The pellets were named after Adam Politzer (1835-1920), an Austrian otologist.

Poultices: Originally spelled "pultes" in sixteenth- century England, the term came from the Latin puls or pultes meaning "a pottage of meal".

Powders: Intimate mixtures of dry, powdered medicinal substances reduced to a fine powder by the processes of comminution and trituration. Bulk powders are divided into two categories: simple, consisting of one substance, and compound, consisting of two or more powders mixed together. One of the most ancient compound powders was *hiera picra* ("sacred bitters"), a mixture of aloe and canella introduced about 500 B.C. as a laxative, the prototype of a large number of bitter powders containing aloe as the principal ingredient and bearing the general title Hiera. This powder was listed in various pharmacopoeias and was last recognized in the 4th edition of the National Formulary (official until 1926). Powders were designed originally as a convenient mode of administrating hard vegetable drugs such as roots, barks, and woods; powders were also found to be convenient for the dispensing of insoluble chemical compounds such as calomel, bismuth salts, mercury, and chalk.

Famous compound powders of the past include: Compound Powder of Glycyrrhiza, a variant of Compound Senna Powder recognized by the first London Pharmacopoeia (1618); Powder of Ipecac and Opium, or Dover's Powders, introduced by the English physician Thomas Dover as Pulvis Diaphoreticus in the early eighteenth century; Aromatic Powder of Chalk, a simplified version of a complex confection devised by Sir Walter Raleigh during his imprisonment and introduced into the London Pharmacopoeia of 1721 as Confectio Raleighana; and Antimonial Powder, patented in 1747 as Dr. James's Fever Powder. The United States Dispensatory (1836) noted that "the form of powder is convenient for the exhibition of substances which are not given in very large doses, are not very disagreeable to the taste, have no corrosive property, and do not deliquesce rapidly on exposure." Today, drugs not available in capsule or tablet form can still be conveniently administered in powdered form by placing them on the back of the tongue and swallowing them with water. Flavored powders are particularly useful for children who might have difficulty swallowing a tablet or capsule.

Powders, divided: Intimate mixtures of dry, powdered medicinal substances intended for oral administration, divided into single doses, each of which is folded into a small sheet of glassine paper. One of the oldest of dosage forms, divided powders have largely been replaced by capsules or tablets. Nevertheless, the preparation of powders permits the drugs to be reduced to a very fine state of subdivision, a physical condition which frequently intensifies their therapeutic activity, a factor in increasing the efficacy of homeopathic triturations, calomel and sodium bicarbonate mixtures, and Dover's Powder (Powder of Ipecac and Opium). Divided powders also furnish a convenient means for administering drugs that are not excessively bitter, nauseous, or otherwise offensive to the taste. One of the most durable divided powders is Seidlitz Powders, a saline cathartic originated and patented by Thomas Savory in 1815. Savory claimed that the powders owed their value to the mineral properties of the Seidlitz spring in Germany, which contains magnesium sulfate. The powders consist of sodium bicarbonate (wrapped in blue paper) and tartaric acid and potassium and sodium tartrate (wrapped in white paper), each of which are dissolved separately in water and then mixed. The formula was exposed in a book of recipes for patent medicines published by the Philadelphia College of Pharmacy in 1824, and remained official as Compound Effervescent Powders in the United States Pharmacopeia, and later in the National Formulary through its 12th edition (1965); also known as *Powder papers.*

Precipitates: Drugs prepared by separating particles from a previously clear liquid by physical or chemical means. Precipitation usually occurs when a hot saturated solution of an amorphous substance is allowed to cool or when a liquid in which the dissolved substance is insoluble is added to its solution. Pharmacists formerly used the process of precipitation as a convenient method of obtaining solid substances in fine particles (precipitated calcium carbonate), to purify solids (precipitated calcium phosphate), or to prepare mercury salts. White precipitate (ammoniated mercury) was first described by Beguin in 1632, a soluble double chloride of mercury and ammonium known to the alchemists as sal alembroth and sal sapientiae, respectively. Red precipitate (red mercuric oxide) was known to alchemists as hydra gyrum precipita tum per se or "*precipitate per se*;" yellow precipitate is a synonym for yellow mercuric oxide.

Resins: Solid preparations consisting chiefly of the resinous principles from vegetable bodies. The officially prepared resins differ from alcoholic extracts in that the latter contain all of the alcohol-soluble principles in the drugs, whereas the resins contain only the alcohol-soluble principles that are insoluble in water. The term probably arose from the Greek rheos, "to flow," referring perhaps to the flow of pine resin commonly observed.

Rubificients: Local remedies which produce redness and inflammation of the skin. The word comes from two Latin words: ruber, meaning "red," and facio, meaning "to make."

Saccharures: Preparations made by saturating sucrose with a tincture, drying it, and grinding the mixture to a powder.

Salts: Compounds formed by the union of acids and bases, by the action of alkalies upon metals, or by the direct union of elements. The term is often incorporated in the common name of salts used as pharmaceuticals: bitter salts, epsom salt, or Seidlitz salt (magnesium sulfate), preparing salt (sodium stannate), Preston's salts (ammonium chloride), Rochelle salt or Seignette's salt (potassium and ammonium tartrate), salt of Mars (ferrous sulfate), salt of Saturn (lead acetate), salt of tartar (potassium carbonate), salt of tin (stannous chloride), salt of wisdom (mercury bichloride and ammonium chloride), sore-throat salt (fused potassium nitrate), vinegar salts (calcium acetate), and vomiting salt (zinc sulfate). The term is also applied to some acids, such as salt of lemon or sour salt (citric acid), salt of sorrel (oxalic acid), and spirit of salt (muriatic acid).

Salts, artificial: A mixture of the more important chemical salts naturally present in several of the well- known mineral springs of Europe. "These are properly labeled artificial," notes Remington's Practice of Pharmacy (1926), "and if used to prepare effervescent salts or mineral waters they should be sold only as an imitation of the genuine."

Salts, effervescent: Granular effervescent salts were formerly made by mixing dry powders with dry tartaric acid and sodium bicarbonate and moistening the mixture with strong alcohol. The pasty mass was passed through a sieve, and the granules dried quickly in a hot room, sifted, and filled into bottles, which were hermetically sealed to prevent the access of moisture. This method was greatly improved upon by mixing the powders in a flat enameled dish and heating in an oven to about 100°C or by heating the mixture in a deep jacketed kettle or in a pill- coating pan, heated, as it revolves, by a gas flame. When the mixture becomes moist, it is manipulated with a wooden spatula to make it uniform in consistency, and rubbed through a coarse tinned iron sieve; the granules obtained are dried slowly at a low heat in an oven.

Salts, smelling: Ammonia-based preparations used as a restorative in "*hysterical syncope*" (fainting). Dry smelling salts (or vinaigrettes) are composed of ammonium chloride and potassium carbonate, perfumed with lavender; liquid smelling salts are composed of ammonium carbonate dissolved in stronger ammonia water and alcohol, and perfumed with oils; solidified smelling salts are similar preparations solidified with stearic acid.

Salves: This term probably arose from the Anglo-Saxon sealf or the German salbe, both meaning "ointment".

Scutum: Abbreviation for scutum stomachicum, a large plaster, applied to the breast or stomach.

Serums: Serum therapy came into prominence in the 1890s when Emil von Behring (1854–1917) extended Pasteur's theory of attenuated viruses. Behring demonstrated that the serum of animals immunized against attenuated diphtheria toxins can be used as a preventive or therapeutic inoculation against diphtheria in other animals through a specific neutralization of the toxin of the disease. In 1894, Behring began to produce his new antitoxic serum on a grand scale; it soon became recognized as the specific treatment for diphtheria. Antisera act by combining with the toxin in the blood of the patient, rendering it inert.

Scarlet fever, tetanus, erysipelas, botulism, and gas gangrene have been successfully treated by antitoxic serums prepared in this manner. Antibacterial serum is produced by injection of an animal with successive doses of bacteria. The immune substances thus formed act by enhancing phagocytosis, destroying the bacteria. Pneumonia, streptococcic infection, and spinal meningitis have been aided by the use of this type of serum. Mixed serum contains both antitoxic and antibacterial immune bodies; the serum used to treat scarlet fever is of this type.

Shampoos: A wash for the hair or soap for hair washing. The term comes from the Hindustani word tshampa, meaning "to squeeze" or "to press," probably associated with hot oriental baths.

Silk, oiled: A thin, very soft, and pliable protective dressing made of fine silk, coated with a flexible linseed-oil varnish. Oiled Silk is available in semitransparent or opaque form, made by the addition of talc or starch in the final coating. Oiled Muslin or Oiled Cambric is similar to oiled silk, except that the basic fabric is of glazed cotton; it is thicker and heavier and consequently less pliable than oiled silk.

Soaps: A class of chemical substances which are metal salts of fatty acids. Pliny (first century A.D) records that the ancient Romans learned the preparation of soap from Nordic tribes, who used a pomade prepared from goat fat and the calcined ashes of beech- wood. Sapo, the Latin word for soap, is derived from the Nordic sepe. The chemistry of soaps was elucidated in the early nineteenth century by French chemist M.E. Chevreul (1786–1889). Soaps may be divided into two classes: soluble soaps (or detergent or cleansing soaps), which are compounds of fatty acids with alkali metals, particularly sodium and potassium; and insoluble soaps, which are compounds of fatty acids and metals of any other group, such as Lead Oleate Plaster or Lime Liniment, a calcium soap of lin-seed oil. Soluble soaps include Hard Soap (or Castile Soap), prepared from olive oil and sodium hydroxide, official through the 11th edition of the National Formulary (1960), and Soft Soap (or Green Soap), prepared from linseed oil, glycerin, and dekanormal solutions of sodium and potassium hydroxide; the latter continues to hold official status.

Solutio: Dental preparations official only in the 5th edition of the National Formulary (1926), consisting of a solution of a resinous material dissolved in chloroform.

Solutions: Liquid preparations that contain one or more substances dissolved in a solvent and, by reason of their ingredients or method of preparation, do not fall into some other category of preparation. From the seventeenth century on, the term liquor denoted liquid solutions. By the early twentieth century, *liquor* usually referred to aqueous solutions of nonvolatile substances. For example, Solution of Magnesium Citrate, United States Pharmacopeia, had the Latin title Liquor Magnesii Citratis. Beginning with the United States Pharmacopeia XVI and the National Formulary XI (1960), the term liquor was dropped and solutions were listed by their active ingredient or ingredients.

Solutions, irrigating: A preparation to be applied continuously by means of a special device for the purpose, as in the treatment of wounds with Dakin's solution.

Solutions, nasal: Solutions of drugs for instilling in the nose rather than spraying are generally a modern development. The first nasal solutions were formulated with menthol and thymol dissolved in light mineral oil. Later isotonic aqueous solutions were designed as drops.

Solutions, ophthalmic: Originally called collyria, ophthalmic solutions arose from the eye washes of the ancient world. Early preparations were not the sterile, buffered solutions of today. The term collyrium comes from the Greek kollurion, which Hippocrates used to designate fatty suppositories for gynecological purposes. They were formed into sticks, from which a paste was made with a liquid. Eventually, more liquid was added and the paste thinned to a lotion. This lotion was used as an eye wash, and term came to be used exclusively for this type of preparation. Remington's formula for a collyrium of 260mg (4 grains) of sodium borate in 30mL (1 oz) of camphor water was a standard from 1886 into the mid-twentieth century.

Spasmadraps: Pieces of linen or other cloth dipped in or spread with a medicinal plaster, popularized in 1514 by Giovanni da Vigo (1460–1525), physician to Pope Julius II. From the Latin spasma meaning "healing powder" and the French drap meaning "cloth." Also known as sparadraps or cerecloths.

Spirits: Solutions of volatile substances (usually volatile oils) in alcohol. Some editions of the National Formulary stated that "*spirits of volatile oils*" contained 6.5% of the volatile oil, but that figure was later rejected as too low. Although used internally, several spirits were used medicinally by inhalation or as flavorings.

Sprays: Medicated liquids prepared for dispersal by atomizers or nebulizers, usually on external surface or mucous membranes of the respiratory tract. In the United States Pharmacopoeia, sprays were called inhalatio; the National Formulary referred to them as nebulae. Sprays of the early twentieth century were formulated with aromatics dissolved in light mineral oil. As injuries from inhaled oils became apparent, especially among children, these sprays were displaced by buffered aqueous solutions.

Stilus dissolubilis: Dissolving pencils. Pencils containing a caustic or an astringent.

Suppositories: Conical or ovoid medicated solids intended for insertion into one of the several orifices of the body, excluding the mouth. Suppository use has been known as early as 2600 B.C., and was recom-mended in the works of Hippocrates (ca. 400 B.C.). The term derives from the Latin suppositus, meaning "to place under." Premodern suppositories were made by hand using soap or other semisolid fatty substances as the main vehicles. They were not commonly used until the seventeenth and eighteenth centuries and did not become popular until the mid-nineteenth century and the advent of cocoa butter as vehicle.

In 1766, Antoine Baumé described a suppository mold which used liquefied cocoa butter. This technique was popularized in America by Alfred B. Taylor about 1852 using paper cones as molds. Metal molds were introduced about 1860, although many pharmacists continued to form suppositories by hand without heat. Cold compression of cocoa butter was made possible through the introduction of metal suppository presses about 1868, although the first popular compression mold was not introduced until 1879. After 1870, mixtures of glycerin and gelatin came to be used as vehicles for suppositories, beginning a quest for the perfect vehicle that continues to the present.

Suspensions: Heterogeneous systems containing coarsely dispersed material that settles. A wide variety of pharmaceutical preparations have been used as suspensions, for example, White Lotion, Magma of Bismuth, and Compound Mixture of Opium and Glycyrrhiza (Brown Mixture). In addition, several official ointments are suspensions of solids in a semisolid base. A large number of suspensions are categorized as mixtures in the United States Pharmacopeia and the National Formulary.

Swabs: Ampuls containing Iodine Tincture, United States Pharmacopeia, covered with gauze or other absorbent material, and used for first-aid treatment. Iodine Swabs, later called Iodine Ampuls, were official in the National Formulary through its 13th edition (1970). When iodine tincture is required for first aid, the tip of the ampul is broken and the gauze absorbs the iodine and provides a means of applying it directly to the wound. Some Iodine Ampuls are in the form of fine capillary tubes, which are broken when needed and the tincture applied directly.

Syrups: A nearly saturated aqueous solution of sugar (usually sucrose) with or without medicinal or flavoring ingredients. Syrups are usually divided into flavored, containing a fruit or aromatic substance for a pleasant taste, and medicated, containing a drug. Simple Syrup, United States Pharmacopeia was commonly used in the preparation of pill masses and other mixtures.

Tablets: Dosage forms prepared by molding or compressing medicinal substances in dies. Tablets vary widely in shape, the most common form being discoid, and range from 0.06 to 0.60 g in weight. Jean de Renou applied the Latin word tabella to a special type of troche in 1608; Burroughs Wellcome & Company coined the term "tablet" in 1878 to refer to its brand of compressed pills; the term is derived from the French tablette, meaning "shelf" and the Latin tabula, meaning "board." In 1843, the English apothecary William Brockedon patented a device for compressing medicinal agents commonly employed in pills and lozenges without the use of liquid adhesive agents; the resulting product was known as compressed pills.

The Philadelphia druggist Jacob Dunton invented a similar device in 1864, marketing his own compressed pills in 1869; Joseph Remington devised a similar machine in 1875 to allow the retail druggist to "manufacture his own medication called for on prescription."Each of these devices consisted of a compression cylinder and lower die (to hold the medicinal substance) as well as an upper die which was struck with a mallet to compress the material. More reliable compression was achieved by using the screw devices invented by Germany's Professor Rosenthal (1874) and perfected by Austria's Carl Engler (1907). Another advancement was the lever device introduced by Philadelphia's Bennett L. Smedley (1879). The first rotary tablet machine was developed in 1872 by Henry Bower, an employee of the Philadelphia drug manufacturer John Wyeth; two years later, Joseph A. McFerran received a patent for the first fully automatic tablet machine.

Tablets, hypodermic: Molded tablet triturates intended to be dissolved in water to make a solution to be injected parenterally. The usual weight of hypodermic tablets is about 0.03 g, which distinguishes them from ordinary tablet triturates that weigh about 0.06g. Formerly prepared extemporaneously by pharmacists, modern compressed hypodermic tablets are not intended to be sterile, although they are manufactured under strict conditions as a precaution against contamination.

Tablets, poison: Tablets of mercury bichloride in an angular, not discoid shape, blue in color, each having the word "POISON" and the skull-and-crossbones design distinctly stamped upon it. A unique one-product classification, poison tablets (or Toxitabellae) first became official in the United States Pharmacopoeia IX (1916), and two strengths remained official in the National Formulary through its 10th edition (1950); the larger tablets remained official through the 12th edition (1965). Diluted in a solution of 1:1000 concentration, mercury bichloride is an antiseptic used chiefly for the disinfection of inanimate objects and the unabraded skin.

Tablets, solution: Molded or compressed tablets containing large amounts of potent substances not intended for administration, but rather as a convenience in dispensing; also known as Dispensing Tablets. To lessen the risk of their being dispensed by mistake for other tablets, dispensing tablets are always of angular rather than discoid shape. They are usually scored to facilitate division into more or less accurate fractions. Also tablets to be dissolved in water for external use.

Teas: Coarsely powdered mixtures of dried herbs intended for medicinal teas or poultices; also known as Species. The National Formulary recognized an Emollient Species, used as a cataplasm; a Laxative Species (St. Germain Tea); and a Pectoral Species (Breast Tea) for a "catarrhal condition of the respiratory tract" through its 5th edition (1926). Similar teas from home-grown herbs persist as common household remedies. True tea from China was introduced to England by Christopher Borough in 1379. The English word probably comes from the Dutch thee.

Tinctures: Alcoholic or hydroalcoholic solutions of drugs, usually of plant origins. The term comes from the Latin tingere, "to dye or soak in color." Tinctures, as alcoholic solutions, entered medical practice in the thirteenth century through the efforts of Raymond Lull and Arnald of Villanova. Paracelsus (1493–1541) was a strong advocate for tinctures; inasmuch as he was controversial, his advocacy probably discouraged their incorporation into compendia until the 1700s. In the 1800s, wine remained the prime hydroalcoholic vehicle in the United States. After the turn of the century, however, tinctures (and elixirs) displaced those wines because of their wide variation in strength. Moreover, prohibition impeded their widespread use.

As galenicals declined throughout the twentieth century, tinctures lingered on as an official class, mainly as flavorings, for example, Tincture of Orange Peel. Homeopathic Tinctures are generally prepared by long maceration of freshly dried succulent plants or their parts in alcohol, the completed tincture being made to represent one part of the dry crude material in each ten parts of the completed preparation.

Triturates, tablet: Small, disk-like masses of medicinal powders prepared by forcing a moistened tablet mass into a die by manual pressure and allowing them to dry and harden. The basis of tablet triturates is usually a mixture of lactose and sucrose in a 5: 1 ratio, moistened with a volatile liquid such as alcohol. Tablet triturates were introduced in New York in 1878 by Dr. Robert W. Fuller as a palatable and convenient means of administering potent drugs by mouth. Fuller's original triturates consisted of triturations of metallic, mineral, and vegetable matter, mixed into a paste with alcohol or water, and molded into the desired shape. In 1882, Fuller described a perforated, hard rubber tablet triturate mold with a corresponding pegged plate, which was practically identical to those available today. Tablet triturates served the purposes of homeopathic physicians well and undoubtedly helped to further the use of homeopathic doses. Twentieth-century pharmacists also prepared hypodermic tablets (used for preparing hypodermic injections) utilizing the technique. Today, tablet triturates and hypodermic tablets are formed by compression and are termed molded tablets.

Triturations: Dilutions of potent powdered drugs prepared by intimately mixing them with a suitable diluent, usually lactose, in a definite proportion by weight, usually 10%, used as a dispensing aid. Such poisonous substances as strychnine sulfate, arsenic, mercury bichloride, and atropine are much more accurately dispensed using this technique, the pharmacist weighing a multiple of the prescribed drug in a triturated form. Homeopathic triturations were formerly prepared by triturating one part of a drug into 99 parts of lactose over a period of at least one hour; homeopathic tincture triturations were prepared by mixing 10 mL of a homeopathic "strong tincture" with 10g of lactose and triturating the mixture gently until dry. Although such powdered triturations were being replaced by commercially prepared tablet triturates by the mid-1920s, a general formula for triturations, specifying geometric dilution, remained official in the United States Pharmacopeia through its 14th revision (1950).

Troches: Solid dosage forms in the form of small disks, cylinders, or tablets, intended to be placed in the mouth and allowed to dissolve or disintegrate slowly. The term is derived from the Greek trochos, meaning "round" or "circular." They were subsequently called pastils in French and lozenges in English. One of the earliest troches (500 B.C.) was terra sigillata, or sealed earth, a product composed of clay from the island of Lemnos and goat's blood, rolled and cut into disks and impressed with a seal; by the Middle Ages, a variety of troche presses were employed. In 1856, Edward Parrish described an apparatus for rolling and cutting troches consisting of a rolling-board, wooden roller, and cutting punch. During the next two decades, F.L. Slocum (1879), F.E.

Harrison (1880), Wallace Procter (1894), and nearly a dozen others patented similar machines. The 4th edition of the National Formulary (1916) featured nine formulas for troches; by the mid-1930s, troches were being replaced by tablets: the 6th edition (1936) featured only Troches of Elm. Modern troches consist of powdered drugs bound with sugar and tragacanth or incorporated into a hard candy or glycerogelatin base.

Vapors: Steam, plain or medicated, generated by the use of steam or boiling water or by the use of a specially constructed apparatus.

Vesicatories: Local remedies, the application of which produces a serous discharge beneath the skin, forming a blister. Also known as *epispastics*.

Vinegars: Infusions or solutions of drugs in vinegar or acetic acid; one of the oldest methods of drug preparation. Ancients knew that vinegar was often a better solvent than water and a preservative as well. The first preparation listed in the first United States Pharmacopeia (1820) was Acetum Opii, or Vinegar of Opium. The English word derives from the French vin, "wine," and aigre, "sour."

Wafer envelopes: Preformed envelopes of rice flour used to administer bitter or nauseating drugs. Developed by Johann Schmidt as *saccelli amylacei*, wafer envelopes marked an improvement of convenience over wafer sheets. Pharmacists often furnished empty wafer envelopes to their patients, who transferred doses to them from prepared powder papers.

Wafers: Flat sheets of rice flour used to administer nauseating drugs. When dry, wafer sheets are nonadhesive, stiff, somewhat brittle, and slightly thicker than ordinary cardboard. Powders are administered by floating a piece of wafer sheet upon water until it becomes thoroughly softened, passing a tablespoon underneath and lifting it out, and depositing the powder in the center and folding over the corners to thoroughly enclose the powder. If water is poured into the spoon, the concealed powder can be swallowed without any disagreeable taste being perceived. Wafer sheets are made by pouring a mixture of rice flour and water upon hot greased plates or rolling it between two hot, polished, revolving cylinders.

Washes: Aqueous preparations designed to cleanse specific parts of the body. Examples include enemas, eye drops, mouth washes, and nasal washes. General washes made official include Alkaline Aromatic Solution N.F. and Antiseptic Solution N.F.

Washes, mouth: Hydroalcoholic solutions of soap flavored with essential oils for cleansing the oral cavity; they became first official in the National Formulary V (1926).

Waters, aromatic: Saturated solutions usually of volatile oils or similar substances in distilled water. Aromatic waters such as rose water were used in antiquity. Distilled waters containing volatile oils reached their therapeutic peak in the early sixteenth century. Although their therapeutic use declined in modern times, they continued to be used as flavorings. Hamamelis water (witch hazel) has lingered on as an aftershave and astringent.

Wines: Alcoholic liquids prepared from drugs by the process of solution, maceration, or percolation, differing from tinctures only in that wine is used as a solvent or menstruum instead of various strengths of alcohol; one of the oldest liquid preparations, since the alcoholic content of the wine improved its solvent characteristics in many cases. The National Formulary IV (1916) recognized 15 wines, but the passage of Prohibition convinced the revisors to drop red and white wine, as well as all medicated wines, from that compendium.

BIOLOGICAL LIPID

Biologically, lipids function as structural elements in plants and animals, transport vehicles, mediators of chemical reactions, energy sources, and as messenger molecules. It is from an understanding of these functions that lipids have been applied to pharmaceutical applications. Lipids can be appropriately manipulated to conform to various physical states such as solutions, suspensions, emulsions, creams, gels, or solids. Because a combined lipid–drug formulation will present a very different conformation and/or molecular size than the drug alone when introduced into a biological environment, changes in both the distribution and the recognition of the drug by the host defense system will occur.

These changes may be either intended or undesired. Overall, the uses of lipids in pharmaceutical dosage forms can be grouped into four categories: (1) improvement in the processing or stability of the formulation in the preferred physical state; (2) enhancement or reduction in cellular or systemic absorption of the drug from the formulation; (3) more effective drug targeting; and (4) sustained or more controlled delivery of the drug. In this article, the problems and opportunities in utilizing lipids for these applications will be explored from various aspects of pharmaceutical formulation, analysis methods, and other commercial issues.

Clinical Aspects of Lipids

Categories shown are: fatty chain acids, salts, alcohols or amines, oils and waxes, phospholipids, glycolipids, neutral lipids, and non-linear chain compounds, such as sterols.

Fatty acids and derivatives

Fatty acids, salts, alcohols, and amines have been utilized in pharmaceutical formulations. Fatty acids are present in cosmetics, ointments, and suppositories and are used in tablet coating applications and as carriers in inhalant products. Fully saturated fatty acids are solid materials at chain lengths above eight, whereas longer chain polyunsaturated forms may exist as liquids, unless the double bonds are conjugated. Fatty acid salts are used widely in tableting applications and often include magnesium, calcium, and aluminum stearates. The handling properties and physical interaction with other solids make them ideal as conditioning agents to effect even distribution of particles, improve compressibility, and control the eventual release of active agent. The fatty alcohols, including cetyl and palmityl alcohol, are well known and are used extensively in ointments or creams as an emollient or emulsion modifier. Fatty amines are generally utilized as precursor compounds in coupling reactions to produce lipophilic drug derivatives. Fatty amines, like the acids, are solid at higher chain lengths and insoluble in aqueous solution. Finally, aldehydes of fatty chain compounds have application in fragrances and flavorings. The unsaturated liquid forms of fatty aldehydes are most often employed.

Oils, waxes, and neutral lipids

Fatty acids, when esterified to glycerol, form mono-di-and triglycerides. Depending on the number of substituted fatty acids, the carbon chain lengths and the unsaturation level, the resultant product may exist in either liquid state (oil) or solid state (wax). Products in this category may either be naturally derived or synthetically produced and, thus, a wide diversity of products is commercially available. Oils have drug- carrying and solubilization functions in oral, topical, or injectable products. Waxes are presently used in topical and oral preparations to improve desired physical properties and control dissolution of the final product. Common materials in this group that are present in commercialized pharmaceutical preparations include castor oil, hydrogenated vegetable oil wax, paraffin, carnauba wax, white wax, olive oil and olive oil ethyl ester, mineral oil, petrolatum, cetyl ester wax, and beeswax. Neutral lipids, such as steroids, are also of value in pharmaceutical practices. For example, cholesterol and cholesterol esters can be exploited to broaden liposome phase transitions and to allow easier manipulation and or emulsification.

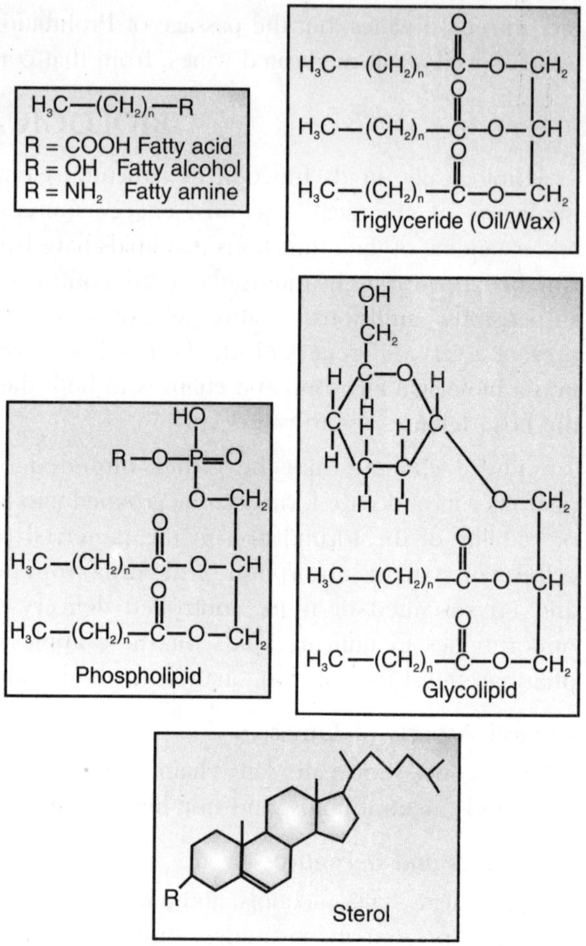

Fig. 1.6. Lipid structures.

Phospholipid compounds

The complexity of glycerides advances by modification of the terminal hydroxyl with phosphate linked head groups to form phospholipids. Common phospholipid head groups include choline, ethanolamine, serine, inositol and inositol phosphates, glycerol, and glycerol esters. As with the triglycerides, numerous species are possible by various combinations of different head- groups and fatty acyl substitution at the 1st and 2nd positions of the glycerol backbone. Fluidity differences are evident as a function of the gel to liquid crystalline transition temperatures. Solubility of phospholipids is intimately linked to the conformation of the aggregate material rather than strictly a chemical function of the molecule. Monoacyl phospholipids, which tend to form micelles, are usually more readily soluble in aqueous solution. Diacyl phospholipids generally form liquid crystalline suspensions as long as the temperature is held at or above the phase transition. Overall, because phospholipids are amphipathic, they function well as emulsifying or dispersing agents. Although most often found in topical products, phospholipids have been employed in oral capsule formulations as well as in liposomal parenteral formulations.

Glycolipid compounds

Glycolipids include compounds formed by the linkage of a sugar moiety to a glyceride backbone. Classical glycolipids such as glycosyl glycerides or ceramides are not common in pharmaceutical preparations primarily due to low abundance and high cost; however, glycosyl modifications can be used to impart targeting mechanisms to aggregate lipid formulations.

Characteristics of Lipid

The colligative and solubilizing properties of lipids have been known since the early part of the 20th century. What had been perceived as clusters of amphipathic molecules (soaps) in aqueous solution are now better understood as micelles. Micellar solubilization of pharmaceuticals is now a familiar classical technique. Additional understanding of drug dissolution and impact of lipid conformations can be gained from studies that utilized macro and microemulsions to solubilize drugs. Beyond the concept of micelles and emulsions, the combination of lipid and aqueous components can result in formulations that exhibit a variety of physical states. These result from lipid orientations, which may be either random, in thermotropic crystalline lattices, or in aggregate structures with long-range order (lyotropic liquid crystals).

Lipid molecules are also capable of packing tightly together to give more ordered states, such as a bilayer. In fact, depending on the components in the formulation, lipids may contribute to distinct phase diagrams containing up to at least six different regions. The "neat" or lamellar (liposomal) phase physically can appear as a suspension to a highly viscous cream, depending on the final volume. Lamellar organizations further include liquid crystalline, gel state, and interdigitated phases. The middle phase (hexagonal) is a cylindrical arrangement with a hydrocarbon core, surrounded by an interfacial layer of hydrated polar groups. This orientation can be reversed to yield the hexagonal II structure, which appears stiff, and suspends as hard particles in water and will not dilute in additional aqueous solution.

Lipids can also organize into cubic structures that appear clear, brittle, and viscous in character. Mixed conformations of lipids have also been observed, as is the case for lipidic particles, with inverse micelles distributed between a lamellar matrix. These different aggregated states are sometimes reached by transitions through other phases simply by changes in temperature or by nature of the external environment (such as solvent concentration). Phase transitions of lipids can be regulated by changes in pH, temperature, cation concentrations, and presence of amphipathic additives. A comprehensive database on lipid phase transitions has been recently compiled by Caffrey. Polymorphic transitions are common in the preparation of lipid-containing formulations and can account for the differences in physical handling properties of the material during manufacture in which varying amounts of solvents, or energy input, are encountered at the different process steps.

Geometric preferences of lipids can be better understood by considering the relative volume occupied by each portion of the molecule. The various macromolecular aggregates of lipids also make it possible to associate with drugs through physical entrapment. In these cases, the concentration of lipid associated drug is dictated by the size or conformation of the drug and the spaces created in the lipid aggregate. Entrapment levels of drug can vary from less than 1% to as high as complete encapsulation. Excluding chemical interactions or solubilization phenomena, drug encapsulation efficiency is linked to both the lipid conformation and the manufacturing method, which juxtaposes the two components. This is particularly true of liposome formulations in which factors of temperature, pressure, solute concentration, solvent volumes, or energy application all contribute to the final amount of drug incorporated within the lipid matrix.

Commercial Production of Lipid-containing Products

Manufacturing issues should be evaluated as part of overall lipid formulation development. Although lipid addition may be desirable for pharmacological purposes there are often commercial issues which present hurdles to eventual marketing. Among those factors are the availability and cost of raw materials, handling of intermediate mixed-phase systems, particulate control, stability of the lipid ingredients, sterilization, packaging interactions, qualification of non-compendial ingredients, and development of validated methods for analysis of either raw material or product.

Commercial formulation

For commercial formulations, raw material sources should be of high purity and quality consistently meeting defined specifications, preferably compendial when available. Fortunately, high-purity lipids have become increasingly available in large supply, and in particular, phospholipid supply has advanced in recent years due to the approval and marketing of parenteral liposome formulations. Large supply of naturally derived lipids can be obtained from plants in which many oils and fatty acids can be readily extracted and purified. Animal sources (e.g., eggs or milkfats) are used to derive complex lipids such as phospholipids and cholesterol. Yield from natural sources is dependent on the weight- percent composition and the efficiency of the extraction procedure. The constitution of fatty acids in vegetable oils varies widely from different sources. For example, oleic acid is present at 64.6% by weight in olive oil but is present at only 0.7% in palm kernel oil. Similarly, castor oil triglyceride is comprised of almost entirely ricinoleic chains.

There are numerous raw material suppliers of oils and oil fractions worldwide. As such, the relative cost of bulk purified fractions and their derivatives such as salt forms and alcohols is quite low (~$50–150/kg). Costs for high purity synthetic phospholipids (>97%) continue to drop when compared to prices 8 to 10 years earlier. This is primarily due to the advent of marketed liposome products. Phospholipids are available worldwide from companies such as Genzyme, Avanti Polar Lipids, Matreya Inc., American Lecithin Co., Lipoid GmBH/Vernon Walden, Lucas Meyer, Nichiyu Liposome/Nippon Oil and Fat and Northern Lipids. Also now available in larger supply are PEGylated phospholipids, lysopholipids, and various cationic lipids used in gene therapeutic applications.

Market value of lipids

Lipids may be supplied either as lyophilized or dried powders, as liquids, or dissolved or suspended in an appropriate solvent. Dried powders are typically waxy or sticky in character, making aliquoting of the product troublesome. Lipid lyophilates can also be hygroscopic, thus presenting some difficulty for fluidized bed operations. Fluid bed applications however, may be successful if the concentration of lipid is held fairly low. Handling of lipid products in a liquid state allows for more accurate compounding. Solubilization vehicles may involve dissolution into either polar or non-polar solvents or surfactant solutions. Aside from safe handling, a primary concern when introducing lipids from solvent environments is the compatibility of the solvent with processing equipment, in particular the gaskets and seals found in the pumps, transfer lines, mixers, and extractors. Elevated temperatures, which are used to dissolve or maintain the phase transition of the lipid or to initiate solvent evaporation, can further induce problems for gaskets and seals. In systems with nonvolatile solvents and detergents, removal of the dissolution vehicle can be achieved with flow centrifugal separation, chromatography or diafiltration.

Introducing a lipid-containing solvent to an aqueous solution of drug will require some means for energy input to mix the components, whether by simple rotating blade stirring, more vigorous homogenization, or milling, by sonication and filtration, or by high-pressure extrusion. Unless cosolvation is achieved a suspension of non-miscible aqueous and non-polar phases will offer more resistance to flow and may present problems for extrusion through filters, which tender a surface that is of opposite character to one of the phases. Binding of active agent, such as proteins, must also be considered. The conformational state of the lipid and the concentration also will govern the workability of the formulation. Lamellar orientations will flow fairly readily at concentrations below about 400 mg/ml whereas cubic or hexagonal phases will present difficulty even at very low concentrations.

Management

For lipids dissolved in solution, particulate control may be necessary, depending on administration route. For parenterals the large- or small-volume parenteral particulate tests of the USP are applicable, as would be the Ph. Eur. clarity test and JP foreign matter insoluble test. For suspended or opaque lipid formulations, compendial particle size tests and specifications for suspensions will apply (i.e., light obscuration or microscopy). Size reduction of lipid formulations can be achieved with a variety of processing equipment including paddle stirrers, homogenization and high shear mixers, Microfluidizer, colloid mills, ultrasonic and piezoelectric emulsifiers, and pressurized filtration or membrane extrusion.

Role of unsaturated lipid

A characteristic of lipid products, particularly those with unsaturated lipids is peroxide formation with oxidation. Free radicals such as $ROO^{\bullet}$, $RO^{\bullet}$, and $OH^{\bullet}$ can damage the drug and induce toxicity. Lipid peroxides may also form due to autoxidation, which increases with unsaturation level. Hydrolysis of the lipid may be accelerated due to the pH of the solution. or from processing energy such as ultrasonic radiation. Antioxidants (i.e., α-tocopherol, propyl gallate, ascorbate, or BHT) may be required.

Heat sterilization

Because of possible oxidation, steam or heat sterilization of lipids may not be an attractive option. Validations at non-standard autoclave cycles or F_0 values may prove valuable. In some instances, as in ophthalmic ointments made with white petrolatum or oleagenous components, the base can be sterilized by dry heat (160–180°C for 1–3 h) and combined aseptically with the sterile drug and additives.

Ointments with lipids such as lanolin, petrolatum, and mineral oil have been terminally sterilized by cobalt-60 gamma irradiation with success. In less viscous solutions, end terminal filtration with aseptic fill is often necessary, requiring justification to regulatory authorities for lower sterility assurance levels. High pressures may be required to force lipid through membranes, thus adequate filter integrity checks are critical. High-pressure extrusion through straight channel membranes (polycarbonates/ceramics) may reduce the pressures and improve flow rates.

Marketing

Lipids will be attracted to hydrophobic surfaces where losses might be expected. Overall binding will be affected if the molecule also contains ionic regions. Generally, lipids are chemically compatible with most plastics or glass used in packaging. Residual solvents that partition into the lipid phase may interact with plastic resins such as polystyrene or polycarbonate or with rubberized seals. Extractables from plastic packaging may be accelerated by the presence of lipid. Extraction of silicone polymer into lipid phases also occurs from silanized surfaces or coated gaskets. Compendial tests and regulations on extractables and impurities will apply here. Presence of metals coming from process or package surfaces are a potential source of catalysis leading to degradation of lipid components. In addition, reducing oxygen headspace should be considered to minimize potential degradation. Inert gas purging can improve lipid stabilities.

Clinical Analysis of Lipid

Because lipids exhibit primary through quaternary conformations, the analysis of lipid-containing pharmaceutical products must therefore involve both chemical and physical determinations to define the product on the molecular level as well as the aggregate state of the product. Stability-indicating assays can be developed from both approaches.

Chemical measurements

Lipids can be measured by standard analytical methods such as wet chemistry, HPLC, thin-layer chromatography, and gas chromatography. Because lipid formulations may have several components and because the formulation may have an overall aggregate structure, it is usually necessary to develop further strategies for isolation and detection of the lipid ingredients. Dissociation of the individual components using temperature, solvents, detergents, reducing or oxidizing agents, or mechanical disruption prior to analysis may be necessary.

Further separation or extraction of the components before analysis is required if interference is encountered. With lipid solvent extraction, there is a greater likelihood for losses of material. More often than not, a combination of techniques will be required for validation of the product. There are worldwide compendial procedures for analysis of various lipids by wet chemistry procedures to determine concentration, end group analysis, acid value, hydroxyl value, iodine value, and saponification value.

HPLC method of analysis

Analysis of non-polar lipids by HPLC is best carried out using normal phase columns. However, for mixed phases with polar drug or drug within an aqueous phase some compromise may be necessary. Good separation of polar and neutral lipids with a C8 column and a four-solvent mobile phase has been reported. Elution of neutral lipids like triglycerides from C18 columns is slow, however, good resolution can been achieved.

Mobile-phase development is usually necessary to effect a high degree of separation and resolution of similarly eluting components. Heated columns are beneficial for increasing temperature above the phase-transition temperature of the lipid and thus minimize clogging. Lipids with predominantly saturated chain lengths will have poor ultraviolet (UV) absorption even at the lower wavelengths (190 nm). Alternative detectors should then be considered.

TLC technology

One- and two-dimensional thin-layer silica gel chromatography remains a cornerstone of lipid analysis. Sensitivity by this method is typically as low as 2 μg/ spot. Lipid visualization can be achieved by many methods, including iodine vapor or charring of plates following exposure to sulfur-dichromic acid, cupric reagents, Phospray, or α-naphthol. Other useful reagents include stains and dyes such as fluorescamine, rhodamine 6G, bromothymol blue, molybdenum blue, phosphomolybdic acid, and silver nitrate, which may be impregnated into the silica. Validation of lipid purity and quantitation is performed using gel or plate scanners.

GC technology

Capillary and packed (GC) columns are of value in the analysis of complex mixtures of lipids. The best capillary column length will depend on the complexity of the material injected, however, 30-m columns are often employed. In packed columns, many types of stationary phases are available for lipid separation, and these include silicone and alkylated or cyanogenated derivatives, polyesters, polyglycol, and carboranes. It is also common to derivatize the fatty-acid side chains to the corresponding methyl esters by reaction in BF_3/methanol prior to chromatographic analysis to achieve more distinct and uniform separations.

Physical measurements

The aggregate states of lipids are discernable from measurements provided by a host of analytical devices, which yield information on physical properties. These techniques may also be used to give clues as to the interaction of the lipid carrier and the drug.

Size of particles

Lipid particulate analysis can be achieved with most commercial laser particle counters. Compendial requirements for suspensions also allow for sizing via electron microscopy determinations, which can provide for qualitative assessments in addition to quantitation. Lipid suspension particles can be detected by negative-stain, freeze-fracture, critical-point drying and scanning techniques.

NMR technology

Nuclear magnetic resonance (NMR) is a valuable technique in the analysis of lipid phases. More specifically, proton, deuterium, carbon-13, fluorine-19, and phosphorus-31 NMR have been utilized for analysis of the dynamic and motional properties of lipids, lipid diffusion, ordering properties, head-group hydration, lipid asymmetry, quantitation of lipid composition, and head-group conformation and dynamics. Cullis et al. and Gruner et al. have shown the importance of P-31 NMR as a tool in the determination of phase properties and lipid asymmetry and the identification of bilayer, hexagonal, and isotropic phases.

ESR technology

Electron spin resonance (ESR) is used to give information on the local environment of a lipid molecule. In normal state these molecules inherently do not exhibit ESR spectrums. However the necessary signal can be generated from reporter labels such as nitroxide or doxyl probes, which can either be linked directly to the lipid molecule or prelinked to a lipid chain and then partitioned into the lipid aggregate formulation. Data from ESR spectrums are valuable for determining molecular properties of the formulation, such as phase transitions and separations, order parameters (anisotropy), polarity, lateral diffusion, segregation and clustering, surface and transbilayer potentials, permeability and internal volumes, surface determinants (antigens), pH gradients, lipid asymmetry, flip flopping, and fusion.

Differential scanning calorimetry

Differential scanning calorimetry is a well known technique in the study of the thermal behavior of lipids and can be used to assess purity and stability of lipids, perturbation of aggregate structures, phase transition temperatures, lipid mixing behavior, and influence of other molecules and ions on structure.

X-ray and neutron diffraction

Diffraction patterns of lipid solutions can yield strong evidence for the presence of specific repeating conformational structures as well as the spacing between lipid molecules in organized films or layers.

Spectroscopic analysis

Appraisals of the optical properties of lipid solutions and dispersions will provide information on concentrations, aggregation and stability, phase transitions, densities, and repeating structures. Measurements of refractive index, scattered light intensity (polarized and depolarized), and birefringence are relatively easy laboratory methods on which certain product specifications may be based. Also, fluorescent techniques can readily provide information on lipid movements and transfer of lipid between particles.

Economic Values

The key purposes of lipid materials in dose forms include: (1) improving the solubility or physical workability of the drug for ease of administration or to enhance stability; (2) augmentation or reduction in absorption of the drug from the formulation; (3) specific drug targeting to maximize response and minimize side effects; and (4) controlled or slow delivery of the drug from the formulation. Currently, the largest use of lipids in pharmaceuticals is in products for oral or topical dosage administration, although there is growing use in parenteral, pulmonary and nasal products.

The mechanics of drug delivery from lipid systems is governed by five structural features: primary structure, that is, chemical or molecular interactions, secondary organization into aggregate structures such as inverse micelles, tertiary organization of the aggregate projected into three dimensions such as three dimensional inverse micelles or hexagonal phase tubules; quaternary associations, agglomeration or interaction of the three dimensional structures such as lateral stacking of hexagonal tubules, and final packing of the molecules in solution (concentration effects) to produce liquids, creams, solids, etc. Mammalian systems have natural degradative pathways for the rapid metabolism of most lipid raw materials. However, the release of drug to a biological system will be controlled at least as much by these structural features as by simple-natural degradative mechanisms.

Prudent tactic

Lipophilic derivatives and prodrugs are a prudent tactic to alter the normal interaction of drug compounds with cells and cellular barriers. The type of modification desired may be a permanent one if the drug compound maintains its activity following conjugation, or a reversible type subject to biological cleavage. A variety of strategies for chemical coupling of lipids to drugs can be developed to produce the desired modified product. Briefly, non-reversible reaction approaches include conjugation to drugs with amine residues either via glutaraldehyde, lipid anhydrides or halides, or succinimidyl derivatives of the fatty acyl chain via carbodiimide activation. Non-reversible conjugations have also been accomplished via carbodiimide-activated carboxyl moieties of drugs reacted with amine containing lipids.

Permanent conjugation to phenolic residues of drugs can be accomplished with diazo derivative of lipids bearing available amines. This approach has been successfully employed to prepare lipophilic derivatives of peptides containing tyrosine residues. Lipid conjugation reactions have also taken advantage of available sulfhydryl groups on drugs, particularly for modification of antibodies. On the other hand, bioreversible conjugates have been made via use of the Schiff base reaction to couple an aldehyde-bearing lipid to amines present on the drug. It is possible to convert the hydroxyl groups of more abundant glycolipids to the corresponding aldehydes using periodate. An extension beyond the Schiff base reaction is the Mannich base condensation, which introduces a nucleophilic reactant such as an enolate anion or amide to the reaction between the aldehyde and amine. In these reactions the amine component is generally a lipophilic primary or secondary amine. Extensive literature is available on the production and biological activities of lipophilic N-Mannich base derivatives and stabilized *a*-acycloalkyl forms.

Enhancing drug permeability

There are a number of excellent reviews on the requirements for enhancing drug permeability across lipophilic biological barriers, with particular reference to the importance of the lipophilic properties of formulations. Administration sites include buccal, oral, nasal, ocular, transdermal, rectal, and pulmonary.

Dispersion

There is fairly extensive use of lipid materials in oral liquid dose forms strictly as vehicles. Primary applications have been as surfactants to promote drug suspension or dissolution (fatty glycols and fatty acids), as flavoring agents (natural or synthetic oils), and as thickening agents (hydrogenated oils). Aspects of pharmaceutical oral suspensions have been discussed in greater depth. Solution based-lipid vehicles have been applied in softgel applications, the primary category being oils that are compatible with gelatins and that that have had applications as both lubricants for processing of the gelatin sheets and as drug vehicles for liquid fill operations into the softgel. High lipid concentrations may allow for higher ethanol content in softgel fills.

Lipids may also be used to create solutions for injectable products, particularly intravenous preparations. Fatty acids, fatty glycols, and fatty alcohols may be used to enhance the dissolution of certain insoluble drugs, act as preservatives, or function as active agents as demonstrated by Scleromate injection, a mixture of fatty acid salts derived from cod liver oil. As another example, benzyl alcohol preservative has been useful in formulations for water insoluble drugs such as etoposide and is a primary active agent in Zilactin gel.

Polyoxyethylated fatty-acid derivative has been used in the dissolution of phytonadione, a lipid-soluble vitamin for subcutaneous or intramuscular injection. Injectable amphotericin B is solubilized with sodium desoxycholate in Fungizone. Polyoxyethylated or PEGylated castor oils are also used in dissolution of injectable drugs such as cyclosporine, paclitaxel, and teniposide and miconazole. Clear colloidal dispersions are also possible using a wide varity of phospholipids, cholesterol esters, and tocopherol esters.

Solid- and liquid-crystalline suspensions

Suspensions and creams are most often developed for aqueous insoluble drugs, as typified by ophthalmic and otic preparations of corticosteroids. In suspensions such as Cortisporin or Pediotic, components such as cetyl alchol, glyceryl monostearate, mineral oil, and propylene glycol are commonly used to effect a homogeneous suspension of drug particles.

The creation of fine suspensions may also be necessary to administer a highly insoluble product by parenteral injection. In these instances lipids may be used as either wetting agents or suspension vehicles. Lecithin (phosphatidylcholine) is a suitable agent for wetting or suspending of drug particles in either aqueous or non-aqueous solutions. This common formulation additive is used for injectable long-acting IM suspensions of penicillin for the bronchodilator inhalation aerosol, Atrovent, and for the otic suspension Cipro HC. Naturally derived and synthetic lecithins, in mixtures with neutral lipids, also serve as the active ingredient in lung surfactants products such as Exosurf, Survanta, and Infasurf.

Lipid materials are used extensively as vehicles in topical creams, ointments, gels, and lotions, and usually serve as the base material for many such preparations. These formulations are principally presented as emulsions and may contain fatty acids, fatty-acid salts, fatty alcohols, petroleum based and natural oils, waxes, fatty glycols, lanolin, and other hydrophobic surfactants. Such emulsions are common in dermal topicals, for which the list of products is too numerous to elaborate. Some of these ingredients also are present as vehicles in suppositories such as semi-synthetic glycerides existing in Nembutal, hydrogenated vegetable oil as found in Dulcolax (Novartis Consumer Health), and glycerides of fatty acids or oils that can be found in vaginal suppositories such as Prostin E2 and Crinone Gel. Mineral oil is frequently present in transdermal products such as Catapres-TTS and Estraderm. In addition, numerous ocular or topical ointments, such as TobraDex and Nitro-Bid, use white petrolatum and mineral oil as a base. Short-chain triglycerides such as triacetin can be found in products such as Prepidil Gel, a cervically administered prostaglandin.

Oil-in-water emulsions have also been applied for intravenous use. Commercial parenteral emulsions include Dizac and Diprivan. Many commercial or experimental parenteral products have been based on vegetable oil (most often soybean, safflower, or cottonseed) stabilized with phosphatides and monoglycerides, which nicely match the *hydrophile-lipophile balance* (HLB) requirements of those oils (~6–7). Further prospects for expanded non-toxic parenteral emulsions may come with use of other phosphatide-based surfactants with high HLB values. Drug delivery from oil/phosphatide emulsions stems from earlier development and marketing of intravenous nutrient emulsion products such as Intralipid and Aminosyn II, which are sterile and non-toxic.

The development of these emulsions as well as distribution profiles of these and similar emulsion compositions have been studied. A cursory scan of the literature on parenteral oil emulsion formulations will obtain studies on amphotericin B, prostaglandin E1, halothane, pregnanolone, paclitaxel, perilla ketone, penclomedine, F-octylbromide, flurbiprofen axetil (Lipfen), lasalocid, lignin, podophyllotoxin, tacrolimus, doxorubicin, epirubicin, menatetrenone, chlorpheniramine maleate, naproxin, cyclosporin A, propranolol, testotsterone, benzocaine, phenylazoaniline, palmitoylrhizoxin, pilocarpine, diazepam, and various peptide and proteins for vaccine delivery.

Liposomes as pharmaceutical dosage forms

Extensive studies on liposomes date back to the 1960s. Many good comprehensive review references exist on compositions and manufacturing of liposomes including the article "Liposomes as Pharmaceutical Dosage Forms" in this volume. The earliest commercial liposomal formulations were developed for veterinary application or over-the-counter cosmetic creams promoted for improved hydration. More recently, parenteral liposome formulations of amphotericin B, doxorubicin, and daunorubicin have been approved and marketed, with others on the horizon for applications in photodynamic therapy. Although the vast majority of liposome preparations are constructed from phospholipids, other nonphospholipid materials can be used either alone or in mixtures to form bilayer arrays.

One such example is Amphotec, which utilizes sodium cholesteryl sulfate as the primary lipid. Other liposome forming materials may include but are not limited to fatty-acid compositions, ionized fatty acids, or fatty acyl amino acids, longchain fatty alcohols plus surfactants, ionized lysophospholipids or combinations, non-ionic or ionic surfactants and amphiphiles, alkyl maltosides, α-tocopherol esters, cholesterol esters, polyoxyethylene alkyl ethers, sorbitan alkyl esters, and polymerized phospholipid compositions.

LDL carriers

Lipoproteins are naturally occurring particulate emulsion carriers for the transport of cholesterol and other lipids such as triglycerides in the blood. Because *low-density lipoprotein* (LDL) particle clearance is receptor mediated, they have been proposed as drug carriers for targeting applications, specifically for targeting of cytotoxic agents to tumor cells, delivery of antiviral agents to parenchymal liver cells, targeting of immunomodulators, antiviral and antiparasitic drugs to Kupffer and endothelial cells, and as gene vectors.

Oral tableting

The main use of lipids in solid form has been for oral tableting applications. Fatty-acid salts such as magnesium and calcium stearates, and various waxes and glycerides are most often used as conditioners and binders during compaction and provide more even, controlled, or slower disintegration of the tablet once administered. Extensive literature is available in which these materials are discussed in context of the preparation of oral tablets or capsules. The use of lipids for solid-implant formulations has also been investigated. Materials such as cholesterol, and high-melting-point fatty-acids, fatty anhydrides, and glycerides have been utilized in compressed implants to prolong systemic delivery of drugs. Lipids are also well suited for suppository and vaginal insert formulations.

Precautions

The safety of a lipid product will in part be a reflection of: (1) the purity of the compounds administered; (2) biological toxicity of the basic chemical ingredients; and (3) reactions to structural presentations of the lipid. Noncompendial lipid materials will necessitate significant toxicology testing.

Adulteration

Sensitive analytical procedures will be required to distinguish contaminants from lipid peaks within the preparation. By-products may be present from the synthetic processes used to produce the lipids or if copurified from the natural source. For example, common impurities in synthetic diacyl chain lipids are the monoacyl forms, which are generally more toxic to biological systems. Endotoxins and pyrogens either may be detectable in a lipid preparation or difficult to detect due to the lipids shielding against the analytical reagents.

Adjuvancy

Consideration should be given to immune reactivities when administering lipids to mammalian systems. Oils are well known for their adjuvancy; different oils will produce varying levels of reactivity. Biological responses such as leukocyte attraction, encysisting of the oil, and edema reactions vary in severity simply as a function of the chemical nature or purity of the oil itself. Importantly, most lipids, including normal endogenous compounds possess antigenic potential to varying degrees. These possible reactivities should be monitored as part of the overall clinical design.

Conformational considerations

There may be biological sensitivities to the conformational presentation and sizing of the lipid formulation. For example the ability of antibodies to distinguish between lamellar organizing lipids and hexagonal-phase lipids are known and may form the basis for certain types of autoimmune dysfunction. Biological factors such as reticuloendothelial cell recognition of particles provide further impetus for control of the particle size and stability.

CURRENT GOOD MANUFACTURING PROCEDURES

Dosage form is a drug delivery system designed to deliver the active ingredient to the body and, upon administration should deliver the drug at a rate and amount that assures the desired pharmacological effect. Such dosage forms are manufactured under *current good manufacturing procedures* (cGMP), using equipment and packaging to ensure product stability. The dosage form must produce the same therapeutic response each time it is administered. To maintain this reproducibility between and within batches, manufacturing procedures are validated under a specific quality assurance program. Non-parenteral dosage forms can be categorized based on the route of administration or physical form. Based on physical form they can be classified as solids, liquids (homogenous and heterogeneous systems), semisolids, and aerosols. Dosage forms can also be categorized based on the route of administration. Solid dosage forms include different types of compressed tablets, granules, troches, lozenges, coated dosage forms, and hard and soft gelatin capsules.

Liquid dosage forms include solutions, suspensions, emulsions, and buccal and sublingual sprays. Topical dosage forms are applied to the skin and include ointments, pastes, creams, lotions, liniments, and transdermal patches. Some dosage forms are formulated for application to body cavities, viz. rectal and urethral suppositories and vaginal pessaries. Inhalation aerosols, using metered dose inhalers (MDIs), dry powder inhalers (DPIs) and nebulizers, are used to deliver drugs to the respiratory tract. Nasal route uses solution and suspension dosage forms. Occular route is used to administer solutions and suspensions to the eye for local and systemic effects.

Solid State

Powders and granules

Powders are intimate mixtures of dry, finely divided drugs and/or chemicals that are intended for oral administration or external use. Powders may also be formulated as larger particle sized, free-flowing granules to aid in handling and administration. Bulk powders usually are packed into a suitable wide-mouth container and contain relatively non-toxic medicaments in large doses, e.g., compounded magnesium trisilicate oral powder. Insufflations are medicated powders blown into ear, nose, or throat.

Tablets

Flowability and compressibility are two important parameters essential for successful manufacture of tablets. Flowability determines ease of material flow from tablet hopper to the press. Inadequate flow gives rise to arching, bridging, or rat-holing in hoppers. Powder flow can be improved mechanically by use of force feeders. Flowability can also be increased by incorporation of glidants like fumed silica and talc. Another method involves the conversion of powder to spherical particles by spray drying or spheronization. Tablets are manufactured by dry and wet methods. Dry methods consist of direct compression, slugging, and roller compaction of drug-excipient blends. Directly compressible excipients may be disintegrants with poor flow, e.g., microcrystalline cellulose (Avicel PH102); free-flowing materials which do not disintegrate, e.g., dibasic calcium phosphate (DiPac); or free-flowing powders which disintegrate by dissolution (e.g., spray-dried lactose, anhydrous lactose, dextrose, sucrose, amylose, etc.).

The drug is mixed with excipients in a blender and then compressed directly on a tablet press. Dry granulation by compression or slugging is used for moisture or heat sensitive actives. The powder blends are compressed into compacts or slugs. An alternative method is to squeeze the powder blends into solid cake between rollers called roller compacts. These slugs or compacts are milled and screened in order to produce granules with improved flow. Granulation is the process of particle size enlargement of homogeneously mixed powder ingredients and simultaneously increasing bulk density, flowability, and compressibility of the system. Wet granulation process involves the massing of the powder mix, using a binder and solvent.

The solvent should be volatile, non-toxic, and removed by drying. This process is not suitable for hydrolysable and thermolabile drugs. The binder is added in the form of a solution, or added dry or its mucilage incorporated with the powder blend. The choice of liquid depends on the properties of the material being granulated. Water is widely used alone or along with a binding agent. Commonly used non-aqueous liquids are isopropanol and ethanol. Massing process is usually performed in a low or high shear granulator where the liquid is poured or sprayed onto a moving powder bed until a moist mass of finely divided material is formed. This is passed through an oscillating granulator with the appropriate screen size to obtain the required granule particle size. Sometimes both intra and extra-granular portions are divided to prevent incompatibility between excipient material hiding or better distribution, tableting or dissolution. Tablets require different functional excipients for their manufacture.

Diluents are inert bulking agents added to actives to make a reasonably sized tablet. Generally, a tablet should weigh about 50-60 mg and therefore very low dose drugs will require these diluents to make at least a 50 mg tablet. Adsorbents such as fumed silica and kaolin are sometimes used for holding large amounts of fluids in an apparently dry state. Binders are used as adhesives to bind powder in wet granulation and give strength to compacts during compression. Binders may be incorporated into the dry blend or added as a solution to the mixed powder during wet granulation. Disintegrants are usually added to promote rapid breakup of tablets to increase surface area and aid drug dissolution. Disintegrants can act by different mechanisms such as like swelling and capillary action. Glidants are materials that are added to tablet formulations to improve flow properties of the granulation. They act by reducing inter-particulate friction (e.g., fumed silica). Lubricants are added to prevent the adherence of granules to the punch and die faces of the tablet press. Many lubricants also facilitate flow of granules. Talc and magnesium stearate are more effective as punch lubricants. Stearic acid works better as a die lubricant.

Specific types of tablets

Lozenges: These are compressed tablets formulated, without a disintegrant and must be allowed to dissolve in the mouth. They are used for local activity (throat lozenges) or for systemic effect (vitamins).

Effervescent tablets: These tablets undergo quick dissolution of actives in water due to internal liberation of carbon dioxide. By combining alkali metal carbonates or bicarbonates with tartaric or citric acid, carbon dioxide is liberated when placed in water.

They are prepared by the heat fusion technique. Usually a water-soluble lubricant is used to prevent scum formation at the water surface. Sweetness is achieved by the addition of saccharin, since sucrose is hygroscopic and increases the bulk of the tablet, e.g., Rochelle Salt.

Chewable tablets: These tablets are preferred for pediatric and geriatric patients who have difficulty swallowing whole tablets. Another advantage is that they do not need water for administration. Mannitol is normally used as the base diluent because of its pleasant taste and texture, and because it can effectively mask the taste of objectionable actives. They are usually prepared by wet granulation and are not compressed very hard. High amounts of flavor are added to increase palatability. Antacids are typically formulated as chewable tablets.

Sublingual and buccal tablets: These tablets are placed under the tongue (sublingual) or the cheek (buccal) and can produce immediate systemic effects by enabling the drug to be directly absorbed through the mucosa by preventing the first pass effect (e.g., isoprenaline sulphate and glyceryl trinitrate). Tablets are small, flat, without a disintegrant, and are compressed lightly to produce soft tablets.

Molded tablets: These are prepared from mixtures of medicinal substances and a diluent usually consisting of lactose and powdered sucrose in varying quantities. The powders are dampened with solutions containing high proportions of alcohol depending on the solubility of the active and filler. The dampened powders are pressed under low pressure in die cavities. Solidification depends upon crystal bridging during the subsequent drying process, and not upon the compaction forces.

Multi-layered tablets: A multilayered tablet consists of several different granulations compressed on top of each other to form a single tablet. They may also be bi-layer when incompatible drug substances are used, e.g., phenylephedrine HCl in one layer and ascorbic acid and paracetamol in another.

Modified release (MR)

Modified release (MR) has been used to describe dosage forms having drug release characteristics based on time, course, and/or location and are designed to accomplish therapeutic or convenience objectives not offered by conventional or immediate release dosage forms. Drugs for chronic conditions with short half-lives, possessing a good therapeutic index and uniform absorption pattern are ideal candidates for such dosage forms. These are either delayed release or extended release (ER) preparations. ER dosage forms allow at least a twofold reduction in dosing frequency as compared to the conventional dosage form. Delayed release dosage forms are designed to release all or a portion of drug at times much later than the time of administration. The delay may be time based or environment specific, as in enteric-coated dosage forms.

Some other MR dosage forms include repeat action and targeted release dosage forms. Most controlled release products are good examples of ER dosage forms. These dosage forms can be classified by their mechanism of release and/or type of formulation. Coated beads, granules microspheres, and other particulate systems are pellet type controlled release dosage forms, where the drug is usually coated onto non-pareil beads (low dose) or made from granules composed of the drug (high dose). These pellets are further coated with functional coating agents (Eudragits, HPMCs, Surelease, etc.) to provide various release characteristics.

These pellets can be used to fill capsules (e.g., Ornade Spansules) or compressed at low pressure into tablets (e.g., Theo-Dur). In some cases small mini-tablets of about 3–4 mm diameter can be compressed. These tablets function like pellets and can be used to fill capsules. Microencapsulation is a process of encapsulating microscopic drug particles with a thin wall of coating material. Several coating materials have been used including gelatin, ethylcellulose, and polyvinyl alcohol, e.g., Micro-K-Extencaps. Matrix systems are dosage forms where drug substance is combined with hydrophilic cellulose polymers (excipient material), which slowly erode in the presence of body fluids. On hydration, the polymers behave like a gel and prevent the fast disintegration of the tablet. Diffusion from the gel controls the drug release (e.g., Oramorph SR tablets). A multi-layered tablet consists of several different granulations compressed on top of each other to form a single tablet composed of two or more layers. Each layer is fed from a separate feed frame with individual weight control.

Precompression tamping helps in good binding of layers. Also, reduced pressures prevent intermixing of granules during compression. They may be bilayer where IR/ER combination are used or mainly when incompatible drug substances is used. Sometimes the release from individual layers is controlled to give a drug delivery system, such as, Geomatrix system. In some cases if the bulk density of the tablet is less than one, it floats in the gastric fluids thus extending the residence time in the gastrointestinal tract (GIT). Such dosage forms are called Hydrodynamically Balanced Systems (HBS), an example being Valrelease (Roche). In some cases the drug is embedded inside inert polymeric matrices with materials such as polyethylene, polyvinyl acetate, and polymethacrylates.

The granulations are then compressed into tablets. These inert matrices are excreted in the feces unchanged (e.g., Ferro-Gradumet (Abbott). Some drugs form complexes resulting in slower dissolution and behave as extended release dosage forms, e.g., Rynatan. A slowly eroding tablet may be granulated with hydrophobic excipients (waxy lipophilic material) so that the drug leaches out over an extended period with an outer shell containing the IR dose (buffered aspirin). In some cases a cationic or anionic drug solution can react with an insoluble resin to form a complex. This complex can be tableted, encapsulated or suspended in a vehicle, e.g., Tussionex Pennkinetic Extended Release Suspension (Medeva). Osmotic pump drug delivery systems consist of a core tablet coated with a semipermeable membrane with a fine orifice made by laser beams. The core usually forms two layers containing the active and osmotic agents.

In the gastric fluids water is imbibed by the osmotic agent (pull) and then exerts pressure on the drug (push) in solution, out through the orifice. Such dosage forms are independent of pH of the gastric fluids and are termed as gastrointestinal therapeutic systems (GITS), e.g., Procardia XL (Pfizer). Repeat action tablets consist of slow release inner core and a immediate release (IR) as in Repetabs (Schering) or as bilayer IR/ER tablets. Delayed release dosage forms are used for drugs that are destroyed in the gastric fluids, or cause gastric irritation, or are absorbed preferentially in the intestine. Such dosage forms are enterically coated using materials such as cellulose acetate pthalate, shellac, and waxes.

The coating allows the drug to release at higher pH (pH dependent) or by enzyme catalyzed reactions, e.g., Erythromycin or Aspirin delayed release dosage forms. Several forms of oral controlled release systems are available in the market; however, most of them are dependent on the rate at which the system passes along the gastro-intestinal tract. This can be overcome by regulating the release of the drug by physical chemical means, or by a process related to the environment in which the delivery system is present at the specific time.

Capsules

The word "*capsule*" is derived from the Latin word *capsula* meaning a small box. Gelatin, a substance of natural origin with unique properties, is the major component of capsules. Gelatin is used because it is non-toxic and readily soluble in biological fluids at body temperature. It has good film forming properties and, as a in water and water-gylcerol systems, undergoes reversible phase change from a solution to gel at only a few degrees above ambient temperature. There are two forms of gelatin (A and B) based on the method of manufacture from animal bone and skin. The properties important for capsule shell manufacture are viscosity and bloom strength.

Hard gelatin capsules are firm and rigid while soft gelatin capsules are soft and flexible. This is because soft capsules contain a larger proportion of plasticizers like glycerol, sorbitol, propylene glycol, acacia, and sucrose. Varying proportions of plasticizers are added depending on the intended use of soft gelatin capsules. The colorants used consist of soluble and insoluble dyes. Titanium dioxide and iron oxide pigments are common, although recently aluminum lakes are being used. Preservatives are added to capsules to prevent microbial contamination. Moisture levels are also maintained at low levels to prevent bacterial growth on storage.

Hard gelatin capsules (HGCs)

Hard gelatin capsules are available in sizes ranging from size 000, (the largest) to size 5 (the smallest). The fill weight of capsule and tapped bulk density of the powder blend determines the selection of capsule size. Recently, better techniques for capsule sealing, like, self-locking and have made it possible for a range of materials. The filler material should not react with gelatin. Aldehydes lead to gelatin cross-linking affecting the integrity of the shell and water in the formula can act as a plasticizer. On the other hand, hygroscopic agents can make the capsule shell brittle. Powders filled into hard gelatin capsules should have good flow properties to maintain uniform fill weights during filling operations. Granules and pellets of spherical shape making them free-flowing and non-friable are good candidates for capsule filling using gravitational systems or specialized dosing chambers to maintain uniform fill weight.

In some instances, minitablets (filmcoated, non-friable) can be filled into capsules to produce specialized dosage forms or to separate incompatible ingredients. A recent innovation in hard gelatin capsule filling is a revival of the old practice of filling liquids or semisolids. The main problem encounted is product leakage. This difficulty was overcome by using self-locking capsules and formulation techniques. The use of mixtures of material which are either thermosoftening or thixotropic in nature has become prevalent. These materials are liquefied by heat or shearing force, and revert to solid state within the capsule shell after filling. Filling machines have been developed to handle such formulations with existing powder filling equipment. This system works with solid, liquid, semisolids, and potent drugs.

The application of semisolids filling is also getting prevalent. A more recent innovation in HGCs to fill liquid dosage forms with the use of new machines which heat-seal the caps permanently to prevent leakage as observed with liquid filled HGCs in the past. A good example is the introduction of Licaps (Capsugel) for liquid fills. Capsules made of non-gelatin ingredients for materials not compatible with gelatin are also available, e.g., cellulose (Vegecaps).

Soft gelatin capsules (SGCs)

SGCs or softgels are continuous gelatin shells surrounding a liquid or semi-solid fill. These capsules are formed, filled, and sealed, all in one operation. These capsules are available in different shapes and sizes. SGCs are preferred for drugs with poor compressibility, poor powder flow, mixing problem, unstable or poor solubility in gastric pH, and bioavailability problems. Such drugs can be solubilized or dispersed in a liquid, where dosage uniformity is more accurate. Some drugs that are liquid or that melt during compression are good candidates, if other means of tabletting are expensive. Gelatin used in SGCs has lower bloom strength than HGCs. The plasticizer type and concentration controls the mechanical strength of the shell. In general, plasticizer amounts are larger, making them more flexible than HGC shells.

Preservatives, colorants, and opacifiers are used in the same manner as in HGCs. Sometimes softgel capsules are enteric coated for drugs which are absorbed in the small intestine. Once the capsule shells dissolve in vivo, the drug is available in a liquid or semi-solid form that dissolves or disperses into fine particles with enhanced bioavailability. SGCs can be filled with several materials such as aqueous solutions, non-aqueous solutions, suspensions, pastes, oily solutions of drug, self-emulsifying system, and water-miscible liquids. Materials that cause migration of water or plasticizer from the shell cannot be filled. Surfactants and systems with extreme pH should be avoided.

Liquid Dosage Forms

Solution

A solution is a homogenous single-phase system consisting of two or more components. Solutions are easier to swallow and are acceptable dosage forms for pediatric and geriatric use. The drug in solution is readily available for absorption and therapeutic response is faster. Solutions, however, are bulky and inconvenient to transport. The stability of actives is poorer than in solids and they provide suitable media for microbial growth. Aqueous solutions are preparations made with water as solvent. Purified Water USP is widely used for most preparations. Some drugs are unstable in water or sensitive to the presence of carbon dioxide or oxygen. Not all substances are completely soluble in water and may lead to precipitation. Several other techniques are used to increase solubility of drugs in solution.

Cosolvency is a process of increasing solubility of a drug by using a combination of solvents. Some suitable cosolvents are ethanol, isopropyl alcohol, sorbitol, glycerol, and propylene glycol. If a drug is a weak acid or base, then its solubility in water is influenced by pH. The pH for optimum solubility may not give a stable product. Thus, a compromise must be reached to ensure proper formulation and bioavailability. Suitable buffer systems may be used if necessary. Solubility of insoluble or poorly soluble drugs can also be increased by addition of surface-active agents. Most surfactants are miscible with solvent system and compatible with other ingredients. Hydrophilic surfactants with HLB values >15 are generally preferred. In some cases complexation of a drug with a material may result in formation of soluble molecular complex. However, such complexation needs to be reversible for the active to cross the biological barrier. Chemical modifications of the drug can also result in more water-soluble derivatives. However, these modified drugs are regarded as new chemical entities.

Non-aqueous solutions are used when complete solution is not possible in water or if the drug is unstable. Ethyl alcohol is the most widely used water-miscible solvent for external preparations. Ethyl ether is occasionally used as a co-solvent, in combination with alcohol in the preparation of some colloidons. Other solvents such as isopropyl myristate and isopropyl palmitate are solvents with low viscosity and are ideally used in cosmetics preparations. Xylene is present in ear drops for human use to dissolve ear wax. Liquid formulation additives used include buffers, colorants, flavoring agents, and preservatives. Buffers are dissolved in solvents to resist pH changes. The choice of buffers depends on the pH and the buffering capacity.

Most pharmaceutically acceptable buffer systems include carbonates, phosphates, citrates, gluconates, and lactates. Colors are added for attractiveness and product identification. Flavors are added to solutions to increase their palatability, particularly for drugs with unpleasant taste. This is especially useful in pediatric formulations. Flavors also help in product identification and are of natural or synthetic sources. Fruit juices, peppermint oil, and menthol are some examples of flavors. Some flavors are preferred for specific products, e.g., mint is associated with antacid formulas. Similarly, flavors are preferred by specific patient groups, e.g., children prefer fruity tastes and smell, while adults prefer flowery and acid flavors.

Preservatives help prevent microbial growth. The choice of preservatives should be based on their performance from a microbial challenge test. Care should be taken to ensure there is no adsorption of preservatives onto product containers or packaging material. Antioxidants are added to prevent degradation of the drug in solution; the amount and type can be determined after careful determination of the degradation pathway and stability testing with different agents. Sucrose is widely used as sweetening agent, because it is water soluble, and stable at a wide range of pH. It has a pleasant texture and soothing effect on the throat. There are several other sweeteners that are less widely used. Artificial sweeteners like sugar alcohols and aspartame are used by diabetic patients.

Types of liquid preparations

Draught and elixirs: Draught is a mixture by which one or two large doses of about 50 ml are given. Traditionally, elixirs are solutions of potent or nauseating drugs containing alcohol as a cosolvent (60–70%).

Linctuses: A linctus is a viscous preparation usually prescribed for relief of cough. They usually consist of a simple solution of active in a high concentration of sucrose, often with other sweetening agents.

Mouthwashes and gargles: These are aqueous solutions for prevention and treatment of mouth and throat infections. They usually contain antiseptics, analgesics, and/or astringents. These solutions are used directly or diluted with warm water.

Nasal drops: These are small volume aqueous solutions. They are usually buffered to pH of 6.8 and are isotonic solutions. These drops are used locally as antibiotics, anti-inflammators, and decongestants.

Ear drops: These are simple solutions of drugs in water, glycerol, and propylene glycol for local use in the ear and include antibiotics, antiseptics, cleaning solutions, and wax softeners (xylene).

Enemas: These are available as solutions (aqueous or oily) as well as suspensions for rectal administration of drugs for cleaning, diagnostic, or therapeutic effect.

Lotions: These are available as solutions and suspensions to be applied topically without friction. They may either contain humectant, so that moisture is retained on the skin after application, or alcohol, which evaporates quickly imparting a cooling sensation to the skin.

Liniments: These are intended for massaging the skin. They may contain ingredients such as methyl salicylate or camphor as counter-irritants.

Colloidons: These are prepared from volatile solvents that evaporate quickly leaving a tough, flexible film on the skin that seals small cuts and or holds the active in intimate contact with the skin.

Intermediate solutions: Pharmaceutical solutions are used as intermediates for manufacturing other preparations. Aromatic water is used as a flavoring agent and peppermint and anise waters have some carminative properties. These are manufactured as concentrated waters and are diluted before use. Infusions are prepared by extracting the drug using 25% alcohol without heat. Extracts are similar to infusions, but are concentrated by evaporation. Tinctures are alcoholic or hydro-alcoholic solutions prepared from vegetable materials or from chemical substances. They are relatively weak compared to extracts. Spirits are alcoholic or hydro-alcoholic solutions of volatile substances prepared by simple solution or by admixture of ingredients. These are used as flavoring agents and may have medicinal value. Syrups are concentrated solutions of sucrose or other sugars to which medicaments or flavoring agents are added. These are bacteriostatic by virtue of their osmotic effect, e.g., simple syrup, USP.

Suspensions

These are liquids consisting of insoluble solid particles dispersed throughout a liquid phase. Most suspensions are ready to use while some are prepared as solids to be reconstituted just before use. Ideally, suspension should be homogenous between the time of shaking and dispensing the required dose. The suspended particles should be small, uniformly sized to give a smooth elegant product free from grittiness. Some insoluble solids are not easily wetted by water and thus need wetting agents to be able to disperse readily throughout the medium. Some wetting agents include surfactants, hydrophilic colloids, and solvents. Surface active agents or surfactants possessing HLB value between 7 and 9 are suitable as wetting agents.

Most surfactants are used at concentrations of 0.1%. For oral use Tweens and Spans are commonly used, while sodium lauryl sulphate (SLS) is used for external applications. Some wetting agents may cause foaming and formation of deflocculated systems. Hydrophilic colloids like acacia, bentonite, tragacanth, alginates, and cellulose derivatives function as a protective colloid by coating the surface of the particles and thus imparting hydrophilic character to the solid particles. Solvents such as alcohol, glycerol, and glycols are water-miscible and reduce the liquid/air interfacial tension, increasing wetting.

Flocculation and deflocculation

Flocculation comes from the Latin word *flocculate* meaning loose and woolly. Flocculated systems result in rapid rate of settling because each individual unit is composed of many particles and is therefore larger. However, due to the loose packing of flocs they are easily dispersible on shaking. Deflocculated systems on the other hand are made up of smaller particles whose settling rate is slower, but the settled particles tend to form an irreversible compact and are difficult to redisperse. This phenomenon is called *caking*. For coarse suspensions, a deflocculated suspension will have better uniformity of dose but poorer stability due to formation of cake. Thus, a stable suspension is obtained by preparing a partially flocculated suspension with controlled viscosity so that settling is minimal. Controlled flocculation is achieved by a combination of particle size control, electrolytes to control zeta-potential and by the addition of polymers. Inorganic electrolytes, added to an aqueous suspension alter the zeta-potential of the dispersed particle.

Lowering the zeta-potential sufficiently will result in flocculation. Some of the commonly used electrolytes include sodium salts of acetates, phosphates, and citrates. Use of ionic surfactants may also result in flocculation by neutralization of particle charges. Starch, alginates, tragacanth, and cellulose derivatives are sometimes added to control the degree of flocculation so that the suspension is in a flocculated state and the sedimentation volume is large. Suspensions should exhibit high viscosity at low shear rate and vice versa. Also, the viscosity should be low enough to be poured from the container but should spread evenly, if it is intended for external application. Suspensions for injection should be able to pass through hypodermic needles. Acacia gum is used as a thickening agent for extemporaneously prepared suspensions. Tragacanth forms viscous aqueous solutions, and its thixotropic and pseudoplastic properties make it a better thickening agent than acacia.

Sodium alginate is used as a suspending agent but is incompatible with cationic materials. Several cellulose derivatives, such as methyl-cellulose (Celacol), hydroxy-ethylcellulose (Natrasol 250), sodium carboxymethylcellulose, and microcrystalline cellulose, disperse in water to produce viscous colloidal solutions and are suitable, as suspending agents. Montmorillonite clays or hydratyed silicates like bentonite, veegum, and hectorite readily hydrate and absorb upto 12 times their weight of water. The gels formed are thixotropic and therefore have wonderful suspending properties. Carbopols are synthetic polyacrylic acid copolymers that function as thickening agents at higher pH values. Buffers are included in suspensions to maintain chemical stability and control tonicity. Density modifiers like sucrose and propylene glycol can be added to prevent large differences in densities that could result in sedimentation. Flavors, colors, and perfumes may be added to improve palatability and appearance of the product.

Humectants like glycerol and propylene glycol are added in concentrations of 5% for external applications to prevent the product from drying out after application to the skin. Addition of preservatives is important, particularly when using naturally occuring adjuvants. In some situations sweeteners may be added but their affect on final product viscosity and degree of flocculation should be well understood. Suspensions are normally manufactured using colloidal mills with rotor-stator mechanism to ensure free flowing and evenly dispersed particles. Oral suspensions usually have flavoring agents intended for oral administration. Good examples are milk of magnesia, bentonite, magma, and jellies. All these systems swell and form a gel like consistency with non-Newtonian characteristics. Topical suspensions like Calamine lotion are for external use only.

Emulsions

These are dispersions of one liquid (dispersed phase) in the form of uniformly divided droplets in another liquid (dispersion medium). Depending on which liquid is the dispersed phase oil-in-water (o/w) or water-in-oil (w/o) systems are obtained. To test the identity, emulsion miscibility, staining and conductivity tests are performed. The choice of emulsion depends on the route of administration and the end use. For oral administration, o/w emulsions are used while for external use, both o/w and w/o systems can be employed. Semisolid o/w emulsions are termed as "creams," and are easily washable after application. Water-in-oil emulsions have an occlusive effect and are therefore preferred as moisturizing lotions and cleansing agents. The choice of oil depends on the application with some oils like castor and cod liver oil having a therapeutic value. Thus, w/o preparations are greasy, with high apparent viscosity while o/w emulsions are less greasy and readily absorbed and washable.

Ideally emulsions should exhibit pseudoplasticity and thixotropy, that is, high viscosity at low shear rates and vice versa. They should be dispensable from containers, bottles, and tubes but at the same time should spread on the skin with light pressure. The rheological properties of emulsions are controlled by the concentration, particle size, and viscosity of dispersed phase, concentration of dispersion phase, and the nature and concentration of the emulsifier. The choice of emulsifying system depends on the route of administration, its HLB value and its toxicity. There is no approved list of emulsifiers but pharmaceutical companies employ emulsifiers approved for use in the food industry. Emulsifiers with surface activity reduce the interfacial tension between the phases, thereby decreasing the need for energy to disperse the internal phase. The surfactants used can be anionic like sodium, potassium, and ammonium salts of long chain fatty acids (e.g., sodium stearate), soaps of di- and trivalents metal ions (e.g., calcium oleate and amine soaps, sulphated and sulphonated compounds (e.g., SLS).

Cationic surfactant, amphoteric surfactants like lecithin, and non-ionic surfactants, like glycol and glycerol esters (e.g., glycerol monostearate), sorbitan esters, polysorbates, fatty alcohol polyglycol ethers (e.g., cetyl or cetostearyl alcohol), fatty acid polyglycol esters are also used. Sometimes naturally occurring materials and their derivatives, such as acacia, semi-synthetic polysaccharides (e.g., methylcellulose), sterol containing substances (e.g., beeswax), and wool fat (anhydrous lanolin), can also be employed as emulsifiers. Other ingredients used in emulsions are finely divided solids, antioxidants, such as butylated hydroxy toulene (BHT) and butylated hydroxyanisole (BHA), humectants and preservatives like benzoic acid, parahydroxybenzoic acid esters, chlorocresol, and phenoxyethanol.

Emulsions can be used orally (o/w) with the therapeutic agent included in the internal phase (as for taste masking bad tasting medicaments). Externally, they can be used as lotions either with therapeutic agents or without (as in cosmetics). Emulsions are made using a variety of equipment depending on the stability requirement and the kind of process used. Both batch and continuous processes can be used. Colloid mills have been used traditionally. However, high-pressure homogenizers, microfluidizers, and ultrasonic homogenizers are being used for manufacturing emulsions.

Semisolid Dosage Forms

Ointments, creams, and pastes are semisolid dosage forms intended for topical application. They may be applied to the skin, used nasally, rectally, and vaginally. Most of them contain some form of medicament. Medicated ointments are semi-solid preparations intended for application to skin or mucous membranes. Non-medicated ointments are used as protectants, lubricants, and emollients. Ointment bases used for- ointment preparation are of fourtypes, hydrocarbon bases, absorption bases, water-removable bases, and water-soluble bases. Hydrocarbon bases have emollient properties and are effective as occlusive dressings (e.g., Petrolatum, USP). Absorption bases permit the incorporation of aqueous solutions to form w/o emulsions (e.g., hydrophilic petrolatum and lanolin). Water-removable bases are also o/w emulsions and are water washable (e.g., hydrophilic ointment). Water-soluble bases have no oleaginous component and are referred to as greaseless water-washable bases (e.g., polyethylene glycol ointment).

The potential for absorption depends on the choice of the bases, and intended use of the medicament. Appropriate selection of ointment bases is important fordermal therapy. Ointments for rectal preparation (e.g., Tronolane ointment for hemorrhoidal analgesia) and vaginal preparations (e.g., Mycelex-7 ointment as antifungal) are available in the market. Creams are semisolid emulsions with one or more medicinal agents intended for external use. The so called *vanishing creams* are o/w emulsions with stearic acid and cold creams are w/o emulsions with an oily base. Creams spread more easily than ointments and are preferred by some patients. Gels are semisolid systems with dispersions of small or large molecules in an aqueous vehicle with a gelling agent e.g., high molecular weight Carbopols that are cross-linked polyacrylic acid. Some gels like milk of magnesia or magma has two phases. These behave as thixotropic systems with the viscosity changing due to a gel–sol transition on shaking.

Pastes are ointments with large amount of powder levigated into the base and are intended for application to the skin. Pastes are more hygroscopic than ointments and are used to absorb serous secretions. (e.g., zinc oxide paste). Plasters are solid or semisolid masses spread on backing paper, plastic, or fabric. Non-medicated forms are termed as adhesive plasters while medicated plasters provide a therapeutic effect at the site of application. Some medicated plasters are termed as cataplasm, e.g., ibuprofen and salicylic acid plasters. *Transdermal drug delivery systems* (TDDS) are used for delivery of actives through the skin into the systemic circulation. Several methods have been used to facilitate the transport of active through the barriers of the skin. Certain absorption enhancers are used to temporarily increase permeability of the skin for increased delivery.

More recently, iontophoretic techniques have allowed the delivery of charged chemicals across the skin using an applied electric field (e.g., amino acids and proteins).Sonophoresis or high-frequency ultrasound is also being studied as a means of effectively enhancing trans-dermal delivery of drugs. Transdermal delivery systems are of two main types the monolithic matrix system that contains the excess drug dispersed in the polymeric matrix and cast into a matrix with a backing layer and frontal membrane, e.g., Estraderm (Novartis). Membrane-controlled transdermal patches contain drug reservoir in the form of a gel or saturated solution of drug with a backing adhesive and a rate controlling membrane e.g., Transderm-Scop (Novartis). Currently, the market is flooded with a variety of transdermal systems for smoke cessation and hormonal delivery patches. More recently efforts are on to develop transmucosal delivery patches.

Other Dosage Forms

Suppositories are solid dosage forms intended for insertion in body cavities like rectum, vagina, and occasionally in the urethra for local or systemic effects. The length, shape, and weight of these depend on the body cavity it is used for. These melt, soften, or dissolve after application depending on the type of suppository base applied. Cocoa butter base suppositories usually melt in contact with the body temperature. Other bases, such as polyethylene glycol, glycerin, and soap based suppositories solubilize. Cocoa butter suppositories are hydrophobic, and dissolve oil soluble drugs but absorption is poor in the aqueous rectal fluids. Thus, better absorption is obtained with water-soluble bases. Rectal suppositories can be used for local (hemorrhoids) or systemic effects. Systemic bioavailability is poor and the amount of drug required is more than oral administration.

Vaginal and urethral suppositories are usually used for their local effects (e.g., anti-infectives). Rectal suppositories weigh about 2 g and have varying shapes like bullets and torpedoes. Urethral suppositories are thin pencil- shaped with tapered ends. Vaginal suppositories are globularized and weigh about 4–5 g. Generally absorption from suppository bases rectally is better when the rectum is empty. Particle size of solids plays an important role in absorption. Most importantly rectal absorption allows to bypass the first pass effect. Most suppositories are manufactured by molding from melts or by compression.

Aerosol Dosage Forms

Aerosols are pressurized dosage forms containing one or more active drug dissolved, suspended, or emulsified in a propellant or a mixture of solvent and propellant, which is released on actuation of the valve as a fine dispersion of liquid or solid in a gaseous medium. Aerosols are intended for topical administration; for administration into body cavities; for administration orally or nasally as fine solid particles or liquid mists through the pulmonary airways, nasal passages, or oral cavity (buccal or sublinual). Those that provide an airborne mist are called space sprays; those intended for carrying actives to surface are termed as surface sprays and other are termed foam aerosols. Aerosol consists of product concentrate and the liquefied propellant. The pressure depends on the types and amounts of propellants and the nature and amount of active present. Propellant is a liquefied gas or mixture of liquefied gas, which serves as the solvent, or vehicle.

In some cases non-liquefied gases like nitrogen, and carbon dioxide are used. Space aerosols (85% propellant) operate at 30–40 psig at 70°F. Surface aerosols (30–70% propellant) operate between 22–5 5 psig at 70°F. Foam aerosols operate at a slightly higher pressure. Foam aerosols are emulsions of the propellant and product concentrate. Aerosols can be two-phase systems comprising of a solution of drug in liquefied propellant and a vapor phase of propellant/gas. Some of them are present as three-phase systems, comprising of water-immiscible propellant, an aqueous product concentrate or drug in suspension/emulsion and a vapor phase. Aerosol containers are made from glass (coated and uncoated), tin, aluminum, and stainless steel containers. There are various forms of spray valves and metered valves (for more accurate dosing of potent drug). Oral aerosols are mostly used via the buccal route, e.g., Nitrolingual spray that emits nitroglycerin at a dose of 0.4 mg per metered dose.

The respiratory tract offers several advantages for administration of drugs. Inhalation systems should be capable of producing fine particles, usually <10 μm for effective drug delivery. Inhalation drug delivery has been traditionally used to treat respiratory disease, but in recent times the lung has been used as a portal for administering drugs to the systemic circulation. With their large effective surface area, the lungs offer an attractive route for systemic drugs. Three main dosage forms, viz. metered dose inhalers (MDIs), dry powder inhalers (DPIs), and nebulizers have gained prominence. Metered dose inhalers are pressurized systems consisting of drug suspension in a propellant and may contain other additives such as surfactants, antioxidants, and solvents, although some solution systems are available. Traditionally MDIs have been formulated using chloroflurocarbons (CFCs). These propellants are now being phased out under the terms of the "Montreal Protocol" due to their ozone depleting potential. This has resulted in an increasing urgency to reformulate existing MDI formulations with alternative hydroflurocarbon (HFC) propellants such as HFC-143a and HFC-227. Formulations using the new propellants have recently won FDA approval.

To offer better coordination between actuation and patient inhalation of the emitted dose, several spacer devices are available with marketed MDI formulation. Recent trends in devices include the breath actuated MDI, where the patient inhalation triggers the dose. Due to the phase out of CFCs and extensive difficulties in reformulating using HFCs, dry powder inhalers, and nebulizers are becoming popular. Dry powder inhalers consist of mixtures of micronized drug and a large particle size carrier (usually lactose). These drug-carrier ordered mixtures are packaged in unit doses (capsules or blisters), as in Spinhaler and Rotahaler or as pure drug bulk powder, which can be metered into single doses. In many instances the patients inhalation maneuver causes the active particles to separate from the carrier in the air stream. More recently, to overcome intra-patient variability in dosing, active DPIs have been designed. In these devices, the patient's breath triggers a deaggregation mechanism, which separates the drug particles from the carrier, which are then inhaled.

Newer fine particle generation technologies, such as spray drying and supercritical fluid extraction, are being used to produce fine particles of drug including proteins and peptides for delivery to the deep lung. This can facilitate the use of inhalation delivery systems for systemic drug delivery. Nebulizers generate fine mists from aqueous and non-aqueous drug solutions, using either compressed air or ultra-sonication. Their main disadvantage is that they are bulky and are not portable. Recent developments in this field have focused on the development of battery operated portable devices, which can nebulize aqueous solution containing minimal, or no preservatives.

Pediatric and Geriatric Dosage Forms

Physiology plays an important role in the development and performance of different functions in the body. It is important to determine their consequence on dosage form development. Pediatric dosing is usually determined by weight and age. FDA classifies the pediatric groups into neonates, infant, child, and adolescent. There are several excipients that have been reported to cause adverse reactions, e.g., azo dyes cause bronchoconstriction, lactose may cause prolonged diarrhea and intolerance, and sweeteners such as saccharin are weak carcinogens. Alcohol is a common solvent for most pediatric OTC liquid products. However, limits on the alcohol content of OTC products have been set to minimize toxicity in children. Oral administration is the preferred route for children. However, children younger than 5 years have difficulty swallowing solid tablets. Thus, oral liquid is the most preferred dosage form in pediatric patients. Liquids are often unstable and have short expiration and accurate dosing is difficult. Recently, there has been increased interest in chewable tablets and "*sprinkle*" powders in capsule formulations as they are well received by children with dentition.

Rectal administration is not popular because of wide variability in absorption. Pulmonary administration is emerging to be a popular delivery mode for children, but needs to be studied further for systemic effects. Transdermal route may be another area to explore for children, since the stratum corneum is well developed in children as in adults. This may be beneficial as an alternate route for children. Pediatric drug therapy has very few drug delivery systems. There has been some in-roads made with OTC cough and cold products. However, most industries do not have resources to perform separate studies for safety and efficacy in children (smaller consumer than adults) for new chemical entities.

Perhaps FDA should take initiatives to provide pharmaceutical companies with returns like tax break or patent extension and specific market for such developmental work especially for life threatening diseases. Apart from alterations in the pharmacodynamics (PD) and pharmacokinetics (PK), the geriatric population suffers from a number of chronic conditions and physical limitations. Clinical monitoring becomes very important to titer dosing accurately. Most of their PK and PD processes take a down turn. Absorption is slower from the oral cavity. In general, the aged skin is more permeable to water and other chemicals. However, the clearance to the blood stream is lowered thus distribution may not be complete.

Physically, impairment or decline in vision may hinder one's ability for self-medication. Also, swallowing and chewing may be a problem in elderly patients. For example, patients suffering from dry mouth may have difficulty swallowing a tablet or capsule. Similarly, elderly patients who are edentulous (i.e., toothless) are incapable of chewing any tablet dosage form. Although sublingual and buccal tablets are used by the elderly population there is very less emphasis on its effect on bioavailability with aging. Patients with dry mouth condition may feel local irritation with such dosage forms. Capsules like tablets may hinder swallowing and are not advisable for elderly patients. Liquids are easier to swallow, but are usually not packaged as unit dosages.

Patients with impaired vision and dexterity may not be able to accurately self-administer the required dose. The trans-dermal route seems to offer better compliance with elderly patients but bioavailability needs to be determined before using this route. Several alternative dosage forms and packaging techniques are emerging to improve compliance of dosage form in elderly patients. Several types of packaging aids like dosett tray, calender-packs and med packs are used to remind the dosing schedule for elderly patients.

Granules of drug may be easy to swallow. They can be mixed with water or food and swallowed easily. Unit dose packs may still be difficult to use for some elderly patients. Effervescent tablets provide an alternative dosage form. These tablets dissolve in water to form a ready-to-use product. The use of an irregular shaped tablet that prevents it from lying flat may be another form that can help patients with impaired dexterity. Similar to granules the drug may also be presented in the form of a small amount of concentrated solution for the entire dose, e.g., 5 ml Rapamune concentrated oral solution. Such products can be mixed with food or drink.

This is similar to the use of a dispersible tablet that forms a uniform stable suspension when dispersed in water. Emerging technology has focussed on newer dosage forms like the use of quick or rapid dissolving technology (RDTs), wherein, the dosage form quickly dissolves in the mouth and rapid absorption of the drug can occur systemically or even from the mouth. Furthermore, compliance in elderly patients has also resulted in the availability of sugar and sodium-free products that are beneficial for such age groups.

New Drug-delivery Technologies

Newer technologies are emerging as we move into the new millenium. These technologies promise to have lots of benefits such as simplifying administration regimens, enhancing compliance, improving clinical benefits, and reducing overall healthcare costs. Rapid-dissolving tablets (RDTs) are designed for patients who have difficulty in swallowing standard tablets/capsules, such as pediatric and geriatric patients. These include the lyophilized foam from Zydis, Flashtab, Orasolv, Wowtabs, and Flashdose. Hydrogel based technology offered by Professor Neil Graham of British Technology group for development of several systems including morphine suppositories. Nanocrystal technology offered by Elan Corporation where the crystalline drug (<400 nm) is thinly coated with a surface modifier to impart physical stability. Liquitard is a liquid taste masking sustained release granules as a suspension and is offered by Eurand America.

Inhale Therapeutics offers proprietary technology for pulmonary delivery of proteins and peptides, using innovations in powder processing to develop formulations for deep lung delivery for systemic and local indications. Jago Pharma developed the Geomatrix systems which involves the use of multilayered hydrophilic matrix systems. Several newer excipients are now available as matrices for controlled delivery. These include polysaccharides from Galactomannan such as guargum and locust bean gum. Considerable research efforts have been towards the development of safe and efficient chitosan-based dosage forms. The development of solid-lipid nanoparticles (SLN) have made it possible for delivering drugs with less side-effects, better targeting, and protection from enzymes. Zambon group in Europe has developed Timeclock technology that involves coating solid drug with hrdrophobic surfactant. Chronotropic drug delivery, which targets delivery to a specific absorption window for local as well as systemic effect.

Emisphere technology uses a carrier that binds to drug molecules non-covalently to form a complex. These complexes easily cross the membranes and then dissociate to release the active. Theratech uses the Theriform microprinting technology to develop oral and implantable dosage form for making Microdose tablets. Delsys Corporation has revolutionized their Accudep technology of electrostatic deposition of dry powder to any surface with great accuracy. Labopharm Inc. has developed Contramid technology obtained by cross-linking of high-amylose starch in three dimensional network which is combined with the active. Once in the stomach, the tablet surface turns into a gel and the active diffuses at an even rate. PORT (Programmable Oral Release Technology) is a technology platform to resolve drugs with biopharmaceutical problems. PORT is a capsule based system with opportunity to provide multiple prolonged release of one or more drugs. This technology is applicable to a wide variety of drug classes and can be used to achieve difficult PK profiles, e.g., zero order with burst system.

Quadrant Healthcare has engineered micron-sized particles with enhanced stability and PK profiles. Solidose technology is based on chemically modifying oiligosaccharides to make them more hydrophobic. In contact with body fluids they undergo phase change and release the drug. Several self-emulsifying and lipid- based systems have been developed that form micro-emulsions of water-insoluble drugs for oral delivery. DanBiosystems has developed a proprietary Targit technology, which involves enteric coating with azo polymers, which degrade only at specific sites by bacterial enzymes. Ethypharm is another company involved with coating of nano- and microparticles, using supercritical fluid (CO_2) technology, suited for fragile water-soluble molecules (peptides and proteins).

Protarga is another company which is involved in the development of dosage form by covalently attaching fatty acids (docosahexaenoic acid or DHA) to actives to create new compounds that can be taken up by cells targeted for treatment. A lot of focus has been directed to tissue engineering that is used to design biological substitutes or regenerate natural tissues for defective or lost tissues and organs through the use of cells and their scaffolds. Fentanyl Oralet (Abbott labs) is a lollipop that has painkillers used as preoperative sedative. Several forms of intra-vaginal drug delivery systems like Progestasert system and Dinoprostone vaginal insert have been developed. Implants like the levonorgestrel Norplant device that is incorporated as contraceptive on the upper arm. Similarly, Gliadel wafers are implanted in the brain tumor cells. Non-parenteral dosage forms can be administered by different mechanisms.

In recent years there has been a plethora of emerging drug delivery technology companies. Most of these technologies are addressing issues related to unique delivery systems and selectivity of these systems for affected organs or diseases. However, rarely it recognized that improvements in drug therapy are a consequence of not only the new chemical entity but also the combination of active and the delivery system (dosage form). Currently the trend is to develop a delivery system and then look for suitable drug candidates to apply. This needs to be changed to gain understanding of what unmet medical need is the active aimed for, the rate, time and site for the active to be delivered and then provide a suitable delivery system. More emphasis should be made on the improvement of drug effect profile. Thus, the future of drug delivery will depend on how they can contribute to drug therapy for unmet medical needs.

INJECTABLE FORM

Parenteral is derived from the two words "para" and "enteron" meaning to avoid the intestine. Parenteral articles are defined according to the USP 24/NF19 "as those preparations intended for injection through the skin or other external boundary tissue, rather than through the alimentary canal, so that the active substances they contain are administered using gravity or force directly into a blood vessel, organ, tissue, or lesion. Parenteral products are prepared scrupulously by methods designed to ensure that they meet pharmacopeial requirements for sterility, pyrogens, particulate matter, and other contaminants, and, where appropriate, contain inhibitors of growth of microorganisms. An injection is a preparation intended for parenteral administration and/or for constituting or diluting a parenteral article prior to administration." Parenteral drug administration is an attractive route of administration when oral administration is contraindicated, and it has been traditionally used in institutional settings.

With an increasing interest in reducing overall health care costs and with the development of new biotechnologically derived compounds and improved and novel infusion-related technologies, parenteral products have become an important component in the care of patients in hospitals and the home health care setting. In the present article, information will be presented on history and the following: the use of parenterals in health care, the advantages and disadvantages of using parenterals, routes of administration, vascular access devices and infusion sets, types of parenteral products, components of parenteral products, parenteral packaging, convenience and needleless systems, needleless injection, extemporaneous compounding of parenteral products, infusion pumps and devices, and future parenteral dosage forms.

Clinical Values

A detailed history of early parenteral medications can be found in the first edition of this encyclopedia. One of the first historical references to the parenteral administration of a compound was in the late 15th century when a blood transfusion from three young boys was given to Pope Innocent VIII, resulting in the death of all four individuals. These deaths led to a ban in the use of this type of medical treatment, namely an infusion, for several centuries. It was not until the 17th century that studies on the parenteral administration of compounds was first studied in animals. The development of the hypodermic needle and the use of parenterally injected drugs in humans is first reported in the mid-19th century. By the end of this century, there was an increased interest in the use of intravenous administration of glucose and normal saline solutions.

Baxter produced the first commercially prepared intravenous solutions in 1931. However, parenteral products and their administration became acceptable and a mainstay in the treatment of patients in the mid 20th century. This could be attributed, in part, to our increased understanding of microbial and viral agents, increased recognition of the dangers associated with parenteral therapy, the development of antibiotics and other drug classes, and the availability of new systems or infusion technologies. In the mid 1960s, many hospitals introduced intravenous admixture services. In the last 20 years, the area of infusion pumps and systems and improved vascular access devices has enabled parenteral therapy to extend beyond the institutional setting to clinics, ambulatory infusion centers, and home health care.

In addition, the administration of parenteral drugs is also frequently utilized in basic and clinical research in animals and humans. In today's health care environment, parenteral products are a key component of therapy for hospitalized patients. Vascular access for parenteral infusion therapy is obtained in the vast majority of hospitalized patients at some point in their therapy. There are very little data as to the use of parenteral products in today's health care environment. It was suggested that 40% of all pharmaceutical dosage forms are administered as a type of injection and that over 350 million units of large volume parenterals and 100 million units of IV admixtures (piggybacks) were used annually in the late 1980s.

These numbers have certainly increased since that time with the advent of new drugs and infusion methodologies. Infusion therapy in home health care continues to have a strong market in the United States. While the annual growth rate of home infusion therapy decreased between 1982 and 1993, 29% of acute care hospitals provided or were developing a home health care program. In the current market, home infusion therapy is being integrated into alternative sites, such as ambulatory infusion centers. Due to the rapid and increasing use of parenteral drugs, it is critical for health care providers and scientists to understand the various available dosage forms, specific products, routes of administration, catheter types, and various infusion devices. In addition, company websites are a valuable source of current information, including available parenteral products, infusion sets and ports, devices and infusion pumps.

Objectives

Generally, parenterally administered drugs are advantageous because they can provide rapid and reliable drug systemic effects, long-term drug delivery, and targeted drug delivery. The disadvantages of parenteral products are predominately associated with safety issues related to infection and thrombosis, tissue damage and/or pain upon injection, and the use of requirements for specific equipment, devices, and techniques.

Direction of Intake

The most commonly used routes are intravenous, intramuscular, subcutaneous, and intradermal. Formulations intended for administration into the central nervous system should not include preservatives. For intramuscular, intradermal, and subcutaneous, a single needle and syringe is generally used to administer the drugs. For intravenous and intra-arterial, drugs are administered using vascular access or port devices. Other parenteral routes of administration (e.g., epidural, intra-articular, and intrathecal) usually require specialized delivery sets. In some cases, the specific drug requires a specialized delivery set be utilized due to the dose of the drug or the physicochemical properties of the drug (e.g., nitroglycerin).

Circulatory Accessment

Vascular access can be achieved through short peripheral, long dwell peripheral, central lines, and ports. These various devices differ in their insertion, characteristics, dwell time or time they should be in place, and usage and safety features. In peripheral access, the distal veins on the hand and arm are often used, with the basilic and cephalic veins in the arms being the most common site for peripheral infusions. Alternatively, the metacarpal veins can also be utilized. The basic devices are a winged infusion device or the over-the-catheter needle, with needle sizes ranging from 16 to 26 G (20- or 22-gauge being the most common size). These catheters are usually utilized only for 48 h. A midline catheter is usually inserted into a large vein and is intended to be used over a 2- to 4- week period (Richardson). Peripherally inserted central venous catheters (PICC) are designed for long-term infusion therapy up to a year. The devices are inserted into a peripheral vein and threaded so that the tip of the catheter is within the central vascular system.

Central venous catheters (CVC) are inserted into the subclavian or jugular vein and threaded until the tip is located in the superior vena cava. The types of central venous catheter are a non-tunneled, Groshong, Hickman, and Brovaic. A non-tunneled CVC can be inserted at the bedside, whereas the Groshan, Hickman, and Brovaic require surgical insertion. Vascular access ports (VAP) are an alternative to central venous access. These devices are surgically implanted usually into the chest wall or arm subcutaneous tissue and are composed of a rigid reservoir with a self-sealing rubber septum, and the tip of the catheter is placed into a central vein. The drug is placed into the reservoir via an injection. A VAP allows repeated, intermittent access and drug delivery (in some cases up to 2000 times), depending upon the size of the needle.

Examples of these types of ports include a P.A.S Port (SIMS Deltec), Vital-Port, and BardPort and CathLink. There are three basic types of intravenous administration: (1) primary set; (2) secondary set; and (3) a volume control set. The basic components of all these sets include a piercing spike (to insert into the bag or bottle), drip chamber and drip orifice, tubing ranging in length from 160 to 250cm (63–98.58 in.), a roller clamp or other flow control device on the tubing, a Y-site for infusion of other components, an in-line filter (ranging from 0.2 to 170 μm) and leuk-lok adapter to attach to the vascular access device. A primary infusion set is designed to deliver solutions from the parenteral container via gravity. If an additional infusion is needed, the secondary

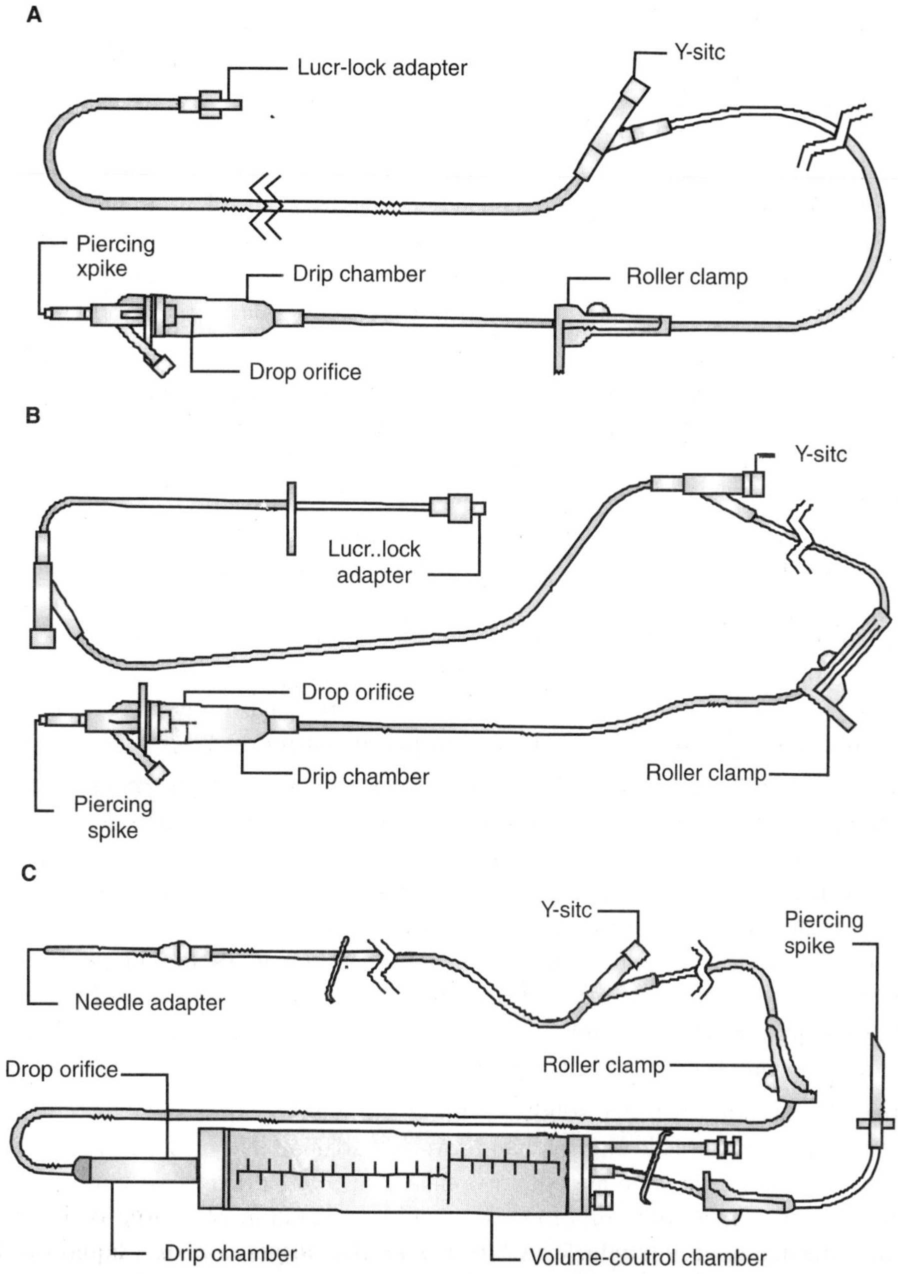

Fig. 1.7. Types of infusion sets used for parenteral therapy: A-primary; B-secondary; and C-volume-controlled infusion sets.

set can be connected to the primary set via one of the Y-sites. A volume control administration set is used to deliver a small amount of solution through the use of the volume chamber. In general, administration sets are classified into macrodrip sets that can deliver 10–20 drops/ml and a microdrip set that delivers drugs at a slower rate of 60 drops/ml.

Nature of Injectable Drugs

Parenteral products can be divided into two general classes according to the volume of the product. All parenteral products are sterilized and must meet all the requirements for sterility and particulate matter and must be pyrogen-free. They must be prepared using strict sanitation standards in environmentally controlled areas by individuals trained to meet these standards. The injections are overfilled with a small excess over the labeled volume to ensure that the required volume can be obtained from the product. *Small-volume parenterals* (SVP) or injections are 100 ml or less and can be provided as a single- or multidose product. In contrast, large-volume parenterals (LVP) are intended to be used intravenously as a single-dose injection and contain more than 100 ml of solution.

SVPs and LVPs are often combined during the extemporaneous preparation of intravenous admixtures, to be discussed later in this article. The U.S. Pharmacopoeia (USP) classifies injections into five different types. The dosage form selected for a particular drug product is dependent upon the characteristics of the drug molecule (e.g., stability in solution, solubility, and injectability), the desired therapeutic effect of the product (e.g., immediate vs. sustained release), and the desired route of administration. Solutions and some emulsions (e.g., miscible with blood) can be injected via most parenteral routes of administration. Suspensions and solutions that are not miscible with blood (e.g., injections employing oleaginous vehicles) can be administered via intramuscular or subcutaneous injection but should not be given intravenously. Parenteral products contain excipients such as buffers, solvents, non-aqueous solvents, antimicrobial preservatives, antioxidants, and chelating agents. Coloring agents are prohibited in parenteral products.

All excipients must meet compendial standards, and the excipients must not interfere with the efficacy of the product (to be discussed more in detail later in this article). Parenterals are packaged in airtight containers using specific, high quality materials so that they do not interact with the product and to maintain the sterility of the product. For example, the type of glass to be used in a specific parenteral drug product is indicated in the monograph. The types of packaging and containers for SVPs and LVPs will be discussed later in this article. A SVP product is available for most of the major therapeutic classes of drug. It is often desirable for a manufacturer to provide both an oral and parenteral dosage form for a specific drug product. A "*drug injection*" is a liquid preparation that is composed of drug substances and or solutions.

A "drug for injection" is a dry solid that upon the addition of a suitable vehicle (usually a vehicle in which the drug is stable and soluble) provides a solution that conforms to the requirements for an injection. Drugs for injection are often lyophilized or freeze-dried to assist in the reconstitution of the solid. A "*drug injectable emulsion*" is a liquid preparation of a drug or drug substances dissolved in a suitable emulsion vehicle. A "*drug injectable suspension*" is a liquid preparation of solids suspended in a suitable vehicle. A "drug for injectable suspension" is a dry solid (often lyophilized) that is intended, upon the addition of a suitable vehicle, to yield a preparation that in all aspects meets the requirements for an injectable suspension. LVPs are often administered via intravenous infusion in a large single-dose container.

The therapeutic goal of these products is to provide electrolytes, body fluids, and nutrition. These solutions may or may not be isotonic with blood depending upon the concentration of the components, which include sodium chloride, dextrose, mannitol, Ringers (sodium, potassium, calcium, and chloride) and Lactated Ringers (calcium, potassium, sodium, and lactate), sodium bicarbonate, ammonium chloride, sodium lactate, fructose, alcohol, dextran, and amino acids. Other drugs (small volume injectables) are often combined to these LVPs, provided that these two products are compatible during the extemporaneous preparation of intravenous admixtures.

Biochemical Nature

Parenteral products are optimized during their development to provide the requisite solubility (per the required dose), stability, and syringeability. In addition, these products must meet the desired requirements for the rate of drug release based upon the dosage form and biopharmaceutical properties. Finally, it is important that parenteral products also be evaluated for their potential to cause tissue damage and/or pain associated with the injection of the formulation. The adjuvants in parenteral products can include solvents, vehicles, cosolvents, buffers, preservatives, antioxidants, inert gases, surfactants, complexation agents, and chelating agents. It is important to understand the various types of waters used in parenteral products. The most frequently used solvent in parenteral products is Water for Injection, USP, which is not required to be sterile but must be pyrogen-free.

In contrast, Sterile Water for Injection, USP is water that has been sterilized, does not include a preservative or antimicrobial agent, is pyrogen-free, and is provided in single containers no larger than 1000 ml. The use of this product is for the reconstitution of other parenteral products, in most cases antibiotics. This product must not be given alone. Bacteriostatic Water for Injection, USP is sterile water that can contain one or more preservative or antimicrobial agent (specified on the label) and is packaged in prefilled syringes or vials that are no larger than 30 ml. It is also used in the reconstitution of SVPs. The limitation with Bacteriostatic Water for Injection, USP is the presence of the antimicrobial agent that is contraindicated in newborns. Other solvents used for parenteral formulations are Sodium Chloride Injection, USP and Bacteriostatic Sodium Chloride Injection, USP, Ringers Injection, USP, and Lactated Ringer's, USP. Other vehicles may be added to parenteral products if the aqueous solubility is limited. However, these vehicles must be non-toxic, non-sensitizing, and non-irritating.

In addition, these solvents must be compatible with the drug and other components in the formulation. Cosolvents often used in parenteral formulations include propylene glycol, ethanol, polyethylene glycols, glycerin, and dimethylacetamide. In addition, fixed vegetable oils, such as peanut, cottonseed, sesame and castor oil, can be used; however the USP provides clear restrictions on their use in parenteral products. Buffers can also be provided in parenteral formulations to ensure the required pH needed for solubility and/or stability considerations. Other excipients included in parenteral products are preservatives (e.g., benzyl alcohol, *p*-hydroxybenzoate esters, and phenol), antioxidants (e.g., ascorbic acid, sodium bisulfite, sodium metabisulfite, cysteine, and butyl hydroxy anisole), surfactants (e.g., polyoxyethylene sorbitan monooleate), and emulsifying agents (e.g., polysorbates). An inert gas (such as nitrogen) can also be used to enhance drug stability. Stability and solubility can also be enhanced by the addition of complexation and chelating agents such as the ethylenediaminetetraacetic acid salts. For a more detailed list of approved excipients in parenteral products, the reader should consult the monographs within the USP.

Packaging Methodology

In general, all parenteral products must be manufactured under strict, current good manufacturing processes (cGMP) to ensure the final product is sterile and pyrogen-free. Sterilization is defined as the complete destruction of all living organisms or their spores or the complete removal from the product. Pharmaceutical products can be sterilized by steam sterilization, dry-heat sterilization, filtration sterilization, gas sterilization, and ionizing-radiation sterilization. The USP provides monographs and standards for biological indicators required to test the validity of the sterilization process. These products must also be tested for pyrogens–fever-producing substances that arise from microbial contamination most likely thought to be endotoxins or lipopolysaccharide in the bacterial outer cell membrane. Injections are provided in either multiple-dose containers or single-dose containers.

A multiple-dose container is often a vial that will allow the withdrawal of successive portions of the contents without a change in the strength of the product and while maintaining the sterility. A single-dose product is intended for a single parenteral administration. These products can be an ampul, vial, or a syringe. For some drugs, there are specific double-chambered vials that contain the reconstitution solvent and the powdered drug (e.g., Mix-O-Vial–to be discussed later). Types I, II, and III glass are required for parenteral products and are specified in the individual monograph for a given drug. Ampuls are utilized for a single dose and, as such, do not require a preservative. However, in many cases, the manufacturer will include a preservative, as the drug formulation is the same for both the ampul and multiple-dose vial.

The disadvantages of ampuls are that these containers become contaminated with glass particles when opened and require the use of a syringe to remove the drug solution. A filter needle must be used sometime during the withdrawal of the solution or delivery of the drug solution to a flexible bag or other intravenous solution to ensure the glass is removed from the solution. Ampuls are opened via breaking the neck at a prescored position. Vials can be used for single or multiple doses. The glass containers are sealed with rubber closures that permit the withdrawal of the drug solution via a syringe. The disadvantage of these systems is associated with ensuring that the drug solution is compatible with the rubber closure. Furthermore, when utilizing vials in the extemporaneous preparation of sterile intravenous admixtures, the health care practitioner must minimize the potential of coring during the introduction of the needle through the rubber seal.

Furthermore, there is always the concern of contamination of the solution with repeated withdrawals. The potential for contamination can be minimized by the use of single-dose vials. Parenteral solutions can also be packaged in syringe dosage forms for a single-dose use. As such, they can be considered a type of convenience container (to be discussed later). The syringe and needle are sterile until opened. They are ideal for emergency situations or the home health care environment. LVPs are usually provided in glass containers, flexible plastic bags, or semirigid containers. These systems are also classified as open systems (non-vacuum) and closed systems (vacuum). The largest manufacturers of LVPs are Abbott Laboratories, Baxter Healthcare Corporation, and B. Braun. Glass containers are sealed with a thick rubber disk and a target in the center for the piercing spike. Glass bottles can be either vented with a plastic venting tube or non-vented, thereby requiring either a non-vented administration set or a vented administration set, respectively.

The advantages of glass containers for parenterals are that they are easy to sterilize, can be accurately read, and are generally inert and less susceptible to incompatibilities with drugs or leaching of components compared to the plastic flexible intravenous bags. The disadvantage of glass is associated with handling the glass bottles and the potential for breakage. Plastic intravenous fluid containers were first introduced due to the need to start intravenous therapy while transporting soldiers from the battlefield or triage area to the hospitals. These containers are flexible due to the presence of plasticizers, with the bags being composed of polyvinyl chloride. In contrast, semirigid containers are often composed of polyolefin. The most common manufacturers of intravenous solutions are Abbott Laboratories– Lifecare, Baxter Healthcare–Viaflex, and B. Braun Medical–Excel. The major advantages of plastic flexible bag systems for parenterals are that they do not require the use of a vented administration set as they collapse when empty, and they are less susceptible to breakage. It is also easier to store and transport these bags.

The difficulties with these flexible plastic bags for infusions solutions are the potential for incompatibilities of the drug substance with the components in the bag, the potential for the bag to get perforated during its use, thus compromising the sterility of the solution, and the difficulty in reading the volume remaining in the bag. One major concern with the use of flexible plastic bags is the potential for the drug compound to leach out the plasticizers from the systems. Semirigid containers are similar to flexible plastic containers in that they are lightweight and nonbreakable and can be easily transported and stored. These containers are less likely to be perforated during their use. More importantly, these containers do not contain plasticizers and, as such, may be more compatible with drug substances. The disadvantages are related to their similar properties to glass containers, in that they require venting, can be more susceptible to cracking upon extreme changes in temperature, should not be frozen, and do not adapt well for ambulatory care.

Advanced Technologies

Whereas the compounding and administration of parenteral products and intravenous admixtures continues to be a vital and important component in the care of hospitalized and home health care patients, there is continued interest in easing the preparation, storage, and administration of these products with respect to controlling contamination of the finished product and protecting the health care providers from needle-stick injuries. It is estimated that more than 750,000 needle-stick injuries occur every year. Commonly used convenience and/or needleless systems include premixes, bags, and vial systems, prefilled syringe systems and double-chambered vial systems. For drugs with suitable stability in intravenous solutions, premixes provide an alternative to the extemporaneous compounding of admixtures.

These products are ready to administer, reduce the chance for a medication error, reduce the potential for infection, and decrease the chance for needle-stick injuries. In addition, there is an advantage in using these products with respect to the shelf-life of the product. The stability and storage requirements for each product is provided by the manufacturer. For example, the FirstChoice Premix products (Abbott) in the over- wrap have a shelf-life of, typically, 18 months, whereas those products in which the overwrap has been removed and which are stored at room temperature can be stable for up to 30 days, provided no additional drugs or additives have been added to the product.

The diluents in these premix products include 0.45 or 0.9% sodium chloride, 5% dextrose, water for injection, Lactated Ringers, and combinations of these diluents in volumes ranging from 50 to 1000 ml in plastic or glass containers. Drug classes that are currently formulated as premixes include amino acids, dextrans, electrolytes, cardiovascular, anti-infectives, analgesics, and gastrointestinal and respiratory compounds. The Add-Vantage system and the Mini-Bag Plus system are needleless drug delivery systems composed of a drug-containing vial and a diluent in a flexible plastic intravenous bag. The drug in the vial comes in contact with the diluent, followed by the drug being transferred back into the bag with the vial still attached; this system can then be attached directly to the infusion equipment. In the Add-Vantage system, there are specialized vials and intravenous bags and it is necessary for the health care professional to activate the system by removing the vial stopper, thus allowing the diluent to enter the vial.

The Mini-Bag Plus system is designed to allow the simple reconstitution of standard 20 mm powdered drug vials. The Monovial Safety Guard is an integrated, self-contained system that allows the transfer of a reconstitution solution from a flexible intravenous bag or vial into a drug-containing vial, and it is only available for a limited number of drugs. As such, the advantages of these systems are that the product can be easily stored and quickly prepared without the need for calculations, specialized equipment (such as laminar airflow hoods) or needles and syringes. They also allow for a quicker turnaround time for the first dose. These products enhance safety for both the patient, by reducing the chance for medication errors (e.g., the wrong drug added to the vial, the incorrect amount of drug to be added to the bag, or a product being incorrectly labeled), and for the health care practitioner, by minimizing the chance for a needlestick injury.

In addition, these products can help to reduce costs associated with unused doses because the unwrapped products can be redistributed for short periods of time. At present, a variety of therapeutic classes of drugs from anti-infectives to cardiovascular agents to pain management at various doses are available for reconstitution in bags containing 0.9 or 0.45% sodium chloride or 5% dextrose in 50 to 250 ml bags. Prefilled syringes are composed of drug solutions placed in a syringe with a needle and needleless systems. The advantages of these systems are the convenience associated with a standard dose, less chance for medication error associated with extemporaneous compounding of these syringes, usefulness in emergency situations, and ease of storage. The needle and syringe are sterile until opened. Prefilled syringes are available for drugs ranging from anti-infectives, analgesics, and antipsychotics to antiemetics. Doubled-chambered vials are advantageous in that the reconstitution solution is separated from the drug until desired by the health care practitioner.

The Mix-O-Vial system is a combination of a powdered or lyophilized drug in a lower container and an appropriate diluent with a preservative and other active ingredients in an upper container. Following removal of the dust cover and upon pressure on the top plunger, the solution comes in contact with the drug and the vial is shaken until a solution is obtained. The upper plunger can then be swabbed with a disinfectant and the appropriate volume of drug removed with a needle as in the standard preparation of an intravenous admixture using a vial. With an increased interest in eliminating needle-stick injuries associated with parenteral drug administration in health care workers, needleless systems are becoming more common in patient care. For example, the Interlink System is designed for needleless access during intravenous therapy. These types of products are available for injection sites, Y-sites, vial adapters, infusion and vein access, syringe products and catheter extension sets. Other needleless catheters and infusion sites include the Introcan Safety IV catheter and the Sifesite injection caps (B. Braun).

Needleless Injection

The concept of needleless injection is not a new one and has been thought about since the 1940s. Current products utilize either spring action or compressed gas (e.g., helium or carbon dioxide) as a propellant to deliver a drug through the skin. These needleless systems offer several advantages. The first potential advantage is reduced pain and anxiety, an advantage for use in children. The second advantage is that needleless injection causes less tissue damage than conventional needles. Finally, a needleless system results in a diffuse pattern of exposure and, therefore, increases surface area and absorption rate. The main disadvantage of this system is unreliability in reference to pain and discomfort and skin characteristics that can influence the amount of drug entering the body. This system has been used to administer vaccines, insulin, and drugs for topical applications (e.g., penile erectile dysfunction) and as a means to deliver DNA for gene therapy.

Institutinal Products

Whereas the presence of the various convenience parenteral products has assisted health care practitioners in safely and accurately delivering drugs to the patients, the extemporaneous compounding of parenteral products continues to be an important component in institutional settings and home health care. Parenteral intravenous admixtures include the withdrawing of the drug solution from an ampul(s) or vial(s) and placing it into various large volume solutions, syringe dosage forms for patients, total parenteral nutrition solutions and cassettes, or other delivery systems for home health care patients. The American Society of Health-System Pharmacists, the National Association of Boards of Pharmacy, and the USP provide practitioners with useful technical assistance bulletins, rules, and standards for the preparation of parenteral products.

The concerns associated with the extemporaneous preparation of parenteral products are maintaining sterility of the products through proper aseptic techniques, calculating and providing the correct dosage, preventing or reducing drug-drug, drug-solution, or drug-container incompatibilities during the preparation or administration of the product, and maintaining and providing drug stability and quality control. The majority of extemporaneous parenteral products are prepared by pharmacists working in hospitals, home health care, or long-term care facilities. These products must be prepared using aseptic technique and using the appropriate supplies (e.g., syringes, needles, and filter needles, and caps) and equipment (Class 100 laminar airflow hoods enclosed within a class 10,000 clean room).

Aseptic technique which differs from sterilization, is a process by which an individual can manipulate sterile products and containers to prevent microbial contamination. Pharmacists and other personnel involved in the extemporaneous preparation of parenteral products require special knowledge and training and should receive additional training and education on a routine basis to ensure proper aseptic techniques are being followed consistently. In addition, these individuals must be able to perform the required calculations (e.g., dosing, milliequivalents, milliosmoles, and powder volume) needed to prepare intravenous admixtures. Equally important to the safe preparation of intravenous admixtures is an understanding of general principles and concepts related to drug and solution or container incompatibilities and to drug stability and also specific knowledge as to whether a specific drug is compatible or stable with another drug, solution, or container.

New Methods of Drug Delivery

Infusion pumps and devices are an essential component to the delivery of parenteral drugs, particularly those given by the intravenous route. For drugs that are administered via intravenous infusion, there are two forces that control fluid flow: (1) the pressure of an active force of the liquid that can be generated via gravity flow (viz., hydrostatic pressure) or mechanically via a positive pressure pump and (2) resistance, or an opposing force, that is generated via the infusion sets, a vascular access device and/or blood vessels. The maximum flow rate will depend upon the ratio of the change in pressure exerted by the liquid to that of the change in resistance. An infusion control device (ICD) is a device that maintains a constant infusion rate in a gravity flow system (controller) or via a positive pressure pump.

A positive pressure pump is a device that provides mechanical pressure (2–12 psi) to overcome the resistance to flow in the vessels. The types of positive pressure pumps are categorized according to how they deliver the solution and their degree of precision in the flow rate. Positive pressure pumps include peristaltic pumps, cassette pumps, syringe pumps, non-electric or disposable pumps, and patient-controlled analgesic pumps (PCA). Syringe pumps are usually the most accurate pumps, with flow variances at 2% or less. Non-electric or disposable syringe and PCA pumps are useful for ambulatory care. PCA pumps are very useful for the parenteral administration of analgesics (viz., morphine) and can be easily programmed to deliver bolus doses and provide a dosing history.

Non-electric or disposable pumps are lightweight, and the solution is delivered based upon a vacuum or through the generation of a gas in the system. A recent consensus development conference on the safety, cost, simplicity of use, and training of intravenous drug delivery systems, focusing on acute care and non-electronic devices, reviewed the use of manufacturer-prepared (e.g., premixed or frozen products), point-of-care activated systems (manufacturer-prepared products that require the drug and diluent to be mixed at the point of care), pharmacy-based intravenous admixture, intravenous push medications in prepared or premade syringes, augmented iv push systems (syringe pumps), and volume control chambers.

Manufacturer-prepared products, point-of-care activated products, and pharmacy-based intravenous admixture programs were recommended as being superior intravenous drug delivery systems, with the manufactured products being considered the safest systems due to the quality assurance in the preparation of these products.

CONCLUDING REMARKS

Current research is leading to newer types of parenteral dosage forms that will be useful for both immediate and sustained drug delivery, for systemic and targeted drug delivery, and for the delivery of small molecules and macromolecules (e.g., proteins, peptides, and DNA). There has been an increased interest in developing a wide variety of particulate drug delivery systems for parenteral products, which have included liposomes or other phospholipid vesicles, microspheres, microcapsules, nanoparticles, or microemulsions. The development of new biomaterials, such as the linear and branched biodegradable polyesters, has increased the interest in the development of these systems for microsphere formulations for parenteral drugs.

An advantage is that drug molecules can be incorporated into these particulate carriers, and, as such, the rate of drug release can be modified, cellular uptake can be facilitated, or the degree of tissue damage or pain can be reduced. Microemulsions, defined as clear solutions obtained by titrating normal coarse oil-in-water emulsions with a medium chain alcohol and composed of the non-polar phase, surfactant and cosurfactant, appear to be potentially useful for parenteral administration, as these are clear and stable formulations that are able to be filtered and might besuitable for intravenous administration.

In addition, other researchers are investigating in situ forming gel or implants that can be easily injected intramuscularly or subcutaneously and that result in the formation of a depot at the site of injection with the potential to modify or extend the release of the drug or macromolecule. Parenteral products will continue to play a vital role in the treatment of patients when the oral route is contraindicated, when it is necessary to carefully control drug blood levels in response to therapeutic effects, when a prolonged therapeutic effect is needed through a long-acting injectable, or when a drug effect is to be targeted to a specific tissue or organ, to name a few instances. Advances in the technology required for the administration of parenteral dosage forms in the last 100 years have expanded their clinical uses for in-patient and out-patient settings. In addition, there is improved convenience and safety for the health care providers who prepare and administer these products. It seems likely that more parenteral dosage forms will become available in the marketplace in response to the compounds being developed through biotechnology.

2

Pharmacodynamic Biomarkers

The concept of applying *pharmacodynamic* (PD) biomarkers into early clinical development is gaining in acceptance. In the past, early-phase studies (phase Ia and Ib) were typically designed to evaluate safety, tolerability, and *pharmacokinetics* (PK) of new therapeutics. This approach was understandable when the target of the therapeutic was not well defined. However, now that the target(s) of most new therapeutics is known, the addition of PD measures into early-phase clinical studies to determine whether the therapeutic is modulating the desired pathway has become more important. It should be noted that observation of target pathway modulation does not necessarily reflect ultimate clinical benefit; however, lack of target pathway modulation would support early termination of the clinical program. By demonstrating measurable PD effects in early clinical trials, meaningful exposure:response (PK:PD) relationships can be determined that will help guide the dose ranges to be explored in later trials and support continued development of the therapeutic agent. Providing early evaluation of the PD effects of a new therapeutic could positively impact the high cost of drug development.

A large portion of the cost of drug development is due to the expense of late-stage trials because of the large number of patients enrolled and the large number of clinical sites that are typically required to accrue patients to these studies. Currently, greater than 50% of phase III trials fail to meet their efficacy end points. If unambiguous no-go decisions can be made based on biomarkers showing failed mechanism of action or exposure:response relationships that are not efficacious within a predicted safety margin, the execution of more expensive later-phase trials could be eliminated thus resulting in reduced overall drug development costs. An increasing number of phase I studies incorporate biomarkers, some of which are well understood to be direct measures of therapeutic impact on a specific pathway, whereas others are more general measures of physiologic changes.

Some of these measures can be conducted on samples that are collected as part of routine practice (such as plasma or serum). However, for the promise of PD biomarkers to be fully realized, tissue types that are not routinely collected need to be obtained, and all types of samples need to be collected in a manner that is compatible with subsequent analyses. In some cases, the assays to be conducted will be well established, whereas others will be cutting-edge assays, new even to the research field. Performing these types of PD assays poses a number of issues that need to be addressed for the strategy to be successful. This chapter will deal with these issues, from the evaluation of preanalytical variables, to sample collection, to fit-for-purpose assay qualification. All the aspects dealt with in this chapter are critical to the success of early-stage biomarker programs, particularly as an increasing number of novel assays (with respect to the clinic) are being employed.

However, diligence in these issues can provide the degree of confidence necessary for the execution of robust assays, whose results can provide evidence for clear go/no-go decisions. Our overall aim in this chapter is to enhance the discussion of issues that affect biomarker programs by providing a background of some of our experiences in this setting. With the range of biomarkers that can be analyzed, the ideas presented here should not be considered all-encompassing but rather provide a basis for discussion of how individual biomarkers may be measured in the clinical setting.

SAMPLING TECHNIQUE

Choice of Samples

The choice of samples to obtain is dictated by the markers to be tested, the accessibility of the preferred tissue, and the capabilities of the clinical site. Fixed samples on slides, flash-frozen tissue pieces, or fresh (often refrigerated) tissue can be collected in cases where preserving structural relationships is important. Whereas the fresh tissue will have the least damage directly caused by preservation methods, it will undergo the most spontaneous change, and those changes should be evaluated before using fresh tissue as a source for measuring critical biomarkers. One should not forget that the cells/tissue will continue to respond to their perturbed environment until they are rendered physiologically incapable of doing so. Whole blood can be used as a source of viable cells for measuring a biological response that does not require a larger physiologic structure (i.e., cell-cell interactions of differing cell types such as skin inflammation responses).

Consistent staining of different cell types can be achieved more readily in the homogenous environment of blood (in which the concentration of drug in the blood is also maintained within the window of *ex vivo* metabolism of the drug) and can enhance measurements of biomarkers in cellular subsets. For molecular measurements where homogenization is acceptable, other tissue collections are possible, including immediate lysis in guanidinium. For PD assessments in early-phase clinical trials, primary considerations in tissue sample choice are data quality, ease of acquisition, value of the biomarker in informing clinical decisions, and the subjects' willingness to donate tissue. A subject's decision to enroll in a trial with PD sampling may be influenced by a range of factors including frequency and invasiveness of specimen collection. For phase I studies performed in patient populations, specimen collection from the site of active disease is preferred because these tissues provide the best PD readout; however, collection of these tissue samples may not be feasible.

Thus, if the target tissue is not easily accessible, consideration should be given to surrogate tissues that can provide information on drug activity. When the preferred tissue is not readily accessible or plausible, such as for multiple tumor biopsies from patients in a cancer trial, surrogate tissues such as skin may be considered. Less-invasive surrogates can also sometimes be found in urine (e.g., collagen breakdown products as measures of bone turnover), saliva, buccal punch biopsies, or hair follicles. Ideally, the surrogate tissue should process signal through the same pathways as is being targeted in the disease tissue; however, resistance mechanisms can arise in the target tissue that are not present in the surrogate, and compensatory mechanisms may be present in the surrogate that are not present in the tumor. Histologic examination of slides is a standard practice, with a large set of qualitative assessments available such as tumor staging, psoriasis plaque staging, or tumor cell presence in lymph nodes. For more objective quantitative measurements, laser scanning cytometry can also be used on stained slides, and in recent years mass spectrometry has been used to examine the physical distribution of different lower-molecular-mass proteins.

A limitation for use of slides can be loss of molecular signals due to the chemical fixation process. Traditionally, proteins and nucleic acids from fixed cells are cross-linked or degraded and thereby less accessible or are qualitatively different in uncontrolled ways. There are now increasingly better methods for dealing with molecular assays on formalin-fixed cells, and the use of molecular biomarkers from these samples may increase. In some cases, opportunistic sampling may be appropriate. When tissue samples are obtained, they may be under the control of the pathologist for diagnosis, and until their needs are complete, the sample will not be released for biomarker studies. Although sampling lymph nodes as a general practice may present risks to a patient, if a sample from the lymph node has been collected as part of standard patient management or monitoring, subsequent use of a portion of the collected sample to measure PD biomarkers is possible. Likewise, the major portion of prostate biopsy samples must go to the pathologist for diagnostic evaluation; however, removing a small tissue print micropeel may provide an opportunity for biomarker analysis without compromising the required pathology evaluation. Although many different individual tissues may be of value for a particular biomarker program, blood is one of the easiest to sample. Almost all clinical sites are capable of handling blood draws and have established methods for processing and shipping samples.

The circulating hematopoietic cells in blood are actively involved in many aspects of inflammation, and even if the core inflammatory reactions in a disease are isolated in an inaccessible tissue, the circulating cells may carry information from when they transited from or through an inflammatory site. Furthermore, blood can accumulate proteins or other analytes from remote sites, thereby providing surveillance of multiple tissues, such as with the classic liver enzymes used to assess hepatitis and other diseases or toxicities that affect the liver. Assays on blood can be either a snapshot of the analytes present in the draw sample or a measure of the response of the blood cells to some stimulus applied *ex vivo*. Maximizing the information to be gleaned from biomarkers in the blood requires selecting appropriate methods and minimizing intrinsic variations in collection, storage, and downstream processing.

The removal of blood from circulation activates many cellular processes, the most notable of which is the initiation of coagulation. These processes can be exploited (for instance, in the deliberate use of coagulation in serum separation tubes) but must be properly controlled for to have interpretable data. An example of where this has not been rigorously considered is in the mass spectrometric analysis of the so-called low-molecular-weight serum peptidome in which the clotting process generates peptides measured by matrix-assisted laser desorption ionization mass spectrometry (MALDI-MS) or surface-enhanced laser desorption ionization mass spectrometry (SELDI-MS). The complications that arise during this process have been described.

Several blood collection products are on the market, including tubes containing anticoagulant suitable for maintaining whole blood and specialized tubes for isolating subsets of cells or that can be used in processes that yield biochemical products such as serum, RNA, or DNA. A clinical trial may require blood collection into several different tube types for each patient because of the multiple analyte types to be measured. The use of multiple different tube types may present challenges for some clinical sites. The local clinician may not be actively supervising the phlebotomy, and the phlebotomist is often focused on filling the order for the tubes he or she is provided. Although the different tube types come with different colored caps, incorrect tubes can inadvertently be used. It is best to have the tubes prepackaged for the clinical site, with all the tubes, coded labels, and summary instructions in a single kit for each patient. Finally, it is recommended to provide a means for the clinical site to document sample collection, storage time, and temperature, which helps provide information about the preanalytical variables.

Nucleic Acid Biomarkers

Nucleic acid biomarkers include both cellular and free nucleic acids, where the free nucleic acids generally are assumed to represent products from damaged cells, and the cellular RNA and DNA represent the current transcriptional program and genetic and epigenetic components of the cell. Free nucleic acids can be measured in circulation and may serve as biomarkers of systemic lupus erythematosus, other autoimmune diseases, and cancer. Surveillance of specific sequences in circulation can provide fetal or tumor genotypes. Free DNA can be collected from either serum or plasma, with serum having a higher yield. The higher yield of DNA in serum may be due to differences in the centrifugation (as discussed below for some protein analytes) or to DNA release from hematopoietic cells during serum preparation. Cellular RNA has also been used as a biomarker for autoimmune disorders such as lupus and leukemias. RNA quality is essential for generating reproducible measurements for transcriptionally based biomarkers.

Traditional methods of assessing RNA quality were based on 28S to 18S rRNA ratio as determined by agarose gel electrophoresis, as well as the 260 nm/280 nm UV absorption ratio. The premier method for assessing quality of RNA preparations is currently an electrophoretic profile on the Agilent BioAnalyzer, which provides both measurements of 28S to 18S rRNA ratios, as well as concentration estimates on samples of 25 to 500 ng. Agilent has also recently developed RIN (RNA Integrity Number) as a single metric of RNA quality and comparability that is predictive of downstream consistency in gene expression measurements. This can be combined with a UV absorption measurement of concentration and purity with the NanoDrop instrument. We have found that PAXgene tubes from PreAnalytix/Qiagen have consistently generated good-quality RNA from both fresh and fresh-frozen blood samples and provide a means for convenient sample collection. It is however, necessary to evaluate and stipulate the appropriate sample collection and storage when using samples collected in this tube type. In the ideal case, one could draw blood into PAXgene tubes, leave the tubes for hours to days at ambient temperature, and/or freeze the tubes and leave them for weeks to years, after which the expression profile for all transcripts would be unchanged.

As described below, we have tested the effect of extended time at ambient temperature after draw and time in the freezer on the consistency of PAXgene RNA, finding that weeks in the freezer have little effect but that a difference of 24 hours at room temperature can strongly perturb the RNA profile. We use the ratio of donor-specific gene expression differences as a measure of reproducible signal. Changes seen in gene expression between PAXgene tubes collected from the same donor at a single draw but handled differently are a measure of the process-specific noise introduced. There can be remarkable reproducibility in expression profiling from two different preparations from the same draw left for different times in the freezer. Multiple PAXgene tubes were collected from three individuals, incubated for 2 to 3 h at room temperature, transferred to a -70ºC freezer, and stored upright in a rack in a secondary sealed plastic box. Tubes were removed from the freezer at different times, thawed, and RNA prepared. After RNA preparation, the samples were profiled on Affymetrix U133A2 chips (Affymetrix, Santa Clara, CA), and the normalized intensities of the approximately 13,000 sequences with the highest intensity were compared. Some random fluctuation in intensity measurements can be expected from the microarray process, and this can be statistically evaluated by various means.

Using the Rosetta Resolver Error Model, set with a p value detection cut-off of 0.01, we would expect about 130 differences between two identical preparations from this gene set. For each of the three donors, comparing the two paired preparations (from each donor, one at 1 week and 26 weeks) only 69, 104, or 243 interdonor differences can be detected, which indicates very little process-specific differences. In fact, more than 90% of the 13,000 sequences have expression levels measured within a factor of 1.5 of each other when comparing the preparations from the same individual processed 25 weeks apart. The remaining sequences have generally lower intensity and noisier measurements and can be statistically evaluated as having lower probability of being a true change. These results are similar to the intradonor differences in side-by-side preparations suggesting that moderate-term storage at -70ºC does not compromise RNA quality or change the relative levels of the different transcripts. Comparatively, between any two individuals, more than 2000 differences are detected at either time point or across time points.

The magnitude of these interdonor differences are remarkably consistent, and the correlation coefficient is greater than 0.93 for all the sequence expression levels that are detected as changed, in either ratio. When all 13,000 sequences are included, the correlation coefficients are still a minimum of 0.73, indicating broad reproducibility. This high correlation continues when RNA from tubes frozen for up to a year are compared with RNA processed within the first week. A different story emerges when PAXgene tubes are held at ambient temperature for variable lengths of time prior to processing. Although the PAXgene literature from PreAnalytix suggests that the sample is stable at room temperature for days after blood draw and shows consistency of transcript stability for a few sequences, it is recommended that this be tested for the transcripts of interest. We conducted a similar experiment to the one above; however, in this experiment, samples were collected and then left at room temperature for 2, 24, or 48 h before processing. When gene expression profiles are compared between donors at either 2 or 24 h, a consistent set of interdonor differences are still seen, with correlation coefficients greater than 0.95, though the differences somewhat diminish by 48 hours. The problem is that comparisons across different incubation times are now vastly different.

In fact, intradonor differences between 2 and 24 h or 24 and 48 h are of the same order as the interdonor differences observed at a single time point When an interdonor comparison is confounded with an inter-time-point comparison, the true donor-to-donor differences can be masked. For instance, comparing interdonor cross-time-point ratios for the donor A sample that sat at ambient for 2 h versus the 24-h sample for donor B, and the ratio for the donor A 24-h preparation versus the donor B 2-h preparation, the different ambient incubation times have a large confounding effect and the correlation coefficients drop to less than 0.4 for the identified changes or less than 0.2 for all sequences. The differences between 2 and 48 h are even greater, vastly exceeding the interindividual variation and generating a negative correlation coefficient. It should be noted that whereas 70% of the RNA transcripts do not show more than a twofold intradonor change over the first 48 h after blood collection, close to 30% of the transcripts do change. This exceeds the number that would change in many treatment situations, and the confounding noise could swamp out the true signal.

It would be potentially disastrous if the transcripts of interest in a particular PD assessment would be included in the set of process-sensitive transcripts. Complex profiles such as those for lupus involve many transcripts and thus increase the odds that the transcript profile would change because of variable incubation time of the PAXgene tubes at ambient temperature. As such, ensuring that samples are collected and handled in the manner specified in the protocol and defined by experiments such as those described above is critical to obtaining robust data from any samples collected from clinical studies. Just as important is the recording of any deviations from the sample collection protocol so that those samples can be flagged for destruction and not included in any analysis.

Hematological Study

For analytes measured from blood, different results will be detected from the same sample collected in different blood collection tubes. A broad look at protein analytes from normal serum or plasma shows a number of specific differences generated based on the tube type used for collection. *Rules Based Medicine* (RBM), a company that provides multianalyte profiling testing, looked at the differences in results from the same sample collected as serum, and in three different plasma collection tubes, with *Ethylenediaminetetraacetic acid* (EDTA), citrate, or heparin as anticoagulant. As expected, they found that circulating fibrinogen was easily measured in the plasma tubes and absent in serum tubes, presumably consumed in the clotting reaction. However, they also observed that other analytes were more easily detected in serum, such as growth factors EGF (17 × to 44 × higher in serum than plasma), and *brain-derived neurotrophic factor* (BDNF) (8 × to 17 ×).

Some analytes measured were below the detectable range in EDTA plasma, such as *creatine kinase* (CK-MB) and matrix metalloproteinase-3 (MMP-3), whereas others such as *monocyte chemoattractant protein*-1 (MCP-1) and *vascular endothelial growth factor* (VEGF) show of the order twofold reduced detection compared with serum. Conversely, serum glutamic oxaloacetic transaminase (SGOT) was only detectable when the sample is collected in EDTA-plasma. In independent *enzyme-linked immunosorbent assays* (ELISAs), we have also seen MCP-1 concentration reduced even more drastically in *tumor necrosis factor*-α (TNF-α) whole-blood stimulations performed from samples collected in EDTA plasma versus heparin plasma. It is not clear what drives the differences seen between the various collection tubes in all cases, but it is notable that differences exist in the measurements of analytes collected in either EDTA-plasma or citrate-plasma collection tubes.

As sample stability often seems similar, and the anticlotting mechanism is presumed for both to be based on chelation of divalent cations, this would suggest that other interactions of the analytes with the tube buffer are responsible. One candidate to explain the differences in analyte concentrations observed between the different collection tubes could also be differential sedimentation or lysis rates of platelets or other particles when the tubes are centrifuged as is discussed below. With the time constraints normally present when qualifying an assay for an upcoming clinical trial, it may be impossible to fully understand all the factors driving differential results for any particular analyte. Therefore, given that individual analytes perform differently in different settings, one should approach an assay for a particular analyte with the knowledge that sample collection in different tube types is a likely source of variability. A complete readout for multiple analytes may require more than one blood collection tube, which may or may not be feasible. Even then, complete characterization of the process from blood collection to sample measurement is necessary to have confidence in the results. As an example of this, we review our experience with a multianalyte measurement of angiogenic cytokines.

MULTIPLEXED ASSAY

Angiogenesis, the formation of new blood vessels, has been a focus for the development of anticancer therapies. Several antiangiogenic therapeutics have been approved or are in clinical development including bevacizumab (anti-VEGF antibody), sunitinib, and PTK787/ZK 222584 [small-molecule antagonists of KDR (VEGFR-2) and other tyrosine kinases]. Amgen is developing new antiangiogenic therapeutics, and as part of the development of these molecules, we have investigated the use of several angiogenic cytokines and soluble receptors as PD biomarkers of drug efficacy. Angiogenic cytokines have been shown to be elevated in the serum and plasma of cancer patients and have previously been shown to be modulated upon dosing of antiangiogenic therapeutics. The following describes a few of the issues that need to be evaluated for the measurement of these analytes including sample freeze-thaw stability, analyte interference, and identification of conditions to diminish analyte interference.

What are not presented are the other aspects of fit-for-purpose assay qualification that we employ, which include parameters such as assay linearity, assay range (upper and lower limits of quantitation), accuracy, precision (inter-/intra-assay reproducibility), parallelism, interference (drug, matrix, abnormal samples [hemolyzed, icteric, and lipemic]), and sample stability. To measure levels of angiogenic biomarkers, we used two multiplexed assays developed by *Meso Scale Discovery* (MSD). We selected the MSD assay format because multiplexing conserves sample and allows multiple analytes to be assessed in the same assay. Another beneficial feature of using MSD assays is that they have wide dynamic ranges (in some cases 3 to 4 logs), which can reduce the need to reanalyze samples, and this in turn can reduce the need for additional freeze-thaw cycles and minimize volume requirements of precious samples.

One of the MSD assays used to measure levels of angiogenic cytokines is a 4-plex that includes basic fibroblast growth factor (bFGF), placental growth factor (PlGF), VEGF, and soluble Flt-1 (fms-like tyrosine kinase 1; VEGFR-1). The second MSD assay is a 2-plex that includes soluble KDR (kinase insert domain receptor; VEGFR2) and soluble cKit. During assay qualification, bFGF, soluble Flt-1, and VEGF were shown to be unstable over multiple freeze-thaw cycles. Previous studies indicated that sample preparation is important for the evaluation of VEGF and bFGF, as these analytes are known to be sequestered by platelets. Because VEGF is released from platelets during the preparation of serum, plasma is the matrix of choice for measuring this analyte. Several different centrifugation conditions can be used in the preparation of plasma, resulting in plasma containing varying amounts of platelets. We investigated the impact of different centrifugation conditions used in preparing plasma on the measurement of levels of the analytes in our angiogenic panel. Plasma was prepared from blood of a normal donor using three centrifugation conditions: $760 \times g$ for 15 min, $1500 \times g$ for 10 min, or $2800 \times g$ for 30 min.

Table 2.1 Plasma prepared under different conditions and subjected to three freeze-thaw cycles shows altered analyte recovery for bFGF, soluble Flt-1, and VEGF but not PlGF

	Freeze-thaw cycle	*Centrifugation speed (× g)* *760 (15 min)* *pg/mL*	*Recovery*	*1500 (10 min)* *pg/mL*	*Recovery*	*2800 (30 min)* *pg/mL*	*Recovery*
bFGF	0	116	100	76	100	3	
	1	177	153	124	163	2	
	2	185	159	143	189	3	
	3	202	174	142	188	2	
Soluble	0	90	100	79	100	100	100
Flt-1	1	111	124	91	115	97	97
	2	121	134	110	139	101	101
	3	124	138	101	127	116	116
PlGF	0	15	100	14	100	13	100
	1	15	103	15	106	13	102
	2	16	111	16	112	14	102
	3	16	111	15	109	15	113
VEGF	0	242	100	163	100	139	100
	1	362	150	239	146	143	103
	2	473	196	322	197	144	103
	3	504	208	313	191	151	108

The latter preparation conditions result in plasma from which most of the platelets have been removed and is referred to as platelet-poor plasma. The plasma samples were subjected to multiple freeze-thaw cycles, and the levels of bFGF, VEGF, PlGF, and soluble Flt-1 were measured using the MSD 4-plex assays. When plasma samples were prepared from a normal donor using centrifugation at either 760 × *g* for 15 min or 1500 × *g* for 10 min, levels of bFGF, soluble Flt-1, and VEGF increased after a single cycle of freeze-thaw. Levels of these three analytes continue to increase with increasing numbers of freeze-thaw cycles, with the increase in VEGF being particularly notable.

In contrast, the level of PlGF did not change appreciably when subjected to three cycles of freeze-thaw regardless of the centrifugation conditions used to prepare plasma. Similar results were obtained with plasma prepared from three other normal donors (data not shown). We also investigated the effect of centrifugation conditions used for plasma preparation on the recovery of soluble cKit and soluble KDR after multiple cycles of freeze-thaw.

Recovery of these analytes is stable through three freeze-thaw cycles, regardless of centrifugation conditions. The increases in measured levels of VEGF and bFGF observed upon freeze-thaw of plasma prepared by centrifugation at 760 × *g* for 15 min or 1500 × *g* for 10 min are consistent with release of these cytokines from platelets. However, the sequestration of soluble Flt-1 by platelets has not been previously reported. VEGF is a ligand for the Flt-1 receptor, and soluble Flt-1 may bind to VEGF in circulation. One possible explanation for the increase in Flt-1 concentration upon multiple freeze-thaw cycles is that soluble Flt-1 binds to VEGF in circulation and in turn is sequestered with VEGF by platelets.

Table 2.2 Plasma collected under different conditions and subjected to four freeze-thaw cycles showed similar recovery of cKit and KDR

	Freeze-thaw cycle	*Centrifugation speed (× g)*					
		760 (15 min)		*1500 (10 min)*		*2800 (30 min)*	
		pg/mL	*Recovery*	*pg/mL*	*Recovery*	*pg/mL*	*Recovery*
cKit	0	2691	100	2564	100	2447	100
	1	2850	106	2512	98	2423	99
	2	2644	98	2686	105	2493	102
	3	2444	91	2551	99	2519	103
KDR	0	536	100	532	100	508	100
	1	581	108	527	99	542	107
	2	526	98	552	104	514	101
	3	499	93	541	102	547	108

VEGF and PlGF are ligands for the Flt-1 receptor. Maynard et al. reported that soluble Flt-1 interferes with the measurement of VEGF and PlGF by ELISA. We decided to test the effect of soluble Flt-1 on the measurement of VEGF and PlGF in our MSD immunoassays. Using Flt-1 standard obtained from both MSD and R&D Systems, we tested the impact of spiking increasing amounts of standard into plasma samples on the levels of analytes as determined in our MSD 4-plex assay. The detectable levels of VEGF decreased with the addition of increasing concentrations of Flt-1, with the highest spike level resulting in 60% recovery of VEGF.

Table 2.3 The concentrations of VEGF, PlGF, soluble Flt-1, and bFGF measured in presence of increasing levels of spiked Flt-1

		Concentration (pg/mL)			
	Spike	*VEGF*	*PlGF*	*Soluble Flt-1*	*bFGF*
Meso Scale Discovery Flt-1	None	2274 (100%)	25	206	28
	Calibrator diluent	2441 (107%)	25	192	26
	0.5 ng/mL Flt-1	1877 (83%)	20	705	21
	5 ng/mL Flt-1	1799 (79%)	23	4783	24
	15 ng/mL Flt-1	1356 (60%)	24	> ULOQ	26
R&D Systems Flt-1	None	2675 (100%)	30	247	31
	Calibrator diluent	2650 (99%)	26	219	25
	0.5 ng/mL Flt-1	2169 (81%)	24	698	27
	5 ng/mL Flt-1	2192 (82%)	26	5503	28
	15 ng/mL Flt-1	1590 (60%)	28	> ULOQ	31

Spiking Flt-1 into the sample did not effect the recovery of PlGF or bFGF. In a separate experiment, the recovery of PlGF was not affected by a Flt-1 spike in plasma samples with higher intrinsic levels of PlGF under conditions in which VEGF recovery was reduced as before. The lack of Flt-1 interference in the measurement of PlGF contradicts a previous study in which spiked Flt-1 interfered with the measurement of PlGF. The authors of the previous report used an ELISA from R&D Systems to measure levels of PlGF. We obtained PlGF ELISA kits from R&D Systems and evaluated the impact of spiked Flt-1 on PlGF recovery, and again we did not observe interference by Flt-1 in the measurement of PlGF. To further evaluate interference in the VEGF assay, we investigated the impact of KDR on the measurement of VEGF in plasma. The recovery of VEGF is reduced in the presence of spiked KDR, and addition of spiked Flt-1 exacerbates the reduction in the quantitation of VEGF.

To mitigate the interference of Flt-1 and KDR in the measurement of VEGF, several pretreatment methods were evaluated. Pretreatment with acetic acid, SDS, Tween, Triton X-100, or urea did not improve recovery of VEGF, whereas pretreatment with heat (80°C for 10 min) or sodium deoxycholate resulted in poorer recovery of VEGF. Of the pretreatment conditions tested, 200 mM NaCl resulted in the best recovery of VEGF. To further optimize pretreatment conditions, we evaluated pretreatment with different concentrations of NaCl on the recovery of VEGF and PlGF in plasma from a normal donor spiked with Flt-1 and/or KDR. Interference in VEGF quantitation was observed by spiking each VEGFR alone and was increased when both VEGF receptors were spiked together.

Treatment with NaCl increased recovery of VEGF when each VEGFR was spiked alone, but was not able to completely overcome the inhibition by both Flt-1 and KDR spiked together. These spiking experiments are meant to explore the endogenous analyte interactions occurring in the samples, which are likely to impact assay results. Readout of a single analyte may be affected by levels of other analytes, due to expected physiologic binding. However, to obtain accurate measurement of an analyte, these parameters should be considered and ideally the assay developed to eliminate these assay variables. As can be seen from this discussion, there were several aspects to sample handling and cross-analyte interference that were found to affect the results of the assays.

In some cases, the process could be controlled to reduce the interfering noise that was generated (i.e., altering centrifugation conditions). In the case of VEGF-VEGFR interference, we were unable to completely control the effect but were able to minimize it and collect a body of data that defined the nature of the interference, such that it could be considered in analysis. This highlights the need to understand the performance characteristics of the assay under development. Other issues not touched upon in this section that need to be considered in multiplex assays include the placement of samples and standard in the assay plate (due to the close proximity of the analytes in the well) and identification of low, medium, and high controls. All of these parameters must be addressed if the performance characteristics of the assay are to be understood and the assay qualified to the appropriate level (fit-for-purpose).

Preanalytical Considerations

Ex vivo stimulation of whole blood generally encounters all of the same variables observed with direct plasma collection, with additional variables that are introduced from continuing cellular activity and the reaction of the blood with the anticoagulant or other components of the tube. *Ex vivo* stimulation assays can be especially valuable in PD assays of biochemical coverage where the therapeutic is expected to modulate a well-expressed stimulus-response pathway in a common circulating cell type. The stimulus is often the addition of a natural cellular signal such as a cytokine or lipopolysaccharide (LPS). The response can be measured as changes in protein phosphorylation, cytokine release, gene expression, or cell surface marker expression. Such measurements are employed in drug development programs where the aim is to determine the effect of therapeutic intervention on a specific response (represented in whole blood) prior to and at various times after dosing. Informative comparison of the response in the presence or absence of therapeutic requires a minimal level of robustness of the response in the range of clinical settings in which it will be measured, in spite of an underlying inherent dynamic response of the blood to its new environment.

Almost immediately, the blood temperature is lowered in the phlebotomy tubing even before the blood gets to the receiving tube where it encounters the anticoagulant, usually in a dried matrix that dissolves in the blood. The environment within the collection tube is also much different than that of the body (no carbon dioxide, oxygen, nutrient, or byproduct exchange) and is missing all the homeostatic influences of the rest of the body. Unfortunately, the program the blood executes *ex vivo* may seriously influence the readout of an assay in ways that can translate poorly between different laboratory and clinical settings. To get an overview of the magnitude of the collection tube response and differences between anticoagulants, we examined the changes in the transcriptional program over the first 6 h after collection using DNA microarrays. Blood was collected from three healthy donors into tubes with EDTA, citrate, or heparin as anticoagulant, as well as directly into PAXgene tubes. The anticoagulant samples were left at room temperature, with individual samples removed from their collection tubes and transferred to PAXgene at 1-h intervals, for up to 6 h. As observed in microarray experiments using Affymetrix U133A2 chips, there was induction of several early-response genes, including ***FOSB, JUN, EGR1, EGR2,*** and ***EGR3,*** by up to 100-fold in 2 h compared with sample transferred to PAXgene immediately after draw.

There is also a consistent induction of some genes involved in inflammatory reactions, including ***IL8, DUSP2,*** and ***CD83*** as well as ***STK17b,*** the latter of which is associated with induction of apoptosis. Table 2.4 lists some of the modulated genes and their altered expression levels in different anticoagulants to illustrate three (immediate early gene, inflammatory and apoptotic induction responses) of the many processes that appear to be taking place in any plasma blood draw at ambient temperature. These common responses were seen in all three donors, as well as two other donors in a separate experiment. During the first 6 h, more than 2000 detectable transcripts showed changes relative to the baseline, with greater than 80% of these changes similar (to different extents) for all three donors and in all three anticoagulants. In addition to the common time-dependent transcriptional changes observed across donors, there were also some transcript changes that were specific to individual donors or anticoagulants.

Table 2.4 The effect of anticoagulant on the expression of specific genes

	Citrate	*EDTA*	*Heparin*
JUN	33.2	4.9	5.7
FOSB	100.0	50.4	21.0
EGR1	75.0	49.9	39.8
EGR2	18.9	48.2	36.3
EGR3	35.6	10.5	4.7
IL8	91.0	42.3	14.8
CD83	19.6	2.5	1.3
DUSP2	33.0	6.2	5.3
STK17B	2.7	3.4	2.7

However, by comparing within a single lot of a particular collection tube, almost all responses observed were consistent between donors, suggesting that keeping good control on collection conditions will result in a consistent background activity in a whole-blood stimulation assay in plasma. Unsurprisingly, an even stronger inflammatory signal can be seen when the collection tubes contain endotoxin contamination. Whereas the manufacturers strive to produce collection tubes with minimal endotoxin, the products themselves are not often tested for this. We have observed significant lot-to-lot variation in our laboratory, and we and others have occasionally seen extreme endotoxin effects in certain lots of tubes. In fact, retrospective analysis of the tubes in the previous time-course experiment showed detectable endotoxin contamination in the citrate tubes. When present at high levels, the endotoxin induces expression of a wide range of cytokine and chemokine genes, observable within an hour, and predominate the expression profile by 2 h.

Expression of effected cytokine and chemokine genes can be increased more than 100-fold even in only moderate levels of endotoxin. Listed in Table 2.5 are six genes with consistently increased expression as a result of contaminating endotoxin. The table shows the change in gene expression at 2 h relative to immediately after draw for the contaminated citrate tubes and the endotoxin-free EDTA and heparin tubes. This lot of citrate tubes tested could fail well-designed quality-control criteria, such as consistent response in a known sample, or low baseline in a 24-h whole-blood stimulation assay, and implementation of this type of quality-control criteria is recommended to mitigate confounding effects due to endotoxin contamination. Undetected endotoxin contamination could generate the internal production of TNF-α, IL-1β, or other signal could independently stimulate the cells and possibly overwhelm the experimental effect. Endotoxin contamination can vary dramatically between different lots of tubes. For example, we have tested several different lots of heparin tubes from one manufacturer and found variable endotoxin signals from all of them.

Table 2.5 The effect of endotoxin contamination on gene expression in whole blood

	Citrate	*EDTA*	*Heparin*
CCL4	18.3	2.2	0.6
IL1B	10.4	0.8	0.4
CCL3	40.6	5.3	1.5
CCL20	58.9	1.0	0.7
TNF	44.2	1.8	0.7
CXCL2	100.0	3.6	1.0

Laboratory safety considerations have led many people to consider plastic collection tubes as a standard. Unfortunately, we have yet to find a lot of plastic tubes with low levels of endotoxin, whereas most lots of glass tubes do have sufficiently low levels and therefore allow reproducible measurements from *ex vivo* stimulation assays. The danger of endotoxin contamination makes it essential to perform up-front quality assessments for blood collection tubes to be used for a whole-blood assay and, if possible, to perform an entire study with one lot of tubes. The complex reaction of blood in the collection tube can play out in the background of a whole-blood stimulation assay. The reaction can produce interfering signals and also sensitize the blood to other stimuli. Assay qualification of a stimulus release assay should always include testing the time between blood draw and initial stimulation.

The blood was collected into heparin and kept sealed at ambient temperature until being transferred to 96-well plates and diluted to 80% with media containing the TNF-α stimulus. The samples were immediately transferred to a 37°C incubator at 0.5% CO_2. All the samples had a strong response to TNF-α, but the size of the response noticeably increased as the heparin tubes were incubated prior to being transferred to the 96-well plate. In some PD assays, this change could be very misleading if the time dependence was not well characterized. One can imagine a TNF-α stimulation assay for an anti-inflammatory agent that would require a blood draw from a patient prior to administration of the therapeutic and then an additional blood draw 4 h after the first sample. If the predose and postdose samples from a placebo-treated individual were then simultaneously stimulated an hour after the second draw, they would effectively represent the third and first points of the graph, respectively. One would conclude that donor A had a predose induction of about 13,000 units and a postdose induction of about 10,000, whereas donor B dropped from 10,000 to 5000.

The variation introduced by sample collection will outweigh that observed effect within the study and negatively impact the ability to interpret the data. The major conclusion to be drawn from these examples is the absolute necessity of qualifying an assay to cover the variables that could be encountered in a clinical trial. The assay range, together with how expected fluctuations in assay conditions can affect the results, must be characterized and considered as a package in deciding if the assay will satisfy the fit-for-purpose requirements.

Again, not all these effects can be eliminated, but characterization of the effects during assay qualification can give one the confidence that the assay is fit-for-purpose or can identify its insufficiency. It is also worth considering the transfer of the assay from its laboratory setting into the clinical setting.

In a whole-blood stimulation assay, where stimulus, incubation, and subsequent harvest are to be performed at the site, the nature of the patient visit and standard practices at distant contract sites should be considered. A protocol that specifies sample processing at 6 or 8 h after blood collection may not be feasible, and a protocol with a 4-h or 24-h incubation may give more consistent results. Even a 4-h incubation will work better if patient visits can be restricted to mornings and the harvest step streamlined.

Thus, careful planning of the actual protocol to be executed at the clinical site (including how much an operator can perform in a given time period) is paramount to a successful program. We have discussed how particular details such as protocol timing, centrifugation speed, and blood collection products can affect the results. There can be other unsuspected variables that will affect the results. A pilot study using untreated volunteer donors at each clinical site can test how well these complicated assays can be transferred to the site.

The pilot study should use all the same production lots of reagents and sample collection tubes and all the same equipment as will be used in the actual trial, as well as include all sample collection and all biomarker analyses that will be part of the main study to ensure that the entire protocol is properly tested. For complex assays, it may even be useful for individuals from both the sponsor and the clinical site to visit each other's labs.

Whether the final analyte measurements are taken at the clinical site, at some other contractor site, or at the sponsor's laboratory, the complete analysis of the pilot study should be compared with the development results to confirm successful transfer. This requires that the assay be fully qualified in the sponsor's laboratory in time for such a pilot study to be accomplished and evaluated before the start of the main trial but can be valuable in identifying and resolving minor issues, maximizing the ability to provide the highest quality information from the biomarker program.

CONCLUDING REMARKS

In this chapter, we provided some examples of the various aspects of biomarker analysis that need to be considered so that robust data can be generated in support of early-phase clinical drug development. This is by no means a comprehensive treatise on this topic but rather meant to point to some issues that one needs to consider when embarking on such studies. As described, even one of the simplest of samples to be collected, whole blood and its component parts (e.g., cells, plasma/serum), can change after collection.

This emphasizes the requirement for careful planning of the clinical site capabilities and ensuring that the needs of the biomarker program can be met before the clinical study and, therefore, sampling begins. It also highlights how many parts of the assay under development need to be examined so that the performance characteristics are well understood. If these aspects of biomarker analysis are not examined and understood, the data generated may lead to inaccurate conclusions being drawn. The hope for biomarkers to positively impact drug development costs will only succeed if the data generated are sufficiently robust, and for that to occur, all aspects of the biomarker program need to be carefully considered and tested. The generation of robust data sets can enable decisions on dosing and program progression to be made.

3

Physicochemical Characteristics of Biofluid

The physical and chemical characteristics of a drug and biofluid can be useful predictors of drug transfer into various compartments of the body. The principal mechanism of transfer for most drugs is passive diffusion. The pKa, lipid solubility, protein binding, and body fluid composition largely determine the extent to which the drug is present. Transfer of drugs from the circulating blood (pH 7.4) to another biological fluid involves transport across membranes that are an effective barrier against ionized, highly polar compounds. Following penetration of the membrane and transfer into the biofluid, the pH differential may result in ionization of the drug, restricting further mobility. Accumulation of the drug in this way is commonly referred to as "*ion trapping*." The increasing use of illegal drugs by expectant mothers has led to an increased need for prenatal toxicological testing.

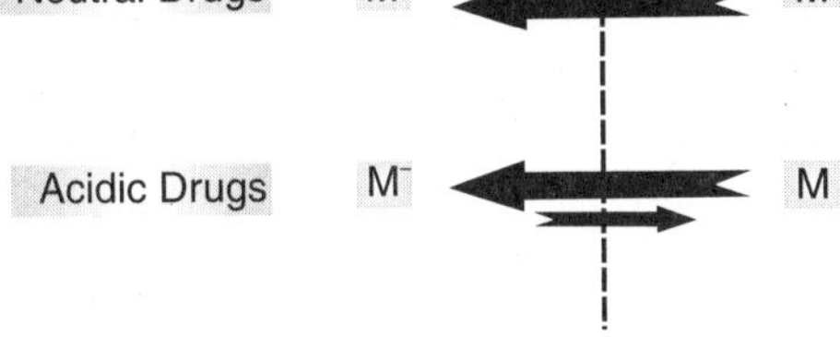

Fig. 3.1. Dissociation of drugs and ion trapping.

Exposure to drugs-of-abuse may result in higher rates of congenital anomalies and neonatal complications. Identification of gestational drug exposure may benefit the newborn in terms of increased vigilance and monitoring of the infant by medical and social services. However, amniotic fluid and breast milk are not routinely used to determine maternal drug use. Other samples of maternal origin such as urine, saliva, blood, or hair can be used for this purpose during pregnancy. More importantly, the detection of drugs or drug metabolites in amniotic fluid and breast milk are essential to our understanding of the pharmacokinetic principles governing intrauterine and prenatal mechanisms of drug transfer.

FREQUENCY OF DRUGS USED

According to self-reported data, it is estimated that almost 4% of pregnant women aged 15-44 years have used illicit drugs. Marijuana was the most widely used drug (2.8%) followed by non-

Table 3.1 Properties of selected drugs

Drug	*pKa*	V_d *(L/kg)*	*Log P*	*Fb (%)*	$T_{1/2}$
Acetaminophen	9.5	1	0.5	25	1-3 h
Alprazolam	2.4	0.7	2.12	70-80	11-15 h
Amitriptyline	9.4	15	4.94	91-97	9-36 h
Caffeine	14.0, 10.4	0.5	-0.07	35	2-10 h
Cocaine	8.6	1-3	2.3	92	0.7-1.5 h
Diazepam	3.3	0.5-2.5	2.7	98-99	20-40 h
Diphenhydramine	9.0	4.5-8	3.3	80	2.4-9.3 h
Fentanyl	8.4	4	2.3	80	3.7 h
Fluoxetine	9.5	27	4.05	95	4-6 days
Ketamine	7.5	4	3.1	20-50	2-3 h
Meperidine	8.7	4	2.7	50-60	3-6 h
Methadone	8.3	4	3.93	92	10-25 h
Methamphetamine	10.1	3-7	2.1	10-20	9 h
Morphine	8.1	3-5	–	20-35	2-3 h
Oxycodone	8.9	–	0.7	87-94	2-3 h
Phencyclidine (PCP)	8.5	6	4.7	65-80	7-46 h
Phenobarbital	7.4	0.5	1.5	50	90-100 h
Phenytoin	8.3	0.5-1.2	2.5	90	7-42 h
Propoxyphene	6.3	16	4.2	70-80	8-24 h
Salicylic Acid	3.0, 13.4	0.1-0.2	2.3	40-80	2-4 h
Sertraline	9.5	20	5.29	98	26 h
Temazepam	1.6	1	2.19	96	8-15 h
Δ-9-tetrahydrocannabinol	10.6	10	7.6	94-9%	2 h
Valproic Acid	4.6	0.1-0.2	2.8	90	6-20 h

Table 3.2 Effect of pH and pKa on acidic, basic, and neutral drugs

Drug type	pH units from pKa				
	-2	-1	0	+1	+2
	% Ionized drug				
Acidic drugs: acetaminophen, ampicillin, barbiturates, NSAIDs, phenytoin, probenecid, and THC metabolites	1	9	50	91	99
Neutral drugs: carbamazepine, glutethimide, meprobamate	0	0	0	0	0
Basic drugs: amphetmines, antiarrhythmias, antidepressants, antihistamines, cocaine, narcotic analgesics, PCP, and phenothiazines	99	91	50	9	1

medical use of prescription drugs (0.9%). Other drugs used included cocaine, heroin, inhalants, and hallucinogens including phencyclidine (PCP) and lysergic acid diethylamide (LSD). Furthermore, the percentage reporting past month use of an illicit drug was only marginally lower for pregnant women aged 15-17 (12.9%) compared with non-pregnant women in the same age group (13.5%). The accuracy of self-reported drug use is questionable because of the stigma of drug use in pregnancy and the associated legal and ethical issues.

Table 3.3 Factors influencing biological specimen selection

Factors
Sample Collection
Invasiveness
Risk of infection, complication, hazards
Protection of privacy
Ease and speed of collection
Training of personnel (medical/non-medical)
Likelihood of adulteration
Contamination
Volume of specimen
Analysis
Qualitative or quantitative results
Window of detection
Drug concentration/accumulation in biofluid
Parent drug or metabolite(s)
Stability of drug(s)
Biofluid storage requirements
Pretreatment of specimen
Limitations of the matrix
Likelihood of interferences
Inter and intrasubject variability of the matrix
Use of existing analytical procedures
Speed of analysis
Personnel training requirements
Appropriate cut-off concentrations
Interpretation
Pharmacologic effects
Indicator of recent drug use (hours)
Short-term drug exposure (days)
Long-term drug exposure (weeks)
Forensic defensibility

Toxicological testing of maternal or neonatal specimens for commonly abused drugs may provide a more realistic estimate of drug use. In these studies, drug prevalence can vary dramatically, depending on whether the population is considered high or low risk. The prevalence of drug abuse among pregnant women throughout the USA is reported to be between 0.4 and 27%. In one study of newborn drug screening among a high-risk urban population, rates of cocaine, morphine, and cannabinoid use were 31, 21, and 12% respectively. However, this study likely overestimates drug use because results were determined by radioimmunoassay and were not confirmed by another technique. More recently, the US National Institute on Child Health and Development undertook a multi-site study of more than 8000 infants. Newborn drug testing of meconium samples revealed that 10.7% of the samples contained cocaine, opiates, or both. In this study, positive immunoassay test results were confirmed by a secondary technique.

Risk of Drugs

The use of drugs during pregnancy may pose a potential risk to both the mother and the fetus. Drug effects on fetal safety are generally evaluated using animal data or available experience from human pregnancies. Based on this approach, the United States Food and Drug Administration established a categorization of drugs according to safety in 1979, which is currently under revision. Although such categorizations provide a rough estimate for adverse fetal consequences, they are often derived from very limited data sets. In a recent study of prescription drug use in pregnancy, an estimated 64% of women were prescribed a drug other than a vitamin or mineral supplement prior to delivery, and as many as 40% received a drug during delivery from category C (drugs for which human safety during pregnancy has not yet been established) according to the FDA classification system. Pre-natal and intrauterine drug exposure remains a major health concern. Numerous effects including intrauterine growth retardation, birth defects, altered neurobehavior, and withdrawal syndromes have been reviewed.

Pre-natal cocaine use is associated with placental abruption and premature labor, whereas intrauterine cocaine exposure is associated with prematurity, microcephaly, congenital anomalies, necrotizing enterocolitis, and stroke or hemorrhage. Amphetamines may lead to complications similar to those of cocaine-exposed infants, including increased rates of maternal abruption, prematurity and low birth weight. Heroin use during pregnancy has been associated with low birth weight, miscarriage, prematurity, microcephaly, and intrauterine growth retardation. Marijuana has been associated with visual anomalies and a number of persistent neurocognitive effects in the latter stages of development. Neonatal abstinence syndrome is associated with in utero exposure to opioids, cocaine, and methamphetamine. Drug effects on the developing organism are dependent on a number of factors including the activity and retention of the drug and its metabolites in the maternal-fetal compartment, as well as the dose and duration of exposure.

FLUID OF AMNION

Amniotic fluid, produced by cells that line the innermost membrane of the amniotic sac (amnion), is the liquid that surrounds and protects the embryo during pregnancy. This fluid cushions the fetus against pressure from internal organs and from the movements of the mother. Production of fluid commences on the first week after conception and increases steadily until the 10th week, after which the volume of fluid rapidly increases. Amniotic fluid, which may total about 1.5 L at 9 months, contains cells and fat that may give the liquid a slightly cloudy appearance. The protein concentration and the pH of the fluid vary with gestational age. Amniotic fluid is constantly circulated, being swallowed by the fetus, processed, absorbed, and excreted by the fetal kidneys as urine at rates as high as 50 mL per hour. This circulation of fluid continuously exposes the fetus to compounds that may be absorbed in the gut or diffused through fetal skin in the early stage of development. The encapsulation of the fetus in this fluid may prolong exposure to harmful drugs and metabolites. The pharmacokinetics of drug disposition in utero varies from drug to drug, and the acute and chronic effects that may result are the topic of continuing research.

Risk of Drug Exposure to Fetus

A number of maternal, fetal, and placental factors have been documented to affect fetal drug exposure. Of these, binding to serum proteins in the maternal and fetal circulation and fetal elimination are particularly important. Conjugated drug metabolites, which tend to be highly water soluble, may accumulate in the fetus or amniotic fluid because of limited placental transfer. The principle routes of drug transfer into amniotic fluid occur through the placenta and the excretion of water-soluble drugs into the fetal urine. The placenta is an extra embryonic tissue that is the primary link between the mother and the fetus. Passive diffusion and to a lesser extent active transport and pinocytosis are responsible for drug transfer.

The physico-chemical properties of the drug, such as pKa, lipid solubility, and protein binding, largely influence the drug's ability to cross the placenta and enter the fetal circulation and the amniotic fluid. The amnion is considered to be a deep compartment, whereby equilibration with adjacent compartments is achieved relatively slowly. Transplacental passage of small lipophilic drugs occurs readily and is limited only by blood flow rates. By comparison, the rate of transfer of a hydrophilic drug may be approximately one-fifth that of a lipophilic drug of similar size. With drugs that tend to be highly protein bound, only the small fraction of free drug may diffuse across the membrane. Small lipid-soluble drugs can rapidly diffuse across the placental barrier, producing similar drug concentrations in both amniotic fluid and fetal plasma.

Larger, water-soluble compounds that are transferred more slowly are incorporated into the amniotic fluid through fetal urine. Basic drugs may accumulate in the amnion due to ion trapping, resulting in drug concentrations in excess of those found in fetal or maternal plasma. Large, lipid-soluble drugs are more readily transferred to the fetus but less readily transferred to the amniotic fluid due to reabsorption in the fetal kidney. The fetal kidney is not an effective route of drug elimination because fetal renal blood flow is only 3% of the cardiac output compared with 25% in an adult. In addition to transplacental passage of metabolites from the mother, biotransformation by the immature fetal liver may also be responsible for the appearance of some drug metabolites in the amniotic fluid. The glucuronidation capacity of the adult liver is estimated to be 6- to 10-fold greater than fetal liver at 15-27 weeks.

Sampling

The collection of amniotic fluid (amniocentesis) usually takes place between the 16th and 20th week of pregnancy. The liquid is usually collected to test for fetal abnormalities or to learn the sex of the child. The presence of illicit drugs or their metabolites in amniotic fluid suggests that the fetus has been exposed to these substances through maternal blood circulation. A maternal serum sample taken at the same time as the test may provide complementary toxicological data and help assess the relative risk to the fetus. Amniocentesis is an invasive procedure. Prior to the test, an ultrasound scan is used to determine the position of the fetus. A needle is inserted through the abdomen into the uterus where there is the least chance of touching the placenta or the fetus. Although complications are rare, miscarriage occurs in approximately 1% of women. Typically, 5-30 mL of amniotic fluid is removed. The pH of amniotic fluid decreases from slightly alkaline to near neutral pH at full term due to fetal urination. Amniotic fluid, which is 99% water, contains dilute plasma components, cells, and lipids. Following amniocentesis, the fluid may be centrifuged and the supernatant layer frozen prior to drug testing.

Drugs present in amniotic fluid can be analyzed using well-established techniques that are routinely used for blood, urine, or serum. Relatively few interferences are encountered with amniotic fluid due to its high water content. Sample pre-treatment prior to an immunoassay screening test may not be necessary; however, confirmation by gas chromatography/mass spectrometry (GC/MS) requires isolation of the drug using liquid/liquid or solid phase extraction techniques. Although amniotic fluid is extremely amenable to routine toxicology tests, it is not routinely used for this purpose because of the invasiveness of the specimen collection procedure. Subsequently, there are considerably fewer toxicological studies compared with other specimens.

Table 3.4 Methods of analysis of drugs in amniotic fluid and breast milk

Methods
Purification/drug extraction
Liquid-liquid extraction
Solid-phase extraction
Supercritical fluid extraction
Drug detection by immunochemical techniques
Cloned enzyme donor immunoassay (CEDIA)
Enzyme-linked immunosorbent assay (ELISA)
Enzyme-multiplied immunoassay technique (EMIT)
Fluorescence polarization immunoassay (FPIA)
Kinetic interaction of microparticles in solution (KIMS)
Radioimmunoasay (RIA)
Drug identification by chromatographic techniques
Capillary electrophoresis (CE)
Gas chromatography (GC)
Gas chromatography/mass spectrometry (GC/MS)
Gas chromatography/mass spectrometry/mass spectrometry (GC/MS/MS)
High-performance liquid chromatography (HPLC)
Liquid chromatography/mass spectrometry (LC/MS)
Liquid chromatography/mass spectrometry/mass spectrometry (LC/MS/MS)
Thin layer chromatography (TLC)

Frequency of Death of Pregnant Woman

Cocaine and benzoylecgonine concentrations in amniotic fluid from known cocaine users ranged from 0.4-5 to 0-0.25 mg/L. Following the death of a pregnant woman, amniotic fluid cocaine and benzoylecgonine concentrations of 3.3 and 1.6 mg/L were reported. After crossing the placental barrier by simple diffusion, the drug distributes between fetal and maternal blood. The amniotic sac and its contents serve as a deep compartment with restricted, slow equilibrium between adjacent compartments. As a result, amniotic fluid inside this protective sac may expose the fetus to potentially harmful drugs or metabolites that are sequestered in the biofluid. Animal studies have shown a three- to fourfold increase in cocaine concentration compared to fetal or maternal plasma.

The concentration of benzoylecgonine in amniotic fluid was also shown to be higher than newborn urine. Limited fetal hepatic function, ion trapping, and the slow pharmacokinetic exchange present in this deep compartment can increase intrauterine drug exposure and thus compound the risk of complications. Other basic drugs, such as PCP and fentanyl, have also been detected. PCP was detected in amniotic fluid at a concentration of 3.4 ng/mL some 36 days after hospitalization of a pregnant woman. Animal studies have shown that PCP readily crosses the placenta and may concentrate in the fetal tissues. Fentanyl was detected in a series of 14 paired maternal serum and amniotic fluid samples taken during the first trimester.

Table 3.5. Drug and drug metabolites detected in specimens of maternal origin

Biofliud	*Drug/drug metabolite*
Amniotic fluid	Benzoylecgonine, caffeine, cocaine, cocaethylene, diazepam, digoxin, ecgonine methyl ester, 2-ethylidine-3,3-diphenyl-pyrrolidine (EDDP), fentanyl, 6-acetylmorphine (6-AM), meperidine, methadone, morphine, morphine-3-glucuronide, nitrazepam, norcocaine, nordiazepam, phencyclidine (PCP), phenobarbital, secobarbital, and valproate (valproic acid)
Breast milk	Acetaminophen, amitriptyline, amphetamine, benzoylecgonine, chloral hydrate, citalopram, cocaine, cocaethylene, codeine, chloral hydrate, desipramine, desmethylcitalopram, diazepam, didesmethylcitalopram, dothiepin, doxepin, fentanyl, flunitrazepam, fluoxetine, hydromorphone, 11-hydroxy THC, imipramine, lorazepam, meperidine, morphine, methadone, nitrazepam, norcocaine, nordiazepam, nordoxepin, norfluoxetine, normeperidine, norsertraline, nortriptyline, oxazepam, oxycodone, paroxetine, salicylate, sertraline, temazepam, Δ9-tetrahydrocannabinol (THC), 11-nor-9-carboxy- Δ9-tetrahydrocannabinol (THCA), and trichloroethanol

The drug was detected in amniotic fluid within 5 min of intravenous drug administration at concentrations that sometimes exceeded those found in maternal serum. Narcotic analgesics are reported to cross the placental barrier rapidly. Following parenteral administration of morphine to the mother, the fetal-maternal ratio of morphine concentration in blood reached unity at 5 min. However, at physiological pH, narcotic analgesics tend to be predominantly charged, so in the absence of other factors, the concentration of the drug in the amniotic fluid is expected to be lower than that of the maternal plasma. In one case study however, fetal-maternal blood concentration ratios for morphine, 6-acetylmorphine (6-AM), and codeine were 4.86, 38, and 3.5, respectively. Concentrations of morphine, morphine-3-glucuronide, and 6-AM in amniotic fluid were 604, 209, and 128 μg/kg compared with 280, 801 and 4 μg/kg in the maternal blood of a 17-year old pregnant female who died of a heroin overdose.

Mean methadone concentrations in maternal serum and amniotic fluid were 0.19 and 0.20 mg/L, respectively, following maternal drug use. Methadone and 2-ethylidine-3,3-diphenylpyrrolidine (EDDP) concentrations of 0.66 and 0.52 mg/L in amniotic fluid were measured full-term in a female who was maintained on 110 mg methadone a day. Benzodiazepines cross the placenta because of their lipid solubility and lack of ionization. However, drug concentrations in the amniotic fluid remain relatively low due to extensive protein binding in the maternal plasma, minimal renal excretion by the fetus, and the absence of ion trapping.

PHYSIOCHEMICAL NATURE OF BREAST MILK

The female breast consists of 15-20 lobes of milk-secreting glands, embedded in the fatty tissue. During pregnancy, estrogen and progesterone, secreted in the ovary and placenta, cause the milk-producing glands to develop and become active. The ducts of these glands have their outlet in the nipple, and by mid-pregnancy, the mammary glands are prepared for secretion. Colostrum, a creamy white to yellow pre-milk fluid, may be expressed from the nipples during the last trimester of pregnancy. The pH of colostrum more closely resembles that of plasma. This difference in composition and pH can influence the drug content. This fluid, which is a rich source of protein, fat, carbohydrate, and antibodies, is replaced with breast milk within 3 days of delivery of the fetus and placenta. Proteins, sugars, and lipids in the milk provide initial nourishment to the newborn infant. The production of between 600 and 1000 mL of milk per day by the milk-secreting cells is stimulated by the pituitary hormone, prolactin. Contraction of the myoepithelial cells surrounding the alveoli allows the milk to be expressed into the duct system.

Transfer of Drugs through Breast Milk

In addition to widespread prescription drug use during pregnancy, more than 90% of women receive medication during the postpartum week. This is accompanied by a near three-fold increase in the number of women who breast-fed their infants in recent years. Fortunately, however, most drugs given to nursing mothers reach infants in smaller amounts compared with drugs given during pregnancy. The transfer of drug into the milk depends on metabolism, protein binding, and the circulation of blood in the mammary tissue. The total protein concentration of plasma (75 g/L) far exceeds that of breast milk (8 g/L), which limits the passage of highly protein-bound drugs into the milk. Passive diffusion is largely responsible for transporting the drug across the mammary epithelium, interstitial fluid, and plasma membranes into the milk. Drugs that are extensively protein bound may not readily pass into the milk, but emulsified fats contained in the milk may concentrate highly lipid-soluble drugs. Drugs with molecular weights less than 200 Da readily pass through small pores in the semi-permeable membrane. The mildly acidic pH of breast milk tends to trap weakly basic drugs.

Analyzing Drugs from Breast Milk

Fluid is collected using a special device such as a breast milk pump, after which well-established analytical techniques may be used to detect drugs-of-abuse. Breast milk, which contains protein (1%), lipid (4%), lactose (7%), and water (88%), is mildly acidic. The average pH is 7.08 although it may range from 6.35-7.35. However, the high lipid content of milk may interfere or decrease the extraction efficiency or recovery of some drugs. Additional washing with non-polar solvents such as hexane may be necessary to remove excess lipids prior to chromatographic analyses. The effect of natural emulsifying agents in breast milk, which have detergent-like activity, may interfere with antibody-antigen reactions that take place in immunoassay screening tests.

The daily variation of breast milk composition, combined with drug dose and time of administration relative to the expression of milk, is likely to affect the amount of drug present and the effect on the infant. The concentration of drug in the breast milk is subject to both within and between subject variation, further confounding attempts to generalize infant risk assessment. Composition is known to vary with time of day and method of sampling. The lipid content of the milk varies not only daily but also during a single feed; the latter portion of expressed milk may contain a several-fold increase in fat. Changes in pH, lipid or protein content throughout the stages of lactation and throughout the feeding interval are expected to influence the rates of drug transfer.

Study of Toxicity

Small, relatively lipophilic drugs may readily diffuse into the breast milk and become concentrated in the lipid-rich fluid. Furthermore, basic drugs may become sequestered because of ion trapping. For this reason, PCP was detected in breast milk 41 days following cessation of maternal drug use. A PCP concentration of 3 μg/L was detected almost 6 weeks after drug use. Therapeutic use of narcotic analgesics during the delivery and postpartum phase is not uncommon. Although transfer of morphine to the infant through the milk was once considered negligible, low oral doses of morphine to nursing mothers produced drug concentrations in milk up to 100 μg/L and were extremely variable between feeds.

The morphine concentration in the infant serum was 4 μg/L, which was considered to be in the analgesic range. Based on clearance and bioavailability, the authors suggested that the infant received between 0.8 and 12% of the maternal dose. In seven patients receiving intravenous morphine following cesarean delivery, morphine and morphine-6-glucuronide concentrations in colostrum were 0-48 and 0-1084 μg/L, respectively. Morphine concentrations were always smaller in the milk compared to the plasma. However, morphine-6-glucuronide concentrations were always higher. Although the low bioavailability of morphine (20-30%) lessens the risk to the infant, it should also be considered that inactive conjugated metabolites such as morphine-3-glucuronide in breast milk may undergo reactivation by deconjugation in the gastrointestinal tract of the infant.

Neonatal abstinence syndrome is largely associated with narcotic analgesics, and the perceived significance and risk associated with maternal opioid use is a matter of debate in the scientific literature. In the past, breastfeeding was discouraged among women who were receiving more than 20 mg methadone a day. However, a recent study suggested that breastfeeding among methadone-maintained patients is generally acceptable and even advocates its use for the abatement of symptoms associated with neonatal abstinence syndrome. Methadone is lipophilic and highly protein bound. Peak methadone concentrations are reported to occur about 4 h after oral administration, with an average of 2.2% of the daily dose being secreted into the milk. Methadone maintenance (50 mg/day) in a drug-dependent nursing mother produced breast milk concentrations between 20 and 120 μg/L in the first 24 h after the dose, substantially lower than maternal plasma concentration.

Although methadone concentrations are typically lower in milk compared with maternal plasma, a concentration of 5.7 mg/L in breast milk was reported in another study. Other opioids, including hydromorphone and oxycodone, have also been detected in breast milk. In one study of eight women who received intranasal hydromorphone, it was estimated that the infant received approximately 0.67% of the maternal dose.

Meperidine and its active metabolite normeperidine were detected in breast milk at concentrations in the range 36-314 and 0-333 μg/L following postpartum analgesia. Although some studies suggest the quantities of drug transferred to the infant are minimal, others have shown decreased neonatal alertness and neurobehavioral outcome compared to morphine. Fentanyl was found to concentrate in lipid-rich colostrum at much higher concentration compared to maternal serum. Benzodiazepines tend to be lipophilic and uncharged, factors that facilitate their transport across membranes into the milk. More water-soluble benzodiazepines like temazepam are less likely to accumulate in breast milk compared with lipophilic analogs like diazepam. Although a number of benzodiazepines have been detected in breast milk, concentrations are typically very much lower than those found in maternal plasma.

In one report of a woman receiving treatment for benzodiazepine withdrawal, maximum concentrations of diazepam, nordiazepam and oxazepam in breast milk were 307, 141 and 30 μg/L, respectively. Although the dose of diazepam delivered to the infant through breast milk was estimated to be 4.7% of the maternal dose, the immaturity of hepatic enzymes, slow metabolism, and elimination of some benzodiazepines in the infant should also be considered. A number of antidepressant medications that are frequently used for postpartum depression have been reported in breast milk although concentrations are generally considered to be low. Transfer of selective serotonin reuptake inhibitors (SSRIs) including fluoxetine, sertraline, fluvoxamine, paroxetine, and citalopram are reported to deliver 0.5-10% of the maternal dose to the infant. Detectable amounts of parent drug in the milk at the nursing infant have been measured. Although short-term adverse effects of SSRI administration through breast milk are rare, more research on the long-term consequences of maternal drug use are needed.

Transfer of fluoxetine into breast milk is not unexpected given its lipophilicity and basic nature. In a study of 10 women, the average dose of fluoxetine administered to nursing infants was estimated to be 10.8% of the maternal dose although as much as 28.6% was delivered to one infant. In this case, a fluoxetine dosing regimen of 0.27 mg/kg/day produced mean fluoxetine and norfluoxetine concentrations of 41 and 96 μg/L respectively in breast milk. Peak concentrations in milk were achieved in approximately 6 h. Less than 10% of the therapeutic dose is often considered as a reference point of safety for nursing infants. Citalopram and fluoxetine appear to deliver a comparable dose to nursing infants. Trough plasma concentrations of citalopram in nursing infants were 64% those of maternal plasma. Furthermore, citalopram metabolites were present in milk a two to threefold higher concentration compared with maternal plasma.

Fluvoxamine, paroxetine, and sertraline, which produce peak concentrations in breast milk at 7-10 h, are reported to deliver less drug to the infant than either fluoxetine or citalopram. Stimulant drugs that are basic are particularly susceptible to ion trapping and accumulation in breast milk. Following a therapeutic dosing regimen for narcolepsy, amphetamine concentrations in four milk samples were in the range 55-138 μg/L. These concentrations exceeded those found in maternal plasma by three to seven-fold. In this report the dose of amphetamine (20 mg/day) was very much lower than those used by the drug abusing population. Animal studies using rats have also confirmed that cocaine preferentially partitions into breast milk. Cocaine and several of its metabolites have been detected in breast milk.

The parent drug, which is the predominant analyte in breast milk, was as high as 12 mg/L in one study of six women who used cocaine. Intoxication of breast-fed infants by cocaine-abusing mothers has been reported. Despite the fact that as many as 34% of pregnant women are reported to have used marijuana, the effect of postnatal drug exposure is not completely understood. The primary active ingredient, Δ-9-tetrahydrocannabinol (THC), is lipophilic and readily transfers into the milk where it may accumulate. Concentrations of THC are reported to be eight-fold higher in breast milk compared with maternal plasma. THC concentrations in breast milk of 105 and 340 μg/L have been reported, the latter in a chronic marijuana smoker.

CONCLUDING REMARKS

The direct impact of specific drugs on the newborn child is difficult to evaluate. Many substance-abusing women use multiple drugs, receive inadequate health care, and may be predisposed to other health problems that may impact both neonatal and maternal outcomes. A number of illicit, prescription, and over-the-counter drugs have been detected in amniotic fluid and breast milk using well-established immunochemical, chromatographic, and spectroscopic techniques. Much of the human data to date is quite limited and predominantly consists of individual case reports. Small animal studies, although more numerous, are subject to biological scaling and possible differences in drug metabolism, distribution and toxicity. Long-term implications of prenatal drug exposure are limited, and many consequences of fetal drug exposure are still unknown. Despite adequate understanding of the maternal consequences of drug abuse, fetal consequences for many drugs are poorly understood and this is a challenging area of maternal-fetal medicine. Issues concerning use of prescription drugs during lactation are clinically important but complex. Much of the information available is based upon short-term or single-dose studies.

The assessment of adverse drug reactions in neonates and infants are difficult to discern, and the effects of long-term drug exposure have not been fully elucidated for many drugs. The use of illicit drugs poses many of the same issues but is also compounded by other complications such as multiple drug use and health issues associated with substance-abusing individuals. Although analytical methodology that is routinely used in the toxicology laboratory can be used to detect drugs or metabolites in specimens of maternal origin, appropriate cutoff concentrations and detection limits must be utilized. Clinical studies reported to date are somewhat limited in scope. However, additional research is needed in order to fully understand the mechanisms that influence the transfer of drugs within the maternal-fetal complex and between mother and infant.

4

Drugs in Largest Gland

By definition, this chapter deals with the use of liver for the postmortem determination of drug exposure or intoxication prior to death. The liver is the largest organ in the body (other than the skin or muscle) and is easy to macerate. It has been used for the isolation and identification of poisons by forensic toxicologists for well over a century. Many drugs, especially those with chemically "basic" character (e.g., alkaloids), tend to concentrate in the liver 10- to 100-fold greater than in the blood, making it easier to isolate extracts that could be used in basic pharmacological tests, or if enough substance was present, to isolate and characterize crystalline material.

Of more relevance today, liver tissue can serve a useful alternate specimen if blood is not available in some types of cases (decomposition, severe fire, and exsanguination). However, another important reason that liver is still widely used as a second or third specimen after blood is that postmortem blood drug concentrations are seldom static or homogeneous after death. For many drugs, concentrations in blood can easily increase twofold and often 10-fold or more over perimortem concentrations due to postmortem redistribution or postmortem diffusion. Therefore, measuring liver drug concentrations provides additional data to assist with interpretation. Drugs-of-abuse that are chemically "basic" may be easily analyzed in the liver, including the amphetamines, ketamine, phencyclidine (PCP), and nonamphoteric analgesics such as methadone, meperidine, codeine, hydrocodone, and oxycodone.

The so-called back-extraction procedures work well for basic drugs that have a relatively high pKa and can provide extracts that are virtually free of endogenous compounds such as lipids or fatty acids. Chemically neutral compounds such as delta-9-tetrahydrocannabinol (THC) are considerably more challenging to accurately quantify in the liver because they are present at low concentrations and difficult to separate from other ionically neutral endogenous components. Morphine and hydromorphone, both amphoteric narcotic analgesics, are typically not analyzed for in liver because it is more difficult to obtain clean extracts, and as for most drugs, the concentrations measured are more difficult to interpret than in blood.

HISTOMORPHOLOGY OF DIGESTIVE GLAND

The liver is a large, encapsulated, highly vascular organ situated predominately in the upper right front portion of the abdomen that typically weighs around 1.5 kg. It is divided into two lobes; the smaller left lobe partially overlays the stomach in the central to left portion of the abdomen; the right lobe occupies the upper right abdomen. The liver performs several critical functions, including the metabolism of food, the formation and storage of glycogen, and release, when required, as glucose. The other primary function of the liver is the metabolism of endogenous and exogenous toxicants.

The liver is the primary site for drug metabolism in the body and contains the largest quantity of the critical cytochrome enzyme system, as well as liver alcohol dehydrogenase and many other enzymes. From the perspective of drug absorption and metabolism, the hepatic portal system is the most important. All substances absorbed into the body through the digestive tract ultimately pass through the liver through the portal vein. The portal vein is fed by mesenteric and other venous systems from the stomach, the large and small intestines, the pancreas, and the spleen. Virtually, all blood returning to the heart and lungs through the venous system passes through the liver through the inferior vena cava; the liver is infused with arterial blood through the descending aorta.

PRINCIPLE OF PHARMACOKINETICS

For decades, toxicologists have sought to find a relationship between blood and liver concentrations. The assumption has often been that the ratio between drug concentrations in the blood and liver approximates to some particular value, dependent largely on the physical characteristics and pharmacokinetics of the drug. However, interpretation of drug concentrations in liver, or for that matter any other organ, is far from straightforward. Consideration of the kinetics involved shows why a blood : liver ratio can vary wildly, not just because of postmortem changes in the blood concentration but due to the kinetics of the drug in the body, especially as it relates to absorption, distribution, and excretion. For example, it is a fundamental principle of pharmacokinetics that the liver concentration of a drug during the absorption phase will vary significantly depending on whether it is administered orally or systemically (e.g., intravenous or intramuscular).

With oral administration, the drug passes into the stomach where some absorption may occur, but for the majority of drugs, most absorption occurs in the small intestine. Thus, virtually all of an orally ingested drug passes through the mesentery vessels of the small intestine through the portal vein and into the liver. However, with intravenous or intramuscular administration, all of the drug initially enters the systemic circulation directly, and only after it has passed through a large portion of tissue in the body, does the remainder of it pass through the liver. Therefore, it is to be expected that the liver concentration of virtually any drug will vary at the time of peak blood concentration depending on the route of administration and dose. In other words, peak liver concentrations of a drug can be expected to be significantly higher after oral administration than after systemic administration during the period of initial distribution.

In relation to the blood concentration of a drug, this gives rise to the "first pass effect" where a proportion of the drug never reaches the general circulation due to substantial metabolism in the liver. In addition to these simple pharmacokinetic principles, it has to be considered that in the case of oral ingestion of a drug, postmortem diffusion through the stomach wall to the overlaid left lobe of the liver will increase the ratio of liver to blood concentration, especially if the mass of drug in stomach is relatively large. As far back as 1971, it was shown that in a rat model, concentrations of barbiturates in the liver increase with time after death due to diffusion from the stomach. This has been demonstrated more recently for amitriptyline, also in a rat model. The same authors have shown that limited redistribution of drug from the lungs to the liver may occur postmortem although the magnitude of this increase is not great compared to that for blood. In other words, the drug concentration in the liver is already high compared to that in the pre-mortem blood and therefore, the magnitude of increase of the liver concentration will not be as great as it is in the blood.

Blood is non-homogeneous, with drug concentrations frequently site dependent and the concentration of a drug in the liver may vary considerably depending on the specific portion of the liver that is sampled (e.g., which lobe and even which part of that lobe). While it has been repeatedly advocated that the site of collection of a blood sample should be documented, on only a few occasions has it been advocated that the exact site of collection of a liver sample be stated. Therefore, the concentration of a drug in the liver after oral ingestion will depend on the dose ingested, the time after ingestion, the rate of passage of the drug through the stomach into the small intestine (including the time for stomach emptying), and the postmortem interval.

One additional consideration is to what extent a drug may continue to be metabolized after death. For example, it is known that cocaine is unstable in unpreserved postmortem specimens due to continued hydrolysis by stable esterases. However, much less is known regarding the extent to which other enzyme systems may continue to function in the early postmortem period. When the variability of drug concentrations in blood is considered (e.g., femoral blood vs. cardiac or pulmonary blood), it is not surprising that for many drugs the ratio of concentration in blood versus liver varies considerably. Figure 4.1 shows the degree of variability that can occur in the ratio of codeine in

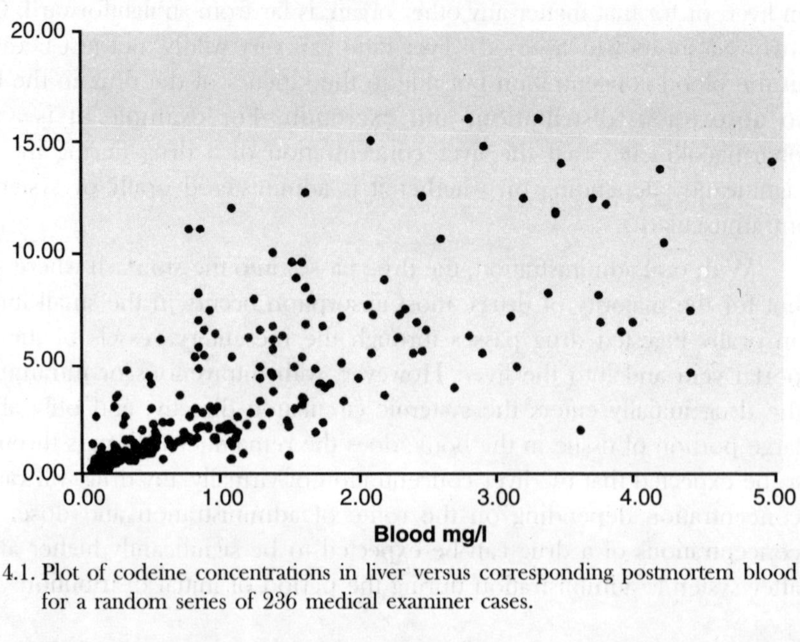

Fig. 4.1. Plot of codeine concentrations in liver versus corresponding postmortem blood for a random series of 236 medical examiner cases.

blood versus liver in a large series of cases where death was attributed to various causes. Although this data includes blood from all sources, the correlation does not improve significantly when blood from only a femoral source is plotted.

Sensitive Methodology

One aspect of toxicology that has received little attention, and indeed is very difficult to study, is the long-term sequestration of drugs in the liver and other tissues. While this issue is not relevant to most death investigation cases, it is relevant to those where sensitive methodologies may detect small amounts of drugs and where those findings may affect the ultimate disposition of a civil case. It is well documented that drug concentrations in the liver are often 50-100 times higher than concentrations in blood. It is also well known that the true terminal half-life of many drugs in the body is considerably longer than the traditional half-life in circulating blood.

There is considerable data regarding the persistence of highly lipid-soluble substances such as chlorinated pesticides, and by extrapolation lipid-soluble drugs such as THC, although hard data for most drugs is anecdotal at best. However, given the lipid-soluble nature of many drugs and the high lipid content of the liver, the persistence of many drugs in the liver is likely to be considerably longer than in blood, and this should be taken into account when interpreting very low concentrations of drugs in the liver.

METHOD FOR DETECTION OF DRUGS

Methods for the detection and quantitation of drugs in liver are not very different than for whole blood once the tissue is homogenized. Traditionally, homogenization of tissue is performed in a blender or homogenizer. Commercial units such as a probe-style homogenizer that combines shearing and ultrasonic action are preferred over food-processor style Waring blenders, because they are considerably more efficient in breaking down cell structures. Liver homogenates may be prepared in water or buffer at dilution from 1:1 to 1:10. If insufficient water or buffer is added, homogenization may be less efficient and the sample may be more difficult to pipette. Over-dilution may cause the concentration of the analyte to drop below the detection capability of the method. Homogenates may be centrifuged in an attempt to remove potentially interfering substances. This is not a problem for qualitative screening tests. However, if this is done for quantitative chromatographic tests, the internal standard must be added and allowed to equilibrate with the homogenate before centrifugation occurs.

Care must also be taken to prevent cross-contamination of liver homogenates with homogenizing devices that can be difficult to thoroughly clean. Most types of immunoassay can be used to screen liver homogenates for drugs-of-abuse although some may be subject to more interference than others. It is therefore important that such assays be investigated for specificity when using liver homogenates and that preferably matrix-matched controls be used. Protease enzymes have been used in the past to liquefy liver homogenates. However, this approach is seldom used today because it can increase the amount of extractable endogenous material and may not offer significant advantages over conventional liquid-liquid or *solid-phase extraction* (SPE) methods that use appropriately chosen internal standards. A major consideration is whether an extraction and analytical procedure is selective enough to allow accurate detection and measurement of the target analyte without interference from the myriad of potentially interfering substances such as lipids or putrefactive amines.

A procedure for basic drugs that incorporates a back-extraction will usually produce less interference and therefore more reliable results than one using a single-step extraction. It also follows that the more decomposed the liver specimen, the more potential there is for interference. Therefore, the more specific the detection method, the more accurate the results are likely to be. However, one caveat to that statement is that non-specific interference that can occur with LC/MS due to ion suppression effects. The majority of drugs-of-abuse, including the amphetamines, cocaine, and most opiates (except morphine and hydromorphone), are moderately strong bases and may be isolated from liver homogenates by extraction into an organic solvent at basic pH and subsequent back-extraction into acid. In a typical procedure, basic drugs can be extracted into a suitable organic solvent after making the homogenate strongly basic (e.g., pH 11-12).

The organic extract is transferred to another tube and strong acid added (e.g., 0.1-1 N sulfuric acid or hydrochloric acid). The aqueous phase containing the back-extracted basic drugs is transferred to another tube and basified with strong alkali (e.g., 2 N sodium hydroxide) and the drug extracted into a suitable organic solvent. This process will remove much of the neutral and acidic interference produced by fatty acids, lipids, and cholesterol. When drying the extracts, care must be taken to protect against loss of the volatile drugs such as amphetamines; some laboratories add methanolic hydrochloric acid to minimize such losses. Morphine and hydromorphone are amphoteric and require a different approach requiring extraction at relatively mild alkaline pH (e.g., 8.0-9.0). Alternatively SPE methods may be used. Kudo et al. have published a method for the determination of free and total morphine in liver and other tissue using acid hydrolysis, extraction with an SPE (Extrelut NT) column, using dihydrocodeine as internal standard, followed by trimethylsilylation.

The derivatized extract was submitted to GC/MS analysis in EI-SIM mode. The limit of detection of morphine was 0.005 mg/kg. Cingolani et al. have demonstrated the successful extraction of morphine from formalin-preserved tissue. The mean levels of recovery of morphine in fixed tissues were 36.29% in liver and 74.93% in the formalin from liver. The major cocaine metabolite, benzoylecgonine, is relatively water soluble and not generally amenable to liquid-liquid extraction. However, benzoylecgonine and related metabolites may be extracted from liver and other tissues using SPE methods. Centrifugation of the liver homogenates after addition of an internal standard will facilitate passage through the SPE column.

CORELATIVE APPROACH

Interpretation of liver drug concentrations is difficult. The kinetics of drug absorption and distribution are such that the relationship between blood and liver concentrations is rarely consistent enough to allow reliable quantitative prediction of a single blood concentration from a liver concentration. The problem is compounded because well-defined reference ranges do not exist and much data for liver concentrations of drugs is typically dotted throughout the literature in a few individual case reports that may be difficult to locate. One well-known book of drug monographs contains numerous references to liver concentrations of drugs. However, those case reports are frequently overdose cases and relatively few are cases where it is known that "*therapeutic*" (or "*recreational*") doses of the drug are involved.

The situation is further complicated because careful review of the original case reports is required to assess the contribution of other drugs, alcohol, or natural disease in the death. In the absence of a reliable database in the literature, interpretation falls to individual laboratories that routinely measure liver drug concentrations in casework - for all types of death. A liver drug concentration can provide useful forensic information when considered with the available history and circumstances of the death, and preferably a blood concentration. For example, a higher than normal liver concentration can corroborate a high postmortem blood concentration and becomes even more useful when interpreted in conjunction with the case history.

As with the interpretation of post-mortem blood drug concentrations, it is critical that consideration should be given to the possibility of drug accumulation due to natural disease or impaired metabolism due to drug interactions or pharmacogenetic factors. Following is a brief summary of the readily available literature regarding concentrations of some drugs-of-abuse in liver. For drugs where data on liver concentrations is sparse in the literature, data are included from the Alberta Medical Examiner's Office in Canada. However, it should be noted that no attempt has been made to differentiate cases where the presence of the drug is the cause of death or merely an incidental finding. Furthermore, the relationship between blood and liver concentrations is not necessarily linear, and indeed, there is a considerable amount of scatter when plotting blood versus liver concentrations for any drug.

Detection of Amphetamines

There are several reports of amphetamine being measured in liver. In one review, a series of 11 amphetamine fatalities were reported with liver concentrations ranging from 4.3 to 74 mg/kg (average 30) with the corresponding blood concentrations 0.5 to 41 (average 8.6) mg/L. A distribution study of several amphetamines in a single fatality reported amphetamine concentrations in liver and femoral blood of 0.857 mg/kg and 0.198 mg/L, respectively.

Detection of methamphetamine

A fatal overdose ingestion of a large amount of methamphetamine resulted in a liver concentration of 206 mg/kg and blood concentration of 40 mg/L. In another methamphetamine-related fatality, concentrations were 4.8 mg/kg in the liver and 2.0 mg/L in the blood. Kojima reported two separate methamphetamine overdose cases with liver concentrations of 174 and 14.1 mg/kg with corresponding blood concentrations of 43 and 8.3 mg/L, respectively. Moore reported the death of a 37-year-old man with postmortem methamphetamine concentrations of 2.2 mg/kg in liver and 0.68 mg/L in blood. Sato reported methamphetamine concentrations in a burned body of 11.7 mg/kg in liver and 2.5 mg/L in blood. Logan reported a massive overdose in a man swallowing a baggie of methamphetamine that produced liver and peripheral blood concentrations of 90.9 mg/kg and 53.7 mg/L, respectively. Bailey reported a large series of deaths involving methamphetamine, with and without concurrent cocaine use. Methamphetamine liver and blood concentrations were 0.13-35.96 mg/kg (average 4.06) and 0.02-3.05 mg/L (average 0.90), respectively, without cocaine and 0-17.1 mg/kg (average 2.83) and 0-7.10 mg/L (average 1.03) with cocaine.

Detection of methylenedioxyamphetamine

Most MDA occurrences today tend to be as the metabolite of methylenedioxy-methamphetamine (MDMA) or co-ingested as an ingredient in illicit tablets. However, in the older literature, methylenedioxyamphetamine (MDA) was more common as a sole intoxicant. In one report of 12 cases, liver concentrations were 8-17 mg/kg (average 12) and the corresponding blood concentrations were 1.8-26 mg/L (average 9.3).

Detection of methylenedioxymethamphetamine

There have been at least three different cases where liver concentrations of MDMA have been reported. DeLetter reported an MDMA overdose with MDMA in liver at 26 mg/kg and in peripheral blood at 3.1 mg/L. Sticht reported concentrations of 29.7 mg/kg in liver and 7.2 mg/L in peripheral blood in an overdose. Dams found MDMA concentrations of 6.66 mg/kg in liver and 1.92 mg/L in femoral blood in a combined MDMA/*P*-methoxyamphetamine (PMA) fatality.

Detection of P-methoxyamphetamine

PMA is a fairly toxic amphetamine compound occasionally detected in forensic cases. Cimbura reported nine cases with liver concentrations ranging from 0.5 to 10 mg/kg and blood 0.3 to 1.9 mg/L. Felgate reported a series of 10 PMA-related deaths with paired liver (femoral blood) concentrations of 11(1.7), 1.4 (0.24), 7.4 (1.3), 7.1 (3.7), 21(4.9), 5.6 (2.2), 6 (1.7), 7.5 (2.2), 6.0 (2.0), and 2.7 (0.53); average 7.8 mg/kg (2.0 mg/L in blood). PMA concentrations have been reported in three other fatalities with liver concentrations of 8.9, 6.8 and 18 mg/kg compared to 1.63, 0.4, and 1.8 mg/L in blood.

Detection of Opiates and Opioids

Felby reported a series of ten fatalities following intravenous use of morphine only; liver total morphine concentrations were 0.4-18 mg/kg (average 3.0) and the corresponding blood concentrations 0.2-2.8 (average 0.7). In a separate case, a man who died of trauma after 2 days in hospital and who received multiple morphine doses had a liver concentration of 0.11 mg/kg and blood concentration of 0.67 mg/L. Cravey also reported blood and liver concentrations of morphine of 8.0 mg/L and 6.0 mg/kg, respectively, in a hospitalized patient. However, it is not clear in these cases whether the morphine reported was unconjugated or total. Chan reported two related morphine fatalities. One case involved oral ingestion with blood concentrations of 0.35 (free) and 1.53 mg/L total morphine and a liver total morphine of 7.0 mg/kg.

The second case was due to intravenous administration and resulted in blood concentrations of 0.07 and 0.42 mg/L of free and total morphine, respectively, and 2.9 mg/kg total morphine in the liver. Moriya determined the tissue distribution of free and conjugated morphine in a man after injection with heroin and methamphetamine. Blood concentrations of free and total morphine varied from 0.462 to 1.350 and 0.534 to 1.570 mg/L depending on the site. A total morphine concentration of 4.20 mg/kg was found in the liver. 6-Acetylmorphine was not measured in the liver.

In a separate in vitro experiment, the same authors found free and conjugated morphine to be stable in the blood and urine at 4, 22, and 37°C for the 10-day study period. In the liver, conjugated morphine was converted almost completely to free morphine at 18-22 and 37°C by the end of the 10 days although it was stable at 4°C. Spiehler analyzed blood and liver for unconjugated and total morphine in 56 cases where the deceased died within 3 h of heroin/morphine use. The unconjugated morphine in blood ranged from 0.08 to 1.65 mg/L, total morphine from 0.10 to 2.2 mg/L, and liver total morphine from 0.41 to 5.5 mg/kg. Using expert systems to analyze the data, Spiehler found blood unconjugated morphine, blood total morphine, and liver total morphine concentrations most useful for interpretation.

Morphine overdoses were characterized by a blood unconjugated morphine greater than 0.24 mg/L and liver morphine concentration greater than 0.50 to 0.75 mg/kg. Kerrigan reported an unusual case involving a morphine pump and acquired tolerance. A 44-year-old male was receiving morphine for pain control. Quantitative analysis of free and total morphine determined unconjugated morphine concentrations in heart blood and liver of 96 mg/L and 88 mg/kg while total morphine concentrations were 421 mg/L and 256 mg/kg, respectively. Records indicate that the infusion pump may have continued to deliver the drug for 15-45 min following death. The cause of death was determined to be complication of adenocarcinoma of the pancreas, and the manner was natural. Lewis reported on the simultaneous analysis of multiple opiates in postmortem tissue, after civil aviation accidents, using SPE, TMS, oxime-TMS derivatives, and GC-MSD. Two cases contained morphine with total blood concentrations of 0.260 and 0.023 mg/L and liver concentrations of 0.87 and 0.083 mg/kg, respectively.

Detection of codeine

Nakamura measured a large series of blood and liver concentrations. In the higher codeine concentrations (11 cases), the liver ranged from 0.6 to 45 mg/kg (average 6.8) with blood concentrations of 1.0 to 8.8 mg/L (average 2.8). A series of 269 deaths associated with codeine in the Alberta Medical Examiner's system in the 6 years from March 1999 to September 2005 were analyzed. The liver concentrations ranged from 0.1 to 96 mg/kg (average 8.0, median 3.0, SD = 13.9), and the corresponding blood concentrations were 0.1 to 83 mg/L (average 2.6, median 1.1, SD = 6.5).

SUICIDAL DOSE OF OXYCODONE

Sedgwick reported on an overdose in a diver with a liver concentration of 12 mg/kg and blood 5 mg/L. Cravey reported two suicidal overdoses with liver concentrations of 22 and 63 mg/kg and corresponding blood concentrations of 4.3 and 14 mg/L. There are two reports of child deaths related to oxycodone. Armstrong et al. reported a 2-year-old girl with oxycodone as the only finding where the heart blood concentration was 1.36 mg/l and liver 0.2 mg/kg. Levine reported the death of a 10-month-old baby boy with only oxycodone present. The blood and liver concentrations were 0.6 mg/L and 1.6 mg/kg, respectively. Lewis et al. reported an oxycodone case after a civil aviation accident with a blood concentration of 0.232 mg/L and a liver concentration of 0.755 mg/kg. A series of 126 deaths associated with oxycodone in the Alberta Medical Examiner's system in the 5 years from March 1999 to December 2004 were analyzed. The liver concentrations ranged from 0.12 to 480 mg/kg (average 10.0, median 1.4, SD = 48.7) and the corresponding blood concentrations were 0.02 to 520 mg/L (average 5.4, median 0.61, SD = 46.32).

Concentration of hydrocodone and hydromorphone

Lewis reported hydrocodone in four aircraft accident victims at liver concentrations of 0.142, 0.02, 0.447, and 0.082 mg/kg and corresponding blood concentrations of 0.022, 0.018, 0.102, and 0.036 mg/L, respectively. Baselt summarized data from eight fatal hydromorphone overdose cases with liver concentrations averaging 1.4 mg/kg (range 0.07-7.7) and blood concentrations averaging 0.3 mg/L (range 0.02-1.2) although data from at least two of the references were not readily available.

Concentration of methadone

In a series of 10 deaths attributed to methadone, Manning et al report liver concentrations of 1.8-7.5 mg/kg (average 3.8) with corresponding blood concentrations of 0.4-1.8 mg/L (average 1.0). A series of 36 deaths due primarily to methadone in the Alberta Medical Examiner's system in the 5 years from June 1999 to August 2004 were analyzed. The liver concentrations ranged from 1.0 to 47 mg/kg (average 6.9, median 4.85, SD = 8.2) and the corresponding blood concentrations were 0.18 to 7.62 mg/L (average 1.0, median 0.505, SD = 1.3).

Concentration of mepheridine (pethidine)

Siek reported the distribution of meperidine and its metabolite, normeperidine, in three oral and three intravenous deaths due to meperidine. For the oral cases, concentrations of meperidine and normeperidine in liver averaged 7 and 31 mg/kg (range 5-10 and 11-66) respectively; for the intravenous cases, liver concentrations average 8.3 and 7.3 mg/kg for (range 2-16 and 0-12) meperidine, and normeperidine respectively. The corresponding blood concentrations averaged 12 and 19 mg/L (range 8-20 and 8-30) for the oral route; blood concentrations average 4.3 and 2.5 mg/L (range 1-8 and 0-7) for the intravenous route. A series of 33 deaths associated with meperidine in the Alberta Medical Examiner's system in the 7 years from March 1999 to March 2006 were analyzed. For 32 paired results, the liver concentrations ranged from 0.23 to 320 mg/kg (average 25.4, median 3.25, SD = 64) and the corresponding blood concentrations were 0.05 to 36.2 mg/L (average 3.2, median 1.2, SD = 6.5). For normeperidine, an active and potentially more toxic metabolite, liver concentrations ranged from 0.26 to 159 mg/kg (average 18.8, median 6.4, SD = 32) and the corresponding blood concentrations were 0.05 to 33.6 mg/L (average 3.3, median 0.72, SD = 6.4).

Detection of Phencyclidine and Ketamine

Baselt summarized 17 PCP cases where the liver concentrations averaged 23 mg/kg (range 0.9-170) and blood 4.8 mg/L (range 0.3-25). Peyton reported two ketamine fatalities with liver concentrations of 6.3 and 0.8 mg/kg and blood concentrations of 7.0 and 3.0 mg/L, respectively. Moore reported a case of a mixed drug fatality involving ketamine and ethanol. The liver and blood ketamine concentrations were 4.9 mg/kg and 1.8 mg/L, respectively, and the blood ethanol concentration was 170 mg/100 mL. Licata et al. described a homicide involving ketamine. The liver ketamine was 6.6 mg/kg and blood 27.4 mg/L.

Detection of Cocaine

Baselt has summarized 19 cases where liver cocaine concentrations were measured as 0.1-20 mg/kg (average 4.2), with corresponding blood concentrations of 0.9-21 mg/L (average 5.3). Bailey reported a series of 66 homicide cases and accidental overdoses. The liver concentrations ranged from 0 to 33 mg/kg for cocaine (average 0.96, SD = 4.59, N = 66) and 0 to 41 mg/kg for benzoylecgonine (average 5.9, SD = 11.59, N = 25). The corresponding blood values were 0-7.4 mg/L (average 0.46, SD = 1.40) and 0-14.4 mg/L (average 2.75, SD = 4.06) for cocaine and benzoylecgonine, respectively.

Spiehler measured cocaine and benzoylecgonine concentrations in blood, liver, and brain in two series of cases, where the death was attributed to cocaine (37 cases) and where it was considered incidental (42 cases). In the cocaine fatality group, liver concentrations ranged from 0 to 393 mg/kg for cocaine (average 6.7) and 1.3 to 87 mg/kg for benzoylecgonine (average 21.3), the corresponding blood values were 0.04 to 31 mg/L (average 4.6) and 0.74 to 31 mg/L (average 7.8) for cocaine and benzoylecgonine, respectively. The second group with incidental cocaine findings had liver concentrations that ranged from 0 to 1.6 mg/kg for cocaine (average 0.08) and 0 to 10 mg/kg for benzoylecgonine (average 1.3), and the corresponding blood values were 0 to 0.5 mg/L (average 0.05) and 0 to 7.4 mg/L (average 0.88) for cocaine and BE, respectively.

Apple and Roe reported six cases with liver concentrations of cocaine and benzoylecgonine ranging from 0 to 5.0 and 0.4 to 24.1 mg/kg, respectively; blood cocaine and benzoylecgonine concentrations ranged from 0 to 5.6 and 0.6 to 18.7 mg/L, respectively. Cingolani has shown that in formalin-preserved tissue, benzoylecgonine can be measured with recovery in fixed tissues averaging 12.31% from liver and 84.47% in formalin from the liver. No cocaine was detected in the liver. Shimomura examined liver, brain, blood, and urine specimens obtained from 15 post-mortem cases for methyl ecgonidine, ecgonidine, cocaine and benzoylegonine. The median concentration, and range, for cocaine in blood was 0.0 12 (0-0.088) mg/L and for liver 0.057 (0-0.503) mg/kg, and for benzoylecgonine in blood 0.458 (0.030-2.071) mg/L and liver 0.821 (0.045-4.98) mg/kg.

Detection of Cannabinoids

Because of the high fat solubility and low doses usually involved, liver is not particularly suitable for the measurement of delta-9-tetrahydrocanabinol and the carboxy-THC metabolite. However, there is a single fatal overdose recorded in the literature with a liver THC concentration of 38 mg/kg! The THC was estimated semi-quantitatively by thin-layer chromatography (TLC) and photoelectric densitometry. No blood concentration was reported.

CONCLUDING REMARKS

Liver is a specimen easily obtained at autopsy in sufficient quantities for comprehensive testing. It is generally considered the preferred specimen when blood is unavailable. Disadvantages of utilizing liver include the necessity to produce an homogenate prior to extraction, matrix effects observed with routine analytical techniques, and lack of a database to aid the interpretation of drug concentrations. The literature is deficient in describing the probable non-homogeneity in liver sampling. However, drug concentrations measured in liver are generally within the detection range of the analytical methods currently utilized in forensic toxicology.

5

Clinical Study of Prodrugs

Most companies prefer to develop drugs that do not require heroic interventions to achieve drug delivery, *pharmacokinetic* (PK) and *pharmacodynamic* (PD) goals. Thus a drug should have adequate water solubility to facilitate dissolution if administered orally and superior solubility if a parenteral solution dosage form is required. The drug should have the ability to cross biological membranes either via passive permeation or via a carrier-mediated process. If it is unable to do so, absorption from the *gastrointestinal tract* (GIT) may be limited and access to intracellular sites a challenge. For passive permeation, having lipophilic properties is ideal, thus this property can conflict with the goal of adequate water solubility. The drug must have adequate chemical stability in its desired dosage forms, once in solution and adequate metabolic stability so that it does not undergo excessive presystemic clearance and has a reasonable *in vivo* residence time.

The drug should have properties such as taste, odor, etc. that allow for ease of formulation and it would be ideal if the drug had greater affinity for the activity receptor site than for sites and receptors that could lead to toxic outcomes. For the last 100 years, many small molecules have been identified with desirable therapeutic benefits. Most have "*drugable*" properties, since the ability to deliver and test drugs without these properties has been limited. That is, earlier *in vivo* screening techniques led to a natural selection process of drugs with drugable properties while discarding otherwise efficacious but poorly deliverable molecules. With combinatorial chemistry increasing chemical libraries to mega-numbers, receptor and cell culture based assays and *high throughput screening* (HTS) assays, and an increase in scientific sophistication, molecules have been identified which should be active, if they are capable of reaching and maintaining adequate concentrations at the appropriate receptor site. Along with this increased sophistication, increased frustration has occurred as receptor or enzyme based assays have not always translated to *in vivo* efficacy.

The negative *in vivo* responses and lack of "*developability*" of many of these molecules have etiologies that include poor "*drugable*" properties. On the positive side, these assays and the novel chemistries have helped identify *new chemical entities* (NCEs) with complex structures, which would have been discarded in earlier times due to apparent lack of apparent *in vivo* efficacy. What if these agents could now be delivered? These "leads" could result in new drug molecules with truly novel structures. However, further analog development of these leads often fall into the category of what has been termed the "*high affinity trap*" whereby the constraints for receptor activity no longer allow the incorporation of drugable properties through analog modifications. A possible solution to this dilemma is the rational design of prodrugs.

DRUG-LIKE CHARACTERISTICS OF PRODRUGS

Many have used the following scheme to describe the prodrug approach to the delivery of problematic molecules. Prodrugs are chemical modifications of drug or potential drug molecules used to overcome barriers to the efficacy, deliverability and utility of problematic molecules. The barrier shown in the scheme can include but is not limited to;

1. Low water solubility leading to compromised oral, dermal. ophthalmic and parenteral delivery
2. High polarity leading to poor cellular permeability
3. Poor chemical and metabolic stability leading to presystemic metabolism and short biological residence times
4. Lack of site of action targeting leading to high side effects and compromised *pharmacodynamics* (PD)
5. Compromised physicochemical properties leading to patient acceptability and formulability issues
6. Poor intellectual property right protection leading to the lack of economic incentive

Ferres (1983) has illustrated the possible advantages of the prodrug solution compared to an analog approach. In this scheme, the delivery problems could be overcome without compromising receptor activity since the active species remains with the analog or what has been referred to in the prodrug literature as the parent molecule or parent drug. A significant number of recently marketed molecules are prodrugs (about 14% at small molecule NCEs approved worldwide in 2001 and 2002) yet the use of prodrugs to address solutions to the formulation and delivery of problematic drugs was often disdained in both big pharma as well as in emerging companies. This appears to be changing. Having consulted for the pharmaceutical industry for the last 32 years on a number of issues including the use of prodrugs, author have had more requests to give talks and to consult on the possible use of prodrugs as

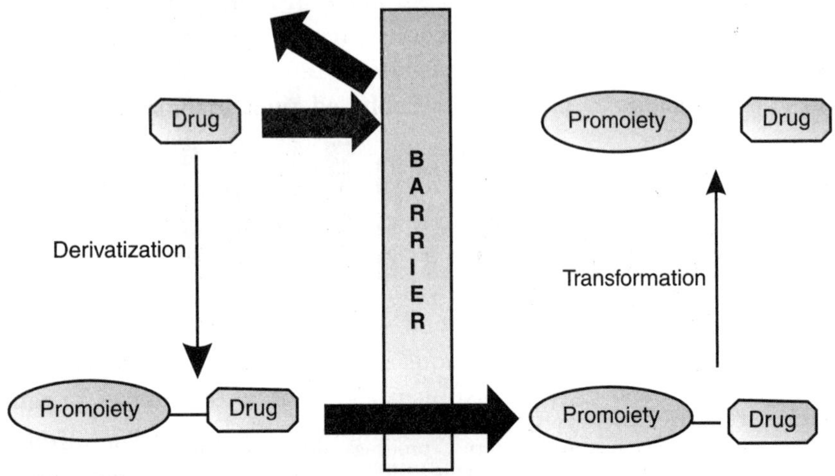

Fig. 5.1. An illustration of the prodrug concept.

a problem solving technique in the last two to three years than in my previous 30 years. The basis for some of this increased interest is discussed in some recent publications.

The number of prodrug patents has increased dramatically in the last 10 years (2000% increase over 1993) and the number of recently approved drugs that are in fact prodrugs strongly suggest that prodrug strategies are becoming an integral part of the drug discovery paradigm. Drug discovery teams seem to be raising the idea of the use of prodrugs earlier rather than as an afterthought. Professor Adrian Albert, the author of the term pro-drug or pro-agent, spoke eloquently of the topic of this workshop in his book, *"Selective Toxicity"* in the 1950s. He recognized the often disconnect between receptor binding and receptor access so suggesting the prodrug approach as a means of overcoming the gap. Have we really come a long way since the 1950s?

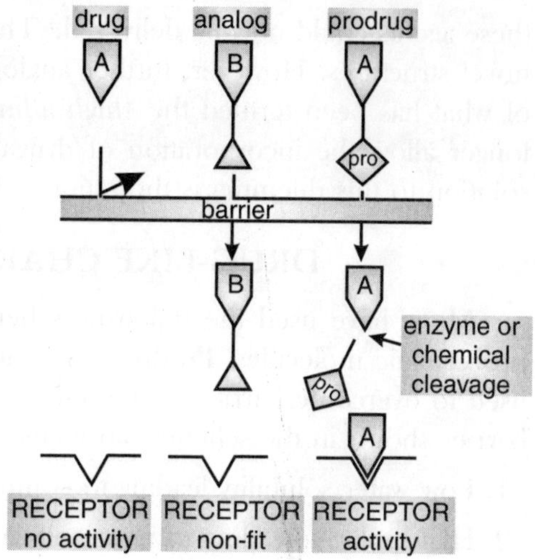

Fig. 5.2. Illustration used to show the potential differences between an analog and a prodrug approach to drug receptor activity.

TARGETING APPLICATIONS

All tell an eloquent tale of the prodrug story while Denny, 2001, 2003; Seddon *et al.*, 2004; and others too numerous to mention discuss how the technique has specific targeting applications in areas such as cancer chemotherapy. The prodrug technique clearly has utility. Of the 43 new drugs approved in 1993, five were prodrugs and an additional two or three had structures indicating the possibility that they might have active metabolites. Bernardelli *et al.* (2002) and Doherty (2003) reported on new approved drugs for 2001 and 2002. In that two-year period there were 49 new chemical entities (non-biologicals) approved, seven, or 14%, of which were clearly prodrugs, one was an acknowledged "soft" drug and one could argue that three other approvals were possibly acting as prodrugs.

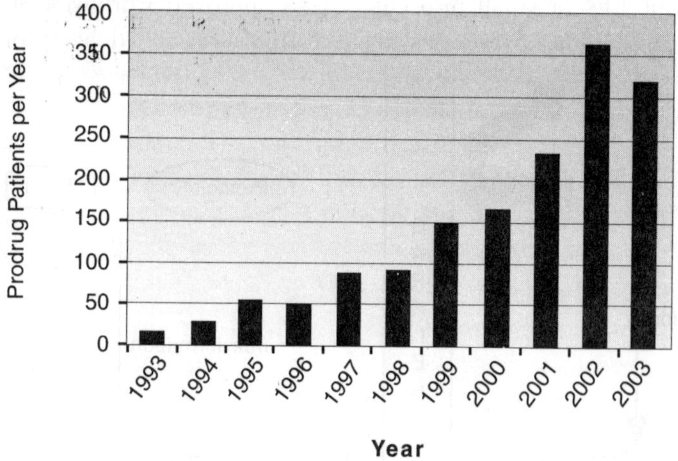

Fig. 5.3. Plot of the number of patents per year over the last 10 years from a search using the terms, prodrugs, drug latentiation, bioreversible derivatives and ADEPT.

Some colleagues are currently completing an extensive new book on prodrugs to be titled "Prodrugs: Challenges and Rewards" that should be published in 2006 by Springer. Included in the book will be numerous case study chapters that will highlight both old and new examples of commercially successful prodrugs. The patent literature also provides some insight into the future, since drugs will only be developed if their exclusivity can be protected. The patent literature on prodrugs has recently been reviewed. Although the search engine used was not comprehensive, the trends were notable. That is, there was an almost exponential increase in prodrug patents over the 10-year period.

Many of these patents were described as being defensive patents. Just as in the past when terms such as "and physiologically acceptable salts thereof" began to be used to cover possible future novel salts from competitors, statements such as "and prodrugs thereof" have begun to appear. This was most prevalent in patents originating from Japan but has also begun to appear in patents from American-based multinationals. A separate search at the US Patent Office site for the period 1976 to the present using just the word prodrug/s resulted in about 6,500 hits. This probably underestimates the real number by at least 50% as many prodrugs are not identified as such. Similar searches in PubMed and SciFinder with the keyword prodrug/s resulted in high publication hit rates confirming the patent trend.

LATEST PRODRUGS

Recently Approved Prostaglandin

Latanoprost (Xalatan), bimatoprost (Lumigan), travoprost (Travatan) and unoprostone isopropyl (Resula) are four recently approved prostaglandin analogs used ophthalmically for the treatment of glaucoma. The use of prostaglandins and their analogs for the treatment of glaucoma derives from the observation in the 1970s that at low doses prostaglandins lowered *intraocular pressure* (IOP) while at higher doses IOP was raised. Prostaglandins themselves were also shown to cause significant hyperemia and irritation. In an effort to separate the activity from toxicity various prostaglandin analogs were prepared, however, the carboxylic acid forms of these agents proved ophthalmically less clinically effective until converted to either their isopropyl ester form (travoprost, unoprostone isopropyl and latanoprost) or their ethanolamine amide (bimatoprost). The FDA approved Latanoprost in 1996 while the other prodrugs were approved more recently. The three isopropyl ester derivatives are clearly prodrugs and are readily converted to the parent prostaglandin analog in the cornea by the action of esterases present in corneal tissue to release isopropyl alcohol and the parent drug.

For a time it was uncertain whether bimatoprost, as an N-ethyl amide, was behaving as a prodrug but studies by Maxey *et al.* (2002) and Hellberg *et al.* (2003) did show hydrolysis in ocular tissues while Sharif *et al.* (2003) showed that bimatoprost acid was a prostanoid FP receptor agonist. The byproducts of bioconversion are isopropyl alcohol for the three ester prodrugs and ethanolamine for bimatoprost. Both of these byproducts appear to be safe and well tolerated. Improved corneal permeation and safety were the non-drugable properties addressed by these prodrugs. In their carboxylate form they are both poorly permeable and more irritating than their prodrug forms. Thus the prodrugs offered an improved therapeutic index compared to their parent molecules.

Recently Approved Antiviral Prodrugs

Many antivirals of the nucleoside type are polar and poorly permeate the GIT. Two recent examples are tenofovir and adefovir. As phosphonates, these antivirals are very polar exhibiting oral bioavailabilities of less than 5% in their non-prodrug forms. Tenofovir was first evaluated as its bis-pivaloyloxymethyl ester but found to have some side effects that were felt to be caused by carnitine depletion due to the pivalic acid promoiety. A series of carbonate bis-esters were found to be superior with tenofovir disoproxil, fumarate salt (Viread), being chosen as the prodrug of choice for further clinical testing. Cleavage of the prodrug results in the formation of two moles each of carbon dioxide, formaldehyde and isopropyl alcohol. The conventional wisdom is that the first phosphonate protecting-group is cleaved via the action of esterase enzymes at the isopropyl carbonate linkage resulting in the monoester intermediate.

Cleavage of the second group is probably via a combination of esterase and phosphodiesterase activity. Animal species differences were seen in tenofovir availability from its prodrugs but human availability appears to be adequate and dose independent. Tenofovir disoproxil was recently approved and is now one of the top-selling anti-AIDs drugs on the market. Many relatively non-polar prodrugs of polar parent molecules often result in less than quantitative oral availability. Many ester type prodrugs of which tenofovir disoproxil is one can be subject to premature cleavage in the GIT resulting in the non-absorbable polar intermediate or parent molecule. Often not considered is the role of cleavage in the enterocytes contributing to less than quantitative availability. The ideal prodrug of a polar non-absorbable parent is one that is chemically and enzymatically stable in the lumen and the enterocytes but is then quantitatively and selectively cleaved on reaching systemic circulation. If cleavage occurs in the entrocyctes diffusion or efflux of the parent back into the lumen is possible and highly probable since there is just as much driving force for the drug to go in that direction as there is for systemic delivery. Adefovir dipivoxil was developed and approved as a bis-pivalate for the treatment of hepatitis B after an initial trial for the treatment of AIDs.

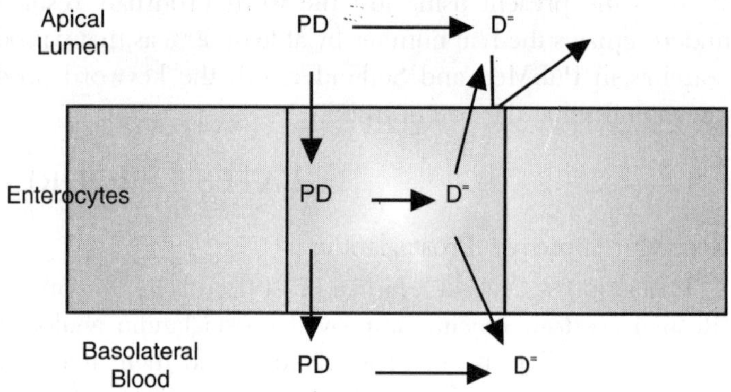

Fig. 5.4. Illustration showing that a non-polar prodrug (PD) of a polar drug (D=) may show less than complete oral availability due to premature cleavage in the GIT or the enterocytes.

Because of the lower dose needed to treat the hepatitis B indication, the amount of pivalic acid formed during the prodrug deprotection did not result in sufficient carnitine depletion to be of concern. Like the tenofovir prodrug, oral availability of the prodrug is good and results in effective therapy. Adefovir dipivoxil and tenofovir disoproxil metabolism results in the production of two moles of formaldehyde. Company toxicologists and regulators often raise concern about formaldehyde production especially since it gives a positive Ames test. There are numerous marketed prodrugs that form formaldehyde, fosphenytoin is a good example. Nevertheless, each time a formaldehyde-producing prodrug is proposed, new concerns are raised. Taking 300 mg of tenofovir disoproxil per day exposes one to 0.035 grams of formaldehyde. To place this into perspective, two glasses of red wine will expose you to about 0.040 grams of methanol/formaldehyde. Yes, one must be concerned with excessive formaldehyde exposure, but by any reasonable estimate most prodrugs will not add significantly greater methanol/formaldehyde/formate exposure than that seen in normal input from diet and the environment. Two earlier marketed nucleoside antivirals are acyclovir and gangciclovir. Both are absorbed from the GIT but on a limited basis. Their limited permeability across the GIT mucosa is the result of their high polarity. Numerous efforts to improve oral availability by the innovators and various other research groups focused on the simple mono-esterification of acyclovir or mono- or bisesters of gangciclovir.

The discovery by Burrows Wellcome scientists that the L-valine ester of acyclovir (valacyclovir, Valtrex) gave superior performance compared to other esters was surprising and was initially attributed to the combination of solubility and lipophilicity enhancement of the prodrug compared to the parent. It was not until later that it was discovered that valacyclovir was actively transported across biomembranes by the di- and tripeptide transporter. Similar results were observed with the L-valine ester of gangciclovir (valgangciclovir, Valcyte). Of note is the better performance of the monoester compared to the bis-ester. Valacyclovir is an example where "*lady luck*" plays a role and illustrates the role that serendipity plays in the drug discovery process. Is the choice of L-valine esters one that someone would make from first principles? It is unlikely, but the discovery and development of valgangciclovir was influenced by the observations made with valacyclovir. Relenza or zanamivir is a neuraminidase inhibitor useful for the treatment of influenza, but, because of its high polarity, is only available in an inhaler form. No oral form is available.

When developing oseltamivir (Tamiflu), another neuraminidase inhibitor, not only did the product have to compete against zanamivir, but the preferred route of administration of most medications in most of the world is oral delivery. As an ethyl ester, oseltamivir is well absorbed orally in man and readily breaks down to its active form, oseltamivir carboxylate (GS-4071), *in vivo*. GS-4071 only showed an oral availability of 4.3 ± 1.6 % in rats. Significant animal species differences were observed in bioavailability after oral dosing of the prodrug, with 30%, 35%, 11% and 73% bioavailability in mice, rats, ferrets and dogs, respectively. The bioavailability in man is around 40-50%. Breakdown of oseltamivir is due to the action of esterases resulting in production of small quantities of ethanol. As an oral product, oseltamivir appears to have had a greater commercial impact than zanamivir, while the discoverers of zanamivir should be credited with pioneering this novel route of influenza treatment.

Recently Approved Drugs

Three additional drugs were recently approved for which the use of a prodrug modification resulted in commercially viable products of parent drugs that were polar. Two that use the same promoiety are olmesartan medoxomil (Benicar), an antihypertensive agent, and prulifloxicin (Sword), an antibacterial. Both these agents use the (5-methyl-2-oxo-1,3-dioxol)methyl promoiety. In one case, it is used in the form of an ester and, in the second case, it is used in the form of an N-alkylation of a secondary amine group. In the case of olmesartan medoxomil, protection of the acidic carboxyl group allows for acceptable absorption of the deprotected antihypertensive active.

The mechanism of deprotection of the (5- methyl-2-oxo-1,3-dioxol)methyl promoiety, first used, to my knowledge, as a promoiety for altering the polarity of penicillins and cephalosporins in the mid-1980s, has been proposed to be first, esterase cleavage of the cyclic carbonate linkage, followed by an elimination reaction. This results in the formation of the parent drug, carbon dioxide and 2,3-butanedione. When attached to the secondary amine group, as in the case of prulifloxicin, some have suggested a more complex mechanism involving N-oxide formation of the tertiary amine formed by the N-alkylation by the promoiety. Ximelagatran (Exanta) is an anticoagulant. Although approved in Europe, it recently was turned down for approval in the USA.

The innovator was asked to supply additional supporting data. Ximelagatran is a double prodrug, whereby a carboxyl group is protected as an ethyl ester and the benzamidine functionality is in the form of an N-hydroxyl moiety. The parent drug, melagatran, has an oral availability of only 3-7%, while ximelagatran results in an oral bioavailability of melagatran of about 20%. Melagatran is zwiterionic, with an anionic carboxylate group and cationic benzamidine group, accounting for its poor permeability from the GIT tract. Gustafasson (2003) has shown, on oral dosing, that ximelagatran produces two intermediates. One is ethylmelagatran resulting from the reduction of the -N-OH to $-NH_2$, but with the ethyl ester moiety intact and melagatran-OH, where the ethyl ester group is removed, presumably via esterase action, but the –NH-OH has yet to be reduced. Neither is present in plasma in high concentration.

Commercialized Prodrugs

There are not many commercialized prodrugs that were designed to overcome the poor oral availability of solubility-limited drugs. A good older example is Merck's non-steroidal antiinflammatory drug (NSAID, sulindac. Sulindac is an S- oxide prodrug of its reduced parent sulindac sulfide. As the S-oxide prodrug, sulindac has superior solubility and lower GIT toxicity than the sulfide and was promoted as a less corrosive NSAID. A more recent example is fosamprenavir, a phosphate ester of the HIV protease inhibitor, amprenavir. Amprenavir is a high dose drug with marginal water solubility. To achieve adequate oral availability, the drug was formulated as a soft gelatin capsule containing 150 mg of drug. Because the dose of amprenavir is 1,200 mg twice a day, eight capsules, taken twice daily, were required for therapy. It can be readily agreed that eight capsules twice a day is not convenient and could lead to patient compliance issues. The innovators synthesized a phosphate ester prodrug, calcium salt, fosamprenavir.

Fosamprenavir has very good water solubility and is readily formulated as 700 mg tablets (molar equivalent of 600 mg amprenavir). Fosamprenavir shows an oral biavailability equivalent to the amprenavir capsules. If fosamprenavir is so water soluble, isn't polarity a problem? The dianionic fosamprenavir is not absorbed intact. The surface of the enterocytes lining the small intestines is very rich in alkaline phosphatases. These embedded enzymes are capable of hydrolyzing fosamprenavir to amprenavir right at the intestinal cell surface. Since amprenavir is readily permeable, it is well absorbed. This coupling of metabolism and absorption is readily seen with the absorption of folic acid from food sources and has been proposed much earlier as a prodrug strategy by others.

Phosphate ester prodrugs appear as some of the better examples of the prodrug technique as a means of increasing the oral availability of solubility-limited drug molecules. The use of water soluble prodrugs of poorly water soluble drugs to effect parenteral dosing has been an accepted solution to solubility limitations since its application in the 1960s to the antibiotic chloramphenicol as chloramphenicol hemisuccinate, sodium salt, and the hemisuccinate and phosphate esters of various steroids. In 1996 fosphenytoin was approved as a water soluble and safe injectable form of phenytoin. More recently, two novel prodrugs are parecoxib sodium (Dynastat), which is the injectable form of valdecoxib (Bextra), and fosfluconazole, a phosphate ester prodrug of fluconazole, both of which were approved for parenteral use. Ueda *et al.* (2003) recently described some novel phosphate ester prodrugs of another clinical antifungal drug. Parenteral formulations of poorly soluble drugs can present some unique challenges to prodrug scientists. First, solubilities greater than 10 mg/ml are often needed, thus for truly poorly soluble parent drugs, orders of magnitude increases in solubility are needed. Consider the phenytoin/fosphenytoin example.

Phenytoin has a solubility of about 25 μg/ml while an injectable dose of 50 mg/ml (as sodium phenytoin) or about 47 mg/ml phenytoin equivalent would have to be matched by a prodrug. Fosphenytoin met that goal without the need for co- solvents. Having met the solubility goal, the second problem was aqueous stability. While fosphenytoin showed excellent chemical stability, it, along with all parenteral prodrugs, suffers from the fact that the most likely degradants of these prodrugs are the poorly soluble parent drugs. Thus the shelf-life of the prodrug in aqueous solution is often dictated, not by the time to lose less than 10% of the prodrug, but by possible precipitation of insoluble degradation products. Therefore, some water-soluble prodrugs of insoluble parent molecules must be formulated, not as ready-to-use solutions, but as freeze-dried powders for reconstitution prior administration. Fosphenytoin is sufficiently stable to allow for a ready-to-use solution, but it does require refrigeration. An alternative room temperature stable formulation has been proposed. Aquavan is a water-soluble prodrug of the anesthetic agent, propofol.

The promoiety, in this case, is identical to that used for fosphenytoin, except that the phosphonoxymethyl group is attached to the sterically hindered phenolic group of propofol. The pharmacokinetic properties of aquavan in rats and humans have been shown to be quite interesting. This drug recently entered phase III clinical trials in the USA. Aquavan was designed to overcome the limitations of the current propofol formulation, where the drug must be administered in a 10% oil/water emulsion. The limitation is low concentration, requiring large volumes of the emulsion to be administered when used to maintain a coma. This can result in lipid overload. Brachial pain on injection, emulsion cracking during long term infusion, and difficulty in maintaining formulation sterility, due to the emulsion being a very good growth media for bacteria, are other limitations. Aquavan is very water-soluble, so no lipids are required and more concentrated solutions are possible. It is rapidly converted to propofol *in vivo*, both in animal models and humans; thus, there is a rapid onset of anesthesia. Since the injection site only sees the inert prodrug, there is no pain on injection. The injection is in the form of a ready-to-use solution and is easy to maintain in a sterile state.

Evolution of Prodrug

Just as it takes a team to efficiently develop analogs, a successful prodrug program depends highly on quick evaluation of prodrug candidates by a team. For big pharma, a prodrug solution to the lack of developability of drug candidates has been an act of last resort, or, when applied to an agent going off patent, an effort to maintain market exclusivity. This was not always the case. In the 1960s, the Parke Davis Company developed a number of very successful prodrugs, as did a concerted effort by the Upjohn Company in the mid-1960s to the mid-1970s. In most companies, a prodrug program is only initiated when a specific problem is identified and the effort is only maintained while the immediate need presents itself. Because the effort is initiated in response to a specific problem area, after more traditional methods of overcoming the barrier to developability are exhausted, the development of the drug candidate is delayed and made more complex by the need for additional toxicology and ADME studies.

Thus, a prodrug solution to a problem is often viewed negatively by management. What if a prodrug solution to a developing problem is proposed early? Imagine the scenario where a discovery team appears to be going down the path leading directly to a series of leads/candidates having clear drugability issues. Would it not make sense to raise the question in the development team of a prodrug solution in the event that other strategies do not work? If implemented early, would a prodrug solution be any more expensive than normal analog development? The answer is probably no. If a prodrug becomes the NCE, some issues in its development may add some additional burden to the drug metabolism and toxicology staff. However, this approach may be less financially burdensome in the long run than the added expense of a significant time delay in development, if the prodrug strategy is implemented six months after the problem/barrier is identified. When the need for a prodrug solution presents itself, who should constitute the "*prodrug team*"? First, it is important to identify the cause or etiology of the drugability problem and to identify the strengths and weaknesses of this team as it relates to the general prevailing knowledge of the prodrug concept.

The views expressed here are based on experience as a prodrug researcher and the knowledge gained from being a consultant to big and small pharma over the last 30+ years and is inherently biased by these experiences. First the etiology of the drugability problem must be identified with some degree of certainty. Is it solubility, permeability (from either the intrinsic properties of the drug candidate or due to the drug being an efflux candidate), presystemic metabolism, or poor pharmacokinetic properties? Or is it a combination of more than one property? Identification of the etiology is not easy and, sometimes, a property such as limited aqueous solubility, can mask other underlying problems and lead to flawed approaches. There is a greater chance that the true cause of the problem is accurately identified if the decision to pursue a prodrug solution is made by a team constituted of medicinal chemists, ADME specialists, toxicologists and pharmaceutical formulation scientists, with input from marketing. If the right experiments are performed and the etiology or causes are identified, the probability of success is greatly enhanced. Once the problem/cause/s are identified, what prodrug solution should be applied? Precedent easily identifies some solutions, while others require greater creativity.

The medicinal chemist brings to the table the ability to synthesize complex molecules as well as highly developed observation skills. However, they are sometimes put off by the perceived trivial nature of some prodrug chemistry and are often naïve when it comes to issues of the safety and reversibility of possible promoieties. The pharmacologist and molecular biologist must test prodrugs with the view that they might not be active per se, so it takes a different mindset in reviewing data from various screens. That is, receptor based screens will not tell the whole story, while cell-based and *in vivo* screens are likely to be more meaningful. The drug metabolism, ADME and early PK specialists often perform *in vitro* metabolism and permeability studies and small animal PK studies on prodrugs. However, they must be conscious of the limitations of the *in vitro* screens and the over-interpretation of studies performed in mice/rats in relation to eventual human studies. Drug metabolism and early PK specialists often help identify the problem/s with the parent molecules and have sufficient knowledge to suggest possible prodrugs, but often do not have the capability to synthesize possible prodrug candidates.

Toxicologists often are not supportive of prodrug programs, because it makes their work more demanding. They not only must be concerned about the active drug, but must also address toxicology questions related to the prodrug and its breakdown products. See earlier questions raised about formaldehyde production and the role of pivalic acid in depleting carnitine levels. The presence of a formulation scientist on the team is critical. Their role is to make sure that any testing of the prodrug uses realistic formulations that can be scaled to human use and that issues, such as chemical stability, are addressed up front. For example, prodrugs proposed and moved into development can be so chemically unstable that they will never be developed as an approvable entity. If properly trained with a strong physical/chemical background, the formulation scientist is critical to the success of a prodrug program. While bench scientists often do not appreciate the role played by marketing, marketing specialists can play a very important role as part of this team. For example, if an injectable prodrug is needed for a specific product, it is important to know if a reconstitutable lyophilized product would be acceptable or if a ready-to-use solution is required. This may be dictated by the disease being treated or by competitor products.

If a ready-to-use solution is needed, then that will dictate what promoiety or chemistry can be applied. Prodrugs, by their very nature, tend to be more or less chemically unstable. This can present a problem for the formulation scientists. Therefore, the type of formulation needed, often dictated by marketing and patient acceptability issues, may require a greater and more creative effort. The formula for a successful prodrug effort is teamwork. Such a team is difficult to put together on an as-needed basis. Big pharma companies have recently shifted toward a paradigm where specialty prodrug teams have been constituted to work with discovery teams to identify prodrug strategies immediately when the need begins to present itself. That is, prodrug strategies are becoming an integral part of the drug discovery paradigm. There are also small companies whose business plan/strategy is to design prodrug solutions, both as a conduit for new drug candidates and/or as CROs, to help big and small pharma solve drugability problems.

6

Drug Withdrawal

Idiosyncratic drug reactions (IDR) remain an important cause of late development program failure or drug withdrawal. They represent a significant economic cost and patient burden. Although our understanding of IDR continues to grow, our ability to predict and avoid these reactions remains unacceptably poor. The intent of this chapter is to review approaches that may be used during lead optimization to minimize the likelihood of producing compounds for development that will have a high IDR liability.

PREDICTION OF IDIOSYNCRATIC REACTIONS

A clear definition and an understanding of how that definition is applied are critical for a rational discussion of prediction of idiosyncratic reactions. The definition of idiosyncratic reactions that will form the basis for the following discussion is: An idiosyncratic adverse drug reaction is an uncommon adverse reaction to a drug that is not related to the pharmacological properties of the drug, that occurs at therapeutic doses and/or serum concentrations, and will not occur in the majority of individuals despite increasing the dose to otherwise toxic concentrations.

The following characteristics are also used to identify adverse drug reactions as idiosyncratic:

1. They are dependent on the chemical, not the pharmacological properties of the drug.
2. They are highly host-dependent. An idiosyncratic reaction is dependent on a specific interaction between the chemical properties of the drug and the host.
3. There is no simple or classical dose-response relationship for the clinical reaction.
4. IDR usually occur at frequencies of less than 1/100, and often at frequencies of 1/1000 or 1/10,000. However, rates of up to 50% in specific populations have been reported.
5. They can not be predictably reproduced in animals. Idiosyncratic reactions do occur in animals, but they can not be reliably reproduced in an experimental setting with a general laboratory population. If the reaction was reliably produced in pre-clinical studies, the compound would

likely not be developed and the reaction would not be identified as being idiosyncratic in humans.

6. The mechanism is unknown. This is certainly true for the majority of idiosyncratic reactions. However, an understanding of the mechanism does not preclude a reaction from being defined as idiosyncratic if the primary characteristics of an IDR are observed.

Table 6.1 Examples of well recognized idiosyncratic drug reactions

Drug	*Predominant manifestation of drug reaction*
Sulfonamides	Systematic hypersensitivity syndrome reaction, including dermatopathy, fever, and hepatitis; Stevens-Johnson syndrome; toxic epidermal necrolysis.
Aromatic anticonvulsants	Systematic hypersensitivity syndrome reactions, including dermatopathy, fever, and hepatitis; Stevens-Johnson syndrome; toxic epidermal necrolysis.
Halothane	Hepatitis with systemic signs, Malignant hyperthermia
Tienilic acid	Hepatitis
Abacavir	Fever, dermatopathy, malaise, gastrointestinal disturbances.
Dihydralazine	Hepatitis
Diclofenac	Hepatitis
Ibufenac	Hepatitis

The clinical presentation of idiosyncratic reactions and the target organ can be highly variable. Common target organs include the skin, liver, kidney, and hematological system. Patients may present with one or more target organs affected. Some drugs target primarily one organ system (e.g. clozapine and agranulocytosis; halothane and hepatitis) while others target multiple organs and have multiple clinical presentations. Idiosyncratic drug reactions commonly present with signs consistent with an immune-mediated pathogenesis, including delayed onset (usually from 1 to 6 weeks after start of therapy), fever, eosinophilia, dermatopathy and multi-organ toxicity. A more rapid onset of clinical signs is often observed on re-challenge, also consistent with an immune-mediated reaction.

Terms such as the idiosyncratic hypersensitivity syndrome reaction, drug-induced immune-mediated disease, or drug-induced lupus, depending on the particular constellation of clinical signs, are used to describe IDR. Demonstration of antibodies or reactive T-cells recognizing the drug, drug-modified proteins, or autologous proteins are required to truly meet the criteria of immune-mediated disease, but many IDR are assumed to have an immunological basis solely on the basis of clinical signs. Many users of the term "*idiosyncratic reaction*" intend to refer specifically to the subset of idiosyncratic reactions that have an immunological basis. It is generally the most important group of idiosyncratic reactions from a drug development perspective. The evidence for an immunological basis of idiosyncratic reactions will be discussed in more detail below. Some authors define the term idiosyncratic drug reaction as a genetically determined abnormal reactivity to a chemical or as a rare toxicity of an unknown mechanism. These definitions share some similarities with the definition of IDR presented above, but they capture different populations of adverse drug reactions. Genetically determined abnormal reactivity to a chemical captures not only immune-mediate idiosyncratic adverse drug reactions, but also drug reactions such as azathioprine hematotoxicity in thiopurine methytransferase deficient individuals and perhexilene hepatotoxicity in CYP2D6 deficient individuals. We prefer to use the term *hypersusceptibility* to characterize these reactions. *Hypersusceptibility* describes those individuals that are particularly susceptible to dose-dependent toxicities.

Individual susceptibility may be determined by genetic factors, but the advers reaction can be produced in more individuals by increasing the dose or serum concentration, and it can be produced in a dose-dependent manner in experimental animals. Environmental exposures and drug interactions can also lead to hypersusceptibility. While we do not consider these reactions to be idiosyncratic, the prediction of potential hypersusceptible individuals in the population, genetic or environmentally-induced, is an important part of drug development and the presence of hypersusceptibility can certainly lead to uncommon adverse events that can result in drug withdrawal. Idiosyncratic reactions to drugs do occur in animals, often with a similar clinical presentation to humans. However, it is rare that they can be reliably induced in the typical laboratory animal population. This has been a significant limiting factor in understanding the molecular events involved in idiosyncratic reactions. Any potential animal models appear to be compound specific and there are no general models that are currently useful. Therefore, animal models will not be further discussed in this chapter.

Incidence of Idiosyncratic Reactions

It is important to differentiate between predicting which *patients* will experience idiosyncratic reactions and predicting which *compounds* will be associated with an unacceptably high incidence of idiosyncratic reactions. Lead optimization occurs prior to human exposure and assessment of individual risk is neither required nor, with our current understanding, applicable at this stage of drug development. Understanding individual risk may be important during clinical development and will ultimately help us in our efforts to predict compounds that will be associated with idiosyncratic reactions but is not an integral part of the lead optimization process. There are many marketed compounds that will cause idiosyncratic reactions but not at an incidence or of a severity that limits their use. While the ultimate goal is to eliminate idiosyncratic reactions associated with new chemical entities, a reduction to a clinically acceptable incidence or severity may serve as an intermediate goal.

The only approach currently available is to identify the characteristics of compounds with a higher probability versus those with a lower probability of triggering idiosyncratic reactions. These characteristics can then be used to optimize lead compounds. We must have a working concept of the pathogenesis of idiosyncratic reactions if we are to develop a paradigm for optimizing lead compounds to minimize the risk of producing a drug that induces IDR. For the purposes of this chapter, we will focus on idiosyncratic reactions in which the clinical manifestations result from cellular damage to an identified organ or tissue. Although not universally accepted, the majority of experts agree that a pathological immune response underlies the clinical manifestations of the majority of idiosyncratic reactions where cellular or organ damage is an integral part of the pathogenesis. Therefore, this chapter will discuss approaches that may be useful in helping to predict and/or eliminate compounds that may be associated with an unacceptably high frequency of immune-mediated idiosyncratic reactions.

IMMUNOPATHOGENESIS

Our understanding of the pathogenesis of *immune-mediated idiosyncratic reactions* (IM-IDR) remains incomplete. The initial hypothesis of an immunological basis for idiosyncratic reactions came from the clinical presentation: delayed onset, fever, and the common occurrence of systemic eosinophilia. Histopathological evaluation of idiosyncratic hepatopathies often reveals infiltrating lymphocytes and plasma cells as well as bridging necrosis reminiscent of viral hepatitis. A skin rash is often part of the clinical presentation and histopathological examination reveals lesions consistent with immune-mediated skin disease. Serum sickness-like reactions, including polyarthritis, have also been reported. On re-exposure, there is often, although not always, a shorter period of time to recurrence of clinical signs. This general clinical picture was consistent with an immunological basis, particularly for the systemic hypersensitivity syndrome reactions associated with drugs such as the sulfonamides and aromatic anticonvulsants. Over the last 25 years, a plethora of studies have identified immunological responses in patients experiencing idiosyncratic reactions to drugs.

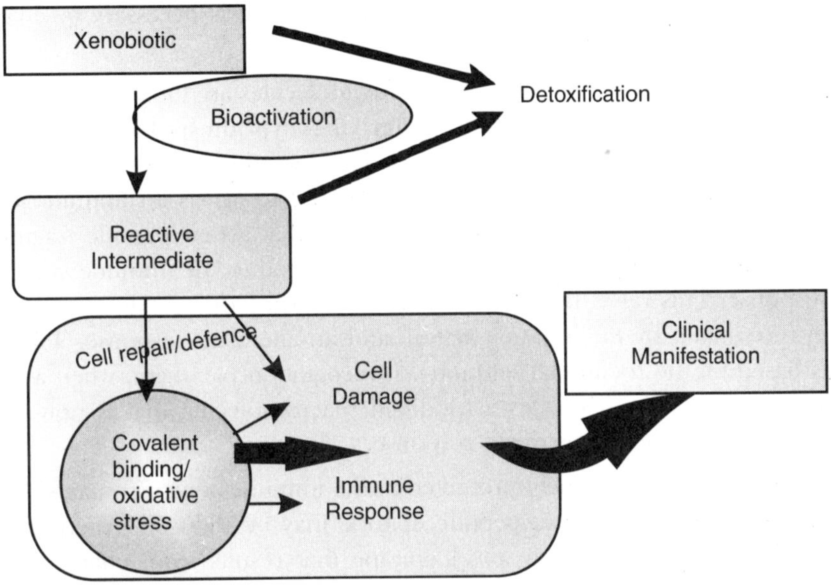

Fig. 6.1. A general paradigm for the pathogenesis of idiosyncratic reactions.

In addition to antibodies against the drug or drug-conjugated proteins, antibodies against native cellular and nuclear proteins have been identified. The key conclusions are: (i) there is very strong evidence for activation of the immune response during idiosyncratic reactions; (ii) the immune response to a drug can be quite varied, with both antibody and cell-mediated responses being documented; and (iii) the nature of the immune response plays a significant role in determining the clinical presentation. Further evidence for an immune response in IDR comes from immunogenetic studies of susceptibility to or severity of idiosyncratic reactions. Specific human major *histocompatibility complex* (HLA) haplotypes and polymorphisms in cytokine or heat shock genes have been identified that influence susceptibility or severity to a variety of idiosyncratic toxins.

The association is most striking with abacavir (approximately 95% of patients have the at risk genotypes compared to 1% in the control population), however in the majority of cases the association is relatively weak. Unfortunately, the specific HLA haplotypes or other markers that appear to confer susceptibility are quite variable and appear to be drug specific. Thus, a specific susceptibility genotype that can be used to identify an "at risk" drug-related immunogen characteristic appears unlikely to be consistent across chemically divergent compounds. Nevertheless, the contribution of immunogenetics to individual susceptibility or severity further supports a role for the immune response. If we accept that an immune response is an important and, in most cases, a required part of the clinical reaction, then we can appreciate that understanding the properties that contribute to a drug acting as an immunogen is necessary to enable us to "dial out" those properties during optimization.

It is generally believed that most drug molecules are too small themselves to induce an immune response. The most common underlying hypothesis to explain how a drug can become immunogenic is the *hapten hypothesis.* This hypothesis states that drugs must be covalently bound to a macromolecule carrier (usually a protein) to act as an immunogen. This macromolecule is then processed by the immune system such that a new peptide fragment or a peptide fragment containing the covalently bound drug is presented to the immune system and induces an immune response. This raises the question then as to when or how does covalent binding of a drug result in recognition by the immune system and an attendant response. The original hapten hypothesis is based on the traditional self/non-self recognition paradigm: when a drug is covalently bound to a macromolecule, it creates a modified macromolecule that is now recognized by the immune system as non-self, triggering a response.

If the peptide fragment presented to the immune system contains the drug, a response directed against the drug or a drug-peptide epitope may be induced. Alternatively, the immune response may be directed against a novel epitope that results from changes in protein confirmation or processing. An alternative explanation, called the *danger hypothesis,* has been proposed by Matzinger to explain why an immune response occurs against immunogens in some patients and not in others. She has postulated that foreignness alone is insufficient to alone trigger an immune response and that context is more important. An immune response will only occur if a second "danger" signal is presented. Potential danger signals include heat shock proteins or cytokines that may be elevated as a result of cell toxicity or concurrent disease causing cell damage (e.g. viral infections). If true, this may help to explain part of individual susceptibility to idiosyncratic reactions, from either a compound perspective (compounds associated with higher cytototoxicity) or a patient perspective (increased incidence of idiosyncratic reactions in HIV-patients). A third possibility that does not depend on haptenization by the drug or drug metabolites has been proposed.

Picher *et al.* have demonstrated that sensitized lymphocytes can recognize drug molecules non-covalently linked to protein, in an MHC-restricted manner and that, at least *in vitro,* free drug can induce lymphocyte activation in some individuals. Whether this "*pharmacological induction*" occurs *in vivo* or induces a sufficiently strong immune response to cause a clinical reaction is unknown. At the moment, the support for this mechanism is not sufficiently strong to use as a basis for lead optimization. The difference between the first two hypotheses may be important for understanding individual susceptibility factors, but for the purposes of lead optimization there is little practical difference. Both hypotheses require the formation of covalently bound drug adducts. Cell damage or stress is required to produce the danger signal in the danger hypothesis, while it is often considered an integral part of the hapten hypothesis as well because adducted proteins must be presented to the immune system. The majority of proteins identified as antibody targets are intracellular proteins. For proteins to be processed, they must be taken up or found within *antigen presenting cells* (APC). One possible mechanism by which modified intracellular proteins can be taken up by an APC is cell damage and death that leads to the release of proteins or engulfment of damaged and dying cells by APC. If protein conjugates are formed in cells that are capable of MHC class I presentation, then intracellular formation of a neo-antigen may lead to processing and presentation for a cell-mediated response.

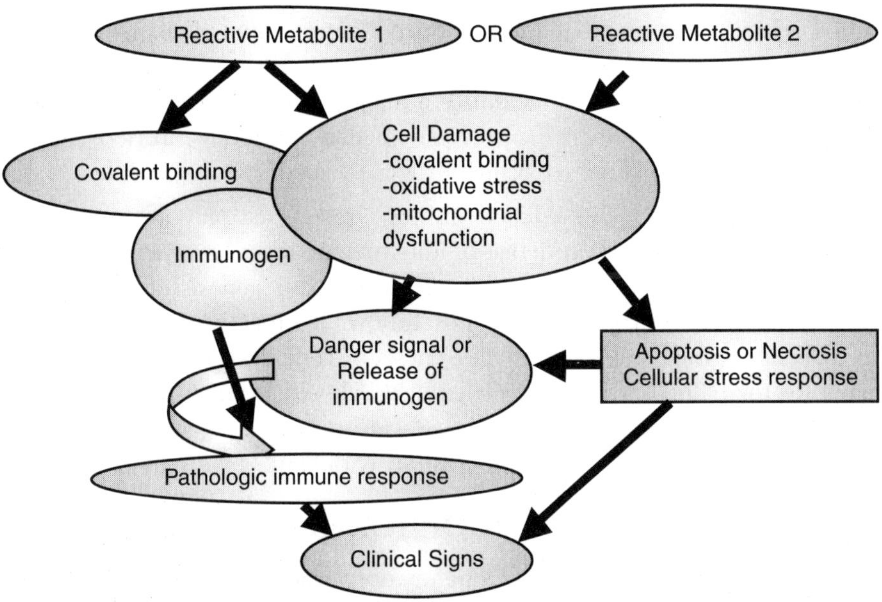

Fig. 6.2. A general scheme for the immunopathogenesis of immune-mediated idiosyncratic reactions.

Possible reasons why some individuals may experience a reaction, and others not, include immunogenetic variation, haptenic density, previous sensitization by immunologically-related epitopes, and degree of cell death (and subsequent release of immunogens). For the purposes of lead optimization, these two competing hypotheses may be considered as equivalent. Once an immune response has occurred, how does the immune system access an intracellular protein? It has now been shown for the majority of proteins identified as immunopathological targets that they are also expressed on the cell surface and that this expression can often be up-regulated. Cell-surface expression presumably allows the immune system to recognize and target cells expressing the antigens. For anaphylactic IgE-mediated reactions, the antibody response against the drug is generally accepted as being critical. The role of the antibodies directed against the drug in other idiosyncratic rections is less clear. The ability of patch or prick testing to identify patients susceptible to idiosyncratic reactions is highly variable between drugs.

Abacavir is a good example of a drug in which patch testing appears highly effective, while for many other drugs skin testing is either unreliable or does not produce a positive response. For some drugs, anti-drug antibodies are specific to hypersensitive patients, but for others, anti-drug antibodies have been found in both patients and non-hypersensitive controls. Further, the clinical reaction may continue after the drug is theoretically cleared from the body, suggesting that the drug-associated antibodies are not the primary cause of the immunopathologic reaction. Animal models exist in which antibodies or an immune response directed against the drug can be induced without associated pathology.

These observations suggest that, with the exception of IgE-mediated drug allergies, antibodies directed against the drug are more a marker for an immune response than an executioner for the clinical disease. Antibodies directed against neo-antigens are not associated with pathology unless the drug is also present, suggesting that a drug-induced conformational change is required, that the drug alters the expression or distribution of the protein such that it becomes accessible to the drug, or that the drug must be present to induce a danger signal. *In vitro* studies have shown that antibodies recognizing drug-protein conjugates can mediate antibody-directed cell cytotoxicity *in vitro* and that cytotoxic T cells directed against drug-associated epitopes can lead to cell death.

These studies suggest that cell-mediated immune responses play an important role in the associated pathology. Our current knowledge of idiosyncratic reactions, then, supports the concept that drug-induced immune responses are involved in the pathogenesis of idiosyncratic reactions. Formation of covalent drug-protein conjugates following bioactivation (reactive intermediate formation) is likely a required step in the pathogenesis of IDR. Immune-mediated cytotoxicity is a required end-point for pathology, but early chemical cytotoxicity may play an important role in triggering the immune response. These three characteristics (bioactivation, covalent binding, and cytotoxicity) may then be used in a lead optimization paradigm.

TOXICOPHORES

It is generally accepted that formation of one or more reactive intermediates is required before an idiosyncratic reaction can be triggered. Avoidance of "*toxicophores*" or chemical structures that are likely to undergo bioactivation is a logical initial approach to lead optimization. A number of readily recognizable chemical structures have been associated with idiosyncratic reactions and avoiding these structures will likely reduce the risk of reactive intermediate formation associated with idiosyncratic toxins. The reader is referred to the chapter by John Walsh in this book for a detailed discussion of reactive intermediate formation and toxicophores associated with bioactivation. There is not, however, a direct correlation between the formation of reactive intermediates and occurrence of idiosyncratic reactions. That is, there are many compounds that form reactive intermediates that are either not associated with idiosyncratic reactions or cause a predictable dose dependent toxicity.

Formation of free radicals is not generally associated with compounds that cause idiosyncratic reactions, presumably because they are not as commonly associated with covalent binding. Many methods have been proposed to identify reactive intermediates: covalent binding studies, screening for suicide inhibition of drug metabolizing enzymes, or screening for glutathione conjugates. These approaches do not identify all possible reactive intermediates either because of complex metabolic pathways (i.e. secondary bioactivation of a metabolite) or because of bioactivation by non-cytochrome

P450 pathways such as myeloperoxidase. They do, however, provide a substantive starting point if reactive intermediate formation is a concern from a structural viewpoint. Eliminating any and all compounds that form or have the potential to form reactive intermediates would be expected to reduce the over-all likelihood of idiosyncratic reactions. However, this approach will also likely result in the loss of potentially useful compounds. Should we optimize simply to minimize the quantitative formation of reactive intermediates or is there a qualitative difference between reactive intermediates that can be used to aid in lead optimization?

A simple quantitative assessment of bioactivation is not a good indicator of likelihood of causing idiosyncratic reactions when comparing across groups of structurally unrelated compounds. However, there are several examples that suggest quantity remains an influence. First, it is now recognized that idiosyncratic reactions are rarely associated with total daily doses less than 100 mg and practically non-existent at dose less than 10 mg a day. Within a structurally related series of compounds, there is evidence that reducing turn-over via metabolic pathways associated with bioactivation is associated with a reduced risk of idiosyncratic reactions. Reduction of metabolic turnover has been achieved either through chemical modification or through individuals that have genetic polymorphisms that influence the rate of bioactivation directly or indirectly.

These examples support the concept that within a structurally-related series, a reduction in the total formation of reactive intermediates will reduce the risk of idiosyncratic reactions. There are also examples where there are not large differences in bioactivation rates for compounds with large differences in idiosyncratic reaction rates (clozapine and olanzapine) or where the rate of metabolite turnover measured does not correlate with risk of IDR (ibuprofen and ibufenac). Therefore, we must look at further properties of the compounds and their associated reactive intermediates, specifically there reactivity with cellular proteins and their propensity to cause direct cytotoxicity.

Characteristics of Covalent Binding

Covalent binding appears to be an integral step in the pathogenesis of idiosyncratic reactions, whether you adhere to the traditional hapten hypothesis or to the danger hypothesis. This would suggest that eliminating covalent binding would eliminate idiosyncratic reactions. However, we know that many effective and safe compounds are associated with covalent binding in *in vitro* systems and *in vivo* in animals or humans. Therefore, for covalent binding to be a more useful marker of risk and hence a useful tool in lead optimization, additional characteristics of covalent binding must be considered. The same arguments that apply to bioactivation to reactive intermediates apply to covalent binding. Compounds that are associated with increased reactive metabolite formation are often also associated with increased covalent binding *in vitro* and *in vivo*. Within a structurally related series, compounds with the lowest covalent binding are likely to have the lowest likelihood of triggering idiosyncratic reactions. However, this does not help resolve comparisons across structurally unrelated compounds.

It is clear that an absolute level of covalent binding has little relevance to prediction of idiosyncratic reactions. Three properties of covalent binding have therefore received attention: targets of binding, association with cytotoxicity, and *in vivo* binding. Considerable effort has been devoted to identifying specific targets or patterns of covalent binding that may be linked to idiosyncratic reactions. The majority of protein targets identified to date have been either endoplasmic reticulum proteins or cell surface proteins. As discussed in the preceding section on immunopathogenesis, the endoplasmic reticulum (ER) proteins that have been identified as targets of the immune response are also found expressed on the cell surface, albeit not in large quantities. This observation also holds for targets of covalent binding. Most of the work has been performed in animals (usually rats) but some comparisons have been performed with either human hepatocytes or in samples obtained from humans receiving the drug.

For several compounds, there have been comparisons of the patterns of covalent bindings based on molecular weights on gel electrophoresis, without attempts to identify specific targets. These studies have not been able to clearly identify any pattern or unique targets associated with idiosyncratic toxins other than a tendency to adduct membrane proteins. It is instructive to view a few specific examples. Tienilic acid and halothane are two compounds that are associated with idiosyncratic hepatitis with limited systemic signs (e.g. dermatopathy is not a major feature of the reaction). Tienilic acid displays a relatively specific binding to predominantly one target, a cytochrome P450: CYP2C11 in rats and CYP2C9 in humans. In contrast, halothane binds to a whole range of ER proteins, including CYP2E1 and a number of ER stress proteins. Thus, the range from specific binding to wide ranging covalent binding is associated with idiosyncratic reactions, and similar observations have been made regarding compounds that are intrinsically toxic.

The most consistent feature of the covalent binding of compounds that are associated with idiosyncratic reactions is that covalent binding is heavily weighted towards membrane-associated proteins. The endoplasmic reticulum, nuclear envelope, and cell membrane proteins are the most common targets of idiosyncratic toxins. Non-idiosyncratic toxins also bind to membrane proteins, but there appears to be higher binding to cytosolic proteins. Our failure to identify one or more critical targets of covalent binding has three possible explanations. The first is that such common, critical targets do not exist. The second is that scientists as a group have inadvertently studied outlier compounds that do not share critical important targets.

The third possibility is that the common, critical targets exist but that we have not been able to identify them because the experimental approaches used have not been sensitive enough or are simply not an appropriate approach. Only further experimentation will resolve this issue. Another way of looking at covalent binding is to assess its link with cytotoxicity. For compounds that are associated with intrinsic toxicity, there is generally a good correlation between covalent binding and intrinsic toxicity. That is, cytotoxicity is apparent shortly after covalent binding appears and increases in association with total covalent binding. For compounds that are associated with idiosyncratic toxicity, there is generally extensive covalent binding in cells (*in vitro* systems) or in tissues (*in vivo*) at concentrations below that causing overt cellular toxicityl, or the covalent binding is dissociated from cytotoxicity. For example, *in vitro* studies have shown that covalent binding of sulfamethoxazole can occur without toxicity. In lymphoid cells, surviving cells after exposure to sulfamethoxazole hydroxylamine have extensive covalent adducts.

Keratinocytes can bioactive sulfamethoxazole to protein reactive compounds without cytotoxicity and circulating adducts of sulfamethoxazole have been observed in people. Studies using isolated cells have shown that the concentration of sulfamethoxazole reactive metabolites required to cause covalent binding sufficient to trigger lymphocyte activation through immune recognition are approximately one-tenth the concentration associated with cytotoxicity. The reactive intermediates responsible for covalent binding and for cytotoxicity are not necessarily the same.

Diclofenac is a good example where it has been shown that the covalent binding is primarily related to acylglucuronide formation while cytotoxicity is primarily the result of cytochrome P450-dependent bioactivation. Early studies on the molecular basis of liver toxicity from acetaminophen established that it was possible to create compounds that were bioactivated to protein-reactive reactive intermediates, but did not cause cell death. The partial or complete dissociation between covalent binding and cytotoxicity may allow the accumulation of covalent adducts such that when they are released or a danger signal is received, a relatively strong immunogen exists within the body as haptenic density is important in inducing an immune response against haptens. In summary, the current data suggest that covalent binding is a risk factor for idiosyncratic reactions particularly when it occurs prior to the onset of intrinsic cellular toxicity. The current data suggest that this property is likely of greater importance than binding to specific cellular targets or the absolute extent of covalent binding in fractionated *in vitro* systems.

Measurable Cytotoxicity

All the idiosyncratic toxins examined to date have been shown to be bioactivated to compounds that cause small but measurable cytotoxicity. As was discussed in the previous section, the cytotoxicity occurs at concentrations higher than those that result in covalent binding or may result from a different metabolite than that responsible for the majority of covalent binding. The ability to measure cytotoxicity from idiosyncratic toxins is dependent on the assay system used and the compound involved. For some compounds, toxicity can be demonstrated in isolated human hepatocytes with intact cytochrome P450 systems while in other cases a bioactivation system (e.g. hepatic microsomes or in some cases a peroxidase-based system) must be combined with a sensitive cell type such as lymphocytes. The utility of such systems early in lead optimization is limited, but they may be used in the late phase of lead optimization to assist in further characterization of compounds or in comparisons of compounds.

The ability of idiosyncratic toxins to cause cytotoxicity suggests that cell damage or stress is a required part of the pathogenesis, whether through the release of immunogens or the generation of a danger signal. The weight of evidence therefore suggests that a small and normally clinically insignificant degree of cytotoxicity is a characteristic of idiosyncratic toxins. Compounds that are intrinsically highly cytotoxic or bioactivated to highly cytotoxic intermediates will more likely be associated with intrinsic than idiosyncratic toxicity. If we accept that a low, but measurable, cytotoxic potential exists for idiosyncratic toxins, how can be exploit this information? The degree of cytotoxicity often parallels the over-all extent of bioactivation, so it may move in parallel with bioactivation and covalent binding. Within the few groups of structurally related compounds studied to date, there appears to be a correlation between the frequency of idiosyncratic reactions and the degree of cytotoxicity.

The volatile anesthetics are a good example of this. Similarly, it has been observed for olanzepine and clozapine that while metabolic bioactivation is similar, cytotoxicity is lower with olanzepine, as is the frequency of idiosyncratic reactions. *In vitro* studies suggest that clozapine bioactivation can result in cytotoxicity at clinically relevant concentrations, while higher concentrations of olanzepine are required. These observations suggest that to a certain extent, increased cytotoxicity increases risk but that eventually a compound will become an intrinsically toxic compound.

Cytotoxicity is a graded response, so one would predict that some compounds will be in a grey zone. This is indeed that case and there are some compounds, such as diclofenac, where clinical presentation is more consistent with intrinsic toxicity in some patients and idiosyncratic immune-mediated toxicity in others. The requirement for an underlying cytotoxic potential may explain the unusual but consistent observation that isolated mononuclear leukocytes from patients are more susceptible to the toxic effects of bioactivation products of drugs associated with idiosyncratic reactions than are controls. The assay is based on exposure of isolated mononuclear leukocytes from patients and controls to the drugs in the presence of a microsomal bioactivating system to generate reactive metabolites. Drugs studied include sulfonamides, cefaclor, aromatic anticonvulsants, sorbinil, and amineptine. The MNL toxicity assay may be used to characterize and screen compounds in development to help optimize leads but only after idiosyncratic or some other unusual toxicity has been observed. It is not an assay that is appropriate for routine use during lead optimization.

Additional Features

The hope remains that array screening will allow us to identify particular patterns of gene activation that will be linked with idiosyncratic reactions. Using proteomics, we may be able to identify critical targets or responses that can be used to guide lead optimization. However, the major limitation is that the animal models that develop the complete idiosyncratic reaction remain limited and not widely applicable, often for practical reasons. It is unclear whether animal models that do not develop the complete clinical profile will ever be predictive. *In vitro* studies with isolated human cells are limited by the lack of cell-to-cell interactions that are likely required for the full manifestation of the response. Paradigms that may identify highly immunogenic compounds have been proposed. These include the popliteal lymph node assay, co-administration of drugs with Freund's adjuvant as a danger signal, and *in vitro* T-cell priming. However, these assays all have limitations and no assays have been developed or validated to the point that they may be useful in lead optimization.

BIOACTIVATION

Many questions remain unanswered regarding idiosyncratic reactions and we do not yet have a clear and consistent marker for compounds that cause idiosyncratic reactions. We can not identify one property of compounds that we can use to eliminate or optimize lead compounds without risking losing valuable compounds. The complete elimination of all idiosyncratic reactions may not be a necessary endpoint. Our goal, however, is to reduce the frequency of idiosyncratic reactions so that the risk associated with a compound is acceptable. Therefore, the current goal for lead optimization is to develop a set of characteristics that identify high risk compounds. The preceding discussion raises several key points that can be used in a step-wise approach in an attempt to minimize the likelihood of compounds causing idiosyncratic reactions.

Step 1: The available evidence strongly supports the contention that protein reactivity is a necessary characteristic of idiosyncratic toxins. Bioactivation may be required for most compounds, although some compounds (e.g. penicillamine, penicillins) may be inherently reactive. Reactive intermediate formation increases the risk of idiosyncratic reactions and minimization of bioactivation potential through elimination of toxicophores is a logical first step. Identification of reactive metabolite formation is a useful approach. The methods for identifying reactive intermediate formation have been reviewed elsewhere and they include demonstration of glutathione conjugates, covalent binding, trapping experiments, and suicide inhibition of metabolizing enzymes. None of these methods are fool-proof. Covalent binding studies *in vitro* and/or *in vivo* are a logical part of any pre-screening program and may be included in either Step 1 or Step 2. If bioactivation with or without covalent binding is identified, the compound may be either modified to remove this property or further characterization to better assess risk undertaken.

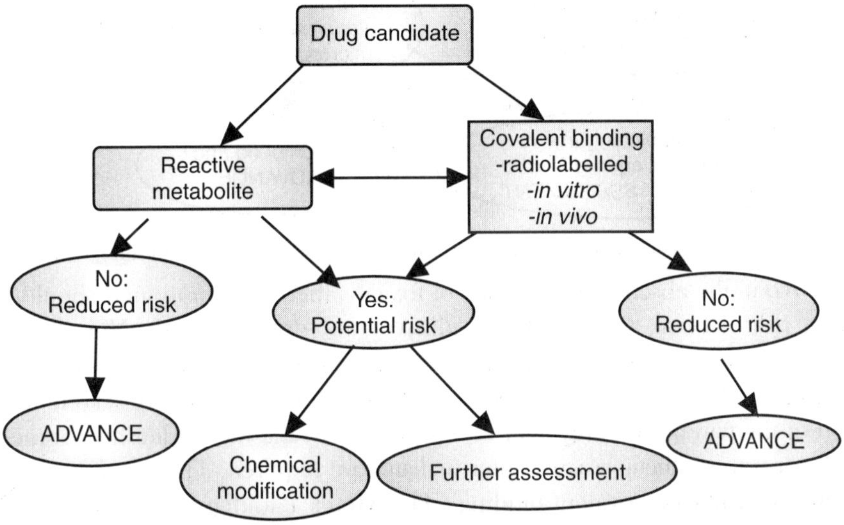

Fig. 6.3. Step 1 in the assessment of risk of a compound being associated with idiosyncratic reactions.

Step 2: Once reactive intermediate formation is identified or predicted, covalent binding *in vitro* (if not previously performed; rat and human systems are both potentially useful) and *in vivo* should be evaluated. *In vivo* studies are generally carried out in rodents. If a series of related compounds is going to be evaluated or re-evaluated, it may be worthwhile to generate antibodies for use in immunochemical detection, otherwise such studies are typically performed with radiolabelled material. There are currently no standardized testing procedures. It has recently been suggested that standard doses of 20 mg/kg *per os* be used for *in vivo* studies and 10 μM for 1 hour be used for *in vitro* studies with human or rat hepatic microsomes. While these are somewhat arbitrary doses or concentrations, if a bank of data in a standardized format is generated by consistent application of these doses/concentrations, sufficient data may be generated to allow for retrospective analysis. However, it is also recommended that studies be carried out at doses or concentrations that are 5 - 10 times the expected peak plasma concentrations to ensure that nothing is missed.

In vivo studies should also include standard assessment for cytotoxicity, such as serum liver-related enzyme activities and histopathological analysis. If significant or extensive covalent binding

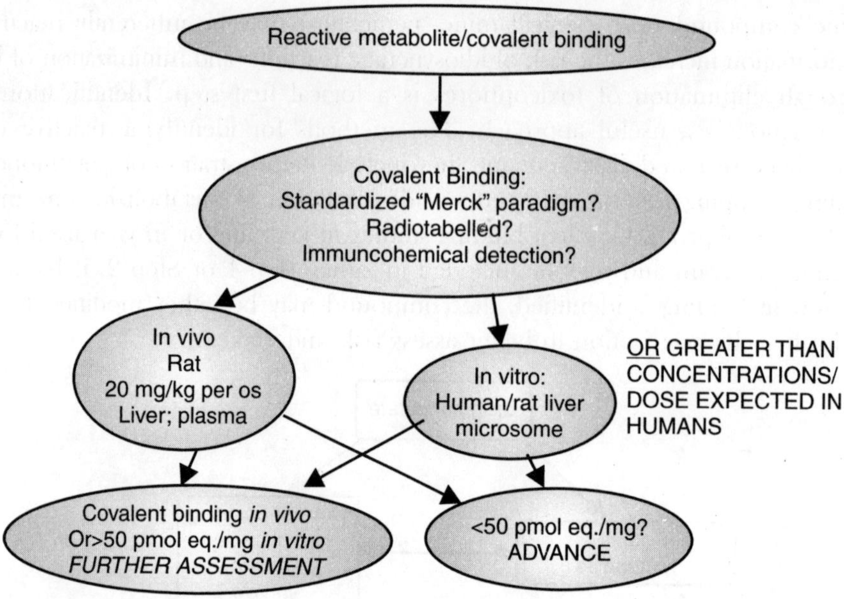

Fig. 6.4. Step 2 in the assessment of risk of a compound being associated with idiosyncratic reactions.

is observed *in vivo* in the absence of evidence of toxicity, then this compound should be considered at an increased risk of being associated with idiosyncratic drug reactions. Not all compounds that are associated with idiosyncratic reactions lead to covalent binding in rodent models when given as the parent compound (e.g. sulfamethoxazole covalent binding is not observed in rodents after administration of the parent, but only after administration of the metabolites). In some cases, studies in isolated cells or with synthetic reactive intermediates are required. The major outstanding question is what constitutes significant covalent binding? The Merck Paradigm suggests 50 pmol/mg protein *in vivo* or *in vitro* as a level of concern. This has not been proposed as a go/no go decision point, but as a level of heightened concern when additional assessment or consideration is required. At the moment, there are insufficient data to support or refute this as a specific level of covalent binding but a review of the literature suggests that is not an inappropriate concentration, provided that doses/concentrations that meet or exceed known or expected human exposures are used.

Step 3: If covalent binding is observed, particularly *in vivo*, assessment of cytotoxicity can be undertaken. When covalent binding is linked to cytotoxicity in Step 2 or in further investigations in Step 3, then the compound should be treated as an intrinsic toxin and standard assessment of risks and benefits of intrinsic toxins undertaken. If the covalent binding is observed at doses or concentrations below those associated with toxicity *in vivo*, then additional studies *in vitro* may be indicated to determine the cytotoxic potential and its association with covalent binding. While this strays closer to lead selection, this information can be part of the feedback loop for lead optimization. In summary, compounds that undergo bioactivation leading to covalent binding at doses or concentrations below those required to cause cytotoxicity or toxicity *in vivo* are likely at increased risk of causing immune-associated idiosyncratic toxicity in the clinic. Available evidence suggests that within a group of structurally related compounds, increased covalent binding and/or

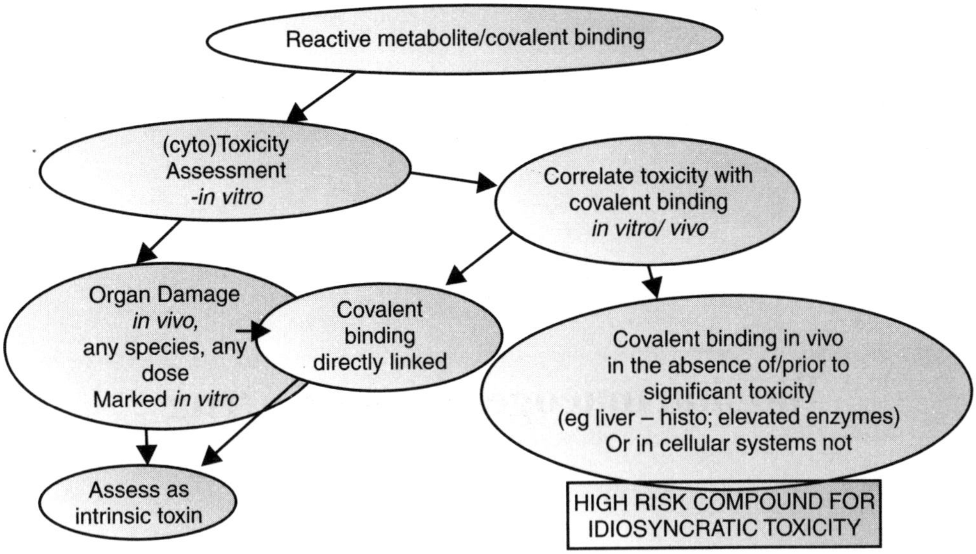

Fig. 6.4. Step 3 in the assessment of risk of a compound being associated with idiosyncratic reactions.

increased cytotoxicity of reactive metabolite increases the risk of idiosyncratic toxicity at some point. Therefore, lead compound structure should be optimized to minimize bioactivation, covalent binding, and cytotoxic potential. Minimizing the total daily dose or duration of therapy further reduces the risk of idiosyncratic reactions. However, dose reduction by increasing affinity for targets must not be undertaken at the expense of drug-like properties. No decision on lead optimization or selection should be taken without giving consideration to the wealth of caveats or modifying risk factors that surround these decisions:

1. What is the potential for chemical modification? If easy without losing other properties, it should be undertaken.
2. If bioactivation is associated with additional risks such as cytochrome P450 inhibition/induction, chemical modification is indicated.
3. The severity of condition treated may warrant advancement of compounds that have characteristics associated with higher risk of idiosyncratic reactions.
4. If alternative therapies with good safety records exist, the risk of idiosyncratic reactions needs to be low.
5. If the expected duration of therapy or expected total daily dose is high, than the risk of idiosyncratic reactions is likely higher for compounds with an "at risk" profile.
6. The intended target population should also be considered.

Our ability to identify compounds that will be associated with a high incidence of idiosyncratic reactions remains unacceptably limited. Continuing investigations are required to develop suitable animal models to evaluate and modify screening and lead optimization paradigms.

7

Biopharmacogenetics of Drug

A number of different strategies are adopted by the pharmaceutical industry in their efforts to identify new drug products. These approaches range from random screening of a wide range of biological materials to knowledge-based drug identification. Once a potential new drug has been identified, it is then subjected to a range of tests (both *in vitro* and in animals) in order to characterize it in terms of its likely safety and effectiveness in treating its target disease. The developer will also undertake manufacturing- related development work, as well as investigating suitable potential routes of product administration. After completing such preclinical trials, the developing company apply to the appropriate government-appointed agency, e.g. the Food and Drug Administration (FDA) in the USA, for approval to commence clinical trials (i.e. to test the drug in humans). Clinical trials are required to prove that the drug is safe and effective when administered to human patients, and these trials may take 5 years or more to complete.

Once the drug has been characterized, and perhaps early clinical work is underway, the drug is normally patented by the developing company in order to ensure that it receives maximal commercial benefit from the discovery. Upon completion of clinical trials, the developing company collates all the preclinical and clinical data they have generated, as well as additional pertinent information, e.g. details of the exact production process used to make the drug. They submit this information as a dossier to the regulatory authorities. Regulatory scientific officers then access the information provided and decide (largely on criteria of drug safety and efficacy) whether the drug should be approved for general medical use. If marketing approval is granted, the company can sell the product from then on. As the drug has been patented, they will have no competition for a number of years at least. However, in order to sell the product, a manufacturing facility is required, and the company will also have to gain manufacturing approval from the regulatory authorities.

In order to gain a manufacturing licence, regulatory inspectors will review the proposed manufacturing facility. The regulatory authority will only grant the company a manufacturing licence if they are satisfied that every aspect of the manufacturing process is conducive to producing a safe and effective product consistently. Regulatory involvement does not end even at this point. Post-marketing surveillance is generally undertaken, with the company being obliged to report any subsequent drug-induced side effects/adverse reactions. The regulatory authority will also inspect the manufacturing facility on a regular basis in order to ensure that satisfactory manufacturing standards are maintained.

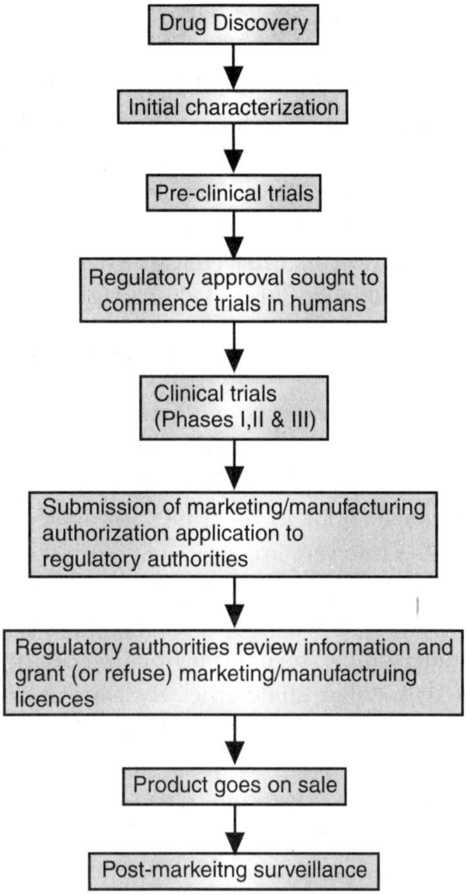

Fig. 7.1. An overview of the life history of a successful drug.

PHYLOGENETIC APPROACH

The discovery of virtually all the biopharmaceuticals discussed in this text was a knowledge-based one. Continuing advances in the molecular sciences have deepened our understanding of the molecular mechanisms that underline health and disease. An understanding at the molecular level of how the body functions in health and of the deviations that characterize the development of a disease often renders obvious potential strategies likely to cure/control that disease. Simple examples illustrating this include the use of insulin to treat diabetes and the use of GH to treat certain forms of dwarfism. The underlining causes of these types of disease are relatively straightforward, in that they are essentially promoted by the deficiency/absence of a single regulatory molecule.

Other diseases, however, may be multifactorial and, hence, more complex. Examples include cancer and inflammation. Nevertheless, cytokines, such as interferons and interleukins, known to stimulate the immune response/regulate inflammation, have proven to be therapeutically useful in treating several such complex diseases. An understanding, at the molecular level, of the actions of various regulatory proteins, or the progression of a specific disease does not, however, automatically translate into pinpointing an effective treatment strategy. The physiological responses induced by the potential biopharmaceutical *in vitro* (or in animal models) may not accurately predict the physiological responses seen when the product is administered to a diseased human. For example, many of the most promising biopharmaceutical therapeutic agents (e.g. virtually all the cytokines), display multiple activities on different cell populations. This makes it difficult, if not impossible, to predict what the overall effect administration of any biopharmaceutical will have on the whole body, hence the requirement for clinical trials. In other cases, the widespread application of a biopharmaceutical may be hindered by the occurrence of relatively toxic side effects (as is the case with *tumour necrosis factor* α (TNF-α).

Finally, some biomolecules have been discovered and purified because of a characteristic biological activity that, subsequently, was found not to be the molecule's primary biological activity.

TNF-α again serves as an example. It was first noted because of its cytotoxic effects on some cancer cell types *in vitro*. Subsequently, trials assessing its therapeutic application in cancer proved disappointing due not only to its toxic side effects, but also to its moderate, at best, cytotoxic effect on many cancer cell types *in vivo*. TNF's major biological activity *in vivo* is now known to be as a regulator of the inflammatory response. In summary, the 'discovery' of biopharmaceuticals, in most cases, merely relates to the logical application of our rapidly increasing knowledge of the biochemical basis of how the body functions. These substances could be accurately described as being the body's own pharmaceuticals. Moreover, rapidly expanding areas of research, such as genomics and proteomics, will likely hasten the discovery of many more such products, as discussed below.

MANIPULATING TECHNOLOGIES

The term '*genomics*' refers to the systematic study of the entire genome of an organism. Its core aim is to sequence the entire DNA complement of the cell and to map the genome arrangement physically (assign exact positions in the genome to the various genes/non-coding regions). Prior to the 1990s, the sequencing and study of a single gene represented a significant task. However, improvements in sequencing technologies and the development of more highly automated hardware systems now render DNA sequencing considerably faster, cheaper and more accurate. Modern sequencing systems can sequence thousands of bases per hour. Such innovations underpin the 'high-throughput' sequencing necessary to evaluate an entire genome sequence within a reasonable time-frame. By early 2006 some 364 genome projects had been completed (297 bacterial, 26 Archaeal and 41 Eucaryal, including the human genome) with in excess of 1000 genome sequencing projects ongoing. From a drug discovery/development prospective, the significance of genome data is that they provide full sequence information of every protein the organism can produce. This should result in the identification of previously undiscovered proteins that will have potential therapeutic application, i.e. the process should help identify new potential biopharmaceuticals.

The greatest pharmaceutical impact of sequence data, however, will almost certainly be the identification of numerous additional drug targets. It has been estimated that all drugs currently on the market target one (or more) of a maximum of 500 targets. The majority of such targets are proteins (mainly enzymes, hormones, ion channels and nuclear receptors). Hidden in the human genome sequence data is believed to be anywhere between 3000 and 10,000 new protein-based drug targets. Additionally, present in the sequence data of many human pathogens is sequence data of hundreds, perhaps thousands, of pathogen proteins that could serve as drug targets against those pathogens (e.g. gene products essential for pathogen viability or infectivity). While genome sequence data undoubtedly harbours new drug leads/drug targets, the problem now has become one of specifically identifying such genes. Impeding this process is the fact that the biological function of many sequenced gene products remains unknown.

The focus of genome research, therefore, is now shifting towards elucidating the biological function of these gene products, i.e. shifting towards '*functional genomics*'. Assessment of function is critical to understanding the relationship between genotype and phenotype and, of course, for the direct identification of drug leads/targets. The term 'function' traditionally has been interpreted in the narrow sense of what isolated biological role/activity the gene product displays (e.g. is it an enzyme and, if so, what specific reaction does it catalyse). In the context of genomics, gene function is assigned a broader meaning, incorporating not only the isolated biological function/activity of the gene product, but also relating to:

1. Where in the cell that product acts and, in particular, what other cellular elements does it influence/interact with;
2. How do such influences/interactions contribute to the overall physiology of the organism.

The assignment of function to the products of sequenced genes can be pursued via various approaches, including:

1. Sequence homology studies;
2. Phylogenetic profiling;
3. Rosetta stone method;
4. Gene neighbourhood method;
5. Knockout animal studies;
6. DNA array technology (gene chips);
7. Proteomics approach;
8. Structural genomics approach.

With the exception of knockout animals, these approaches employ, in part at least, sequence structure/data interrogation/comparison. The availability of appropriate highly powerful computer programs renders these approaches 'high throughput'. However, even by applying these methodologies, it will not prove possible to identify immediately the function of all gene products sequenced. Sequence homology studies depend upon computer-based (bioinformatic) sequence comparison between a gene of unknown function (or, more accurately, of unknown gene product function) and genes whose product has previously been assigned a function. High homology suggests likely related functional attributes. Sequence homology studies can assist in assigning a putative function to 40-60 per cent of all new gene sequences.

Phylogenetic profiling entails establishing a pattern of the presence or absence of the particular gene coding for a protein of unknown function across a range of different organisms whose genomes have been sequenced. If it displays an identical presence/absence pattern to an already characterized gene, then in many instances it can be inferred that both gene products have a related function. The Rosetta stone approach is dependent upon the observation that sometimes two separate polypeptides (i.e. gene products X and Y) found in one organism occur in a different organism as a single fused protein XY. In such circumstances, the two protein parts (domains), X and Y, often display linked functions. Therefore, if gene X is recently discovered in a newly sequenced genome and is of unknown function but gene XY of known function has been previously discovered in a different genome, then the function of the unknown X can be deduced. The gene neighbourhood method is yet another computation-based method. It depends upon the observation that two genes are likely to be functionally linked if they are consistently found side by side in the genome of several different organisms. Knockout animal studies, in contrast to the above methods, are dependent upon phenotype observation. The approach entails the generation and study of mice in which a specific gene has been deleted. Phenotypic studies can sometimes yield clues as to the function of the gene knocked out.

BIOINFORMATIC APPROACH

Although sequence data provide a profile of all the genes present in a genome, they give no information as to which genes are switched on (transcribed) and, hence, which are functionally active at any given time/under any given circumstances. Gene transcription results in the production of RNA, either mRNA (usually subsequently translated into a polypeptide) or rRNA or tRNA (which have catalytic or structural functions). The study of under which circumstances an RNA species is expressed/not expressed in the cell/organism can provide clues as to the biological function of the RNA (or, in the case of mRNA, the function of the final polypeptide product). Furthermore, in the context of drug lead/target discovery, the conditions under which a specific mRNA is produced can also point to putative biopharmaceuticals/drug targets. For example, if a particular mRNA is only produced by a cancer cell, that mRNA (or, more commonly, its polypeptide product) may represent a good target for a novel anti-cancer drug. Levels of RNA (usually specific mRNAs) in a cell can be measured by well-established techniques such as northern blot analysis or by PCR analysis.

However, the recent advent of DNA microarray technology has converted the identification and measurement of specific mRNAs (or other RNAs if required) into a '*high-throughput*' process. DNA arrays are also termed oligonucleotide arrays, gene chip arrays or, simply, chips. The technique is based upon the ability to anchor nucleic acid sequences (usually DNA based) on plastic/glass surfaces at very high density. Standard gridding robots can put on up to 250000 different short oligonucleotide probes or 10000 full-length cDNA sequences per square centimetre of surface. Probe sequences are generally produced/designed from genome sequence data; hence, chip production is often referred to as 'downloading the genome on a chip'. RNA can be extracted

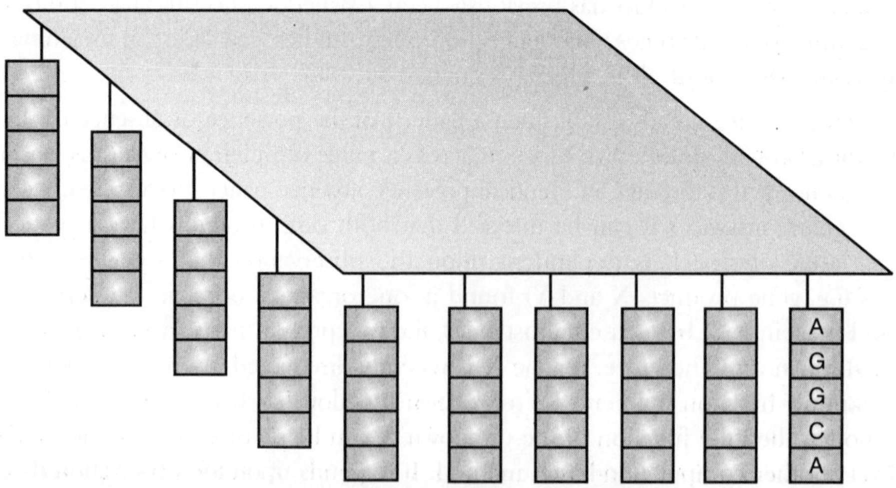

Fig. 7.2. Generalized outline of a gene chip.

from a cell and probed with the chip. Any complementary RNA sequences present will hybridize with the appropriate immobilized chip sequence. Hybridization is detectable as the RNA species are first labelled. Hybridization patterns obviously yield critical information regarding gene expression.

FUNCTIONAL GENOMICS

Although virtually all drug targets are protein based, the inference that protein expression levels can be accurately (if indirectly) detected/measured via DNA array technology is a false one, as:

1. mRNA concentrations do not always directly correlate with the concentration of the mRNA-encoded polypeptide;
2. A significant proportion of eukaryote mRNAs undergo differential splicing and, therefore, can yield more than one polypeptide product.

Additionally, the cellular location at which the resultant polypeptide will function often cannot be predicted from RNA delection/sequences nor can detailed information regarding how the polypeptide product's functional activity will be regulated (e.g. via post-translational mechanisms such as phosphorylation, partial proteolysis, etc.). Therefore, protein-based drug leads/targets are often more successfully identified by direct examination of the expressed protein complement of the cell, i.e. its proteome. Like the transcriptome (total cellular RNA content), and in contrast to the genome, the proteome is not static, with changes in cellular conditions triggering changes in cellular protein profiles/concentrations. This field of study is termed proteomics. Proteomics, therefore, is closely aligned to functional genomics and entails the systematic and comprehensive analysis of the proteins expressed in the cell and their function.

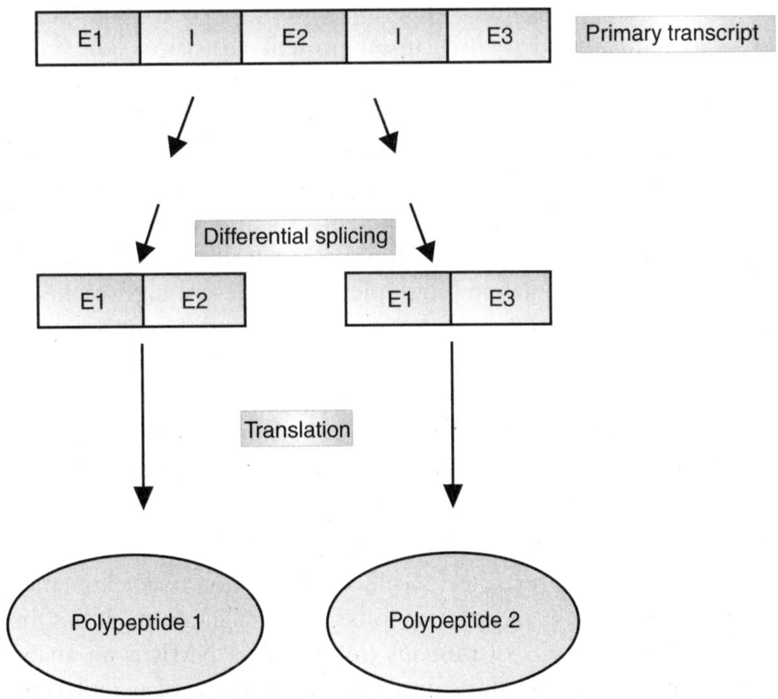

Fig. 7.3. Differential splicing of mRNA can yield different polypeptide products. Transcription of a gene sequence yields a 'primary transcript' RNA.

Classical proteomic studies generally entailed initial extraction of the total protein content from the target cell/tissue, followed by separation of the proteins therein using two-dimensional electrophoresis. Isolated protein 'spots' could then be eluted from the electrophoretic gel and subjected to further analysis; mainly to Edman degradation, in order to generate partial amino acid sequence data. The sequence data could then be used to interrogate protein sequence databanks in order to, for example, assign putative function by sequence homology searches. Two-dimensional electrophoresis, however, is generally capable of resolving no more than 2000 different proteins, and proteins expressed at low levels may not be detected at all if their gel concentration is below the (protein) staining threshold. The latter point can be particularly significant in the context of drug/target identification, as most such targets are likely to be kinases and other regulatory proteins that are generally expressed within cells at very low levels. More recently, high-resolution chromatographic techniques (particularly reverse-phase and ion exchanged-based high-performance liquid chromatography (HPLC)) have been applied in the separation of proteome proteins and high-resolution mass spectrometry is being employed to aid high-throughput sequence determination.

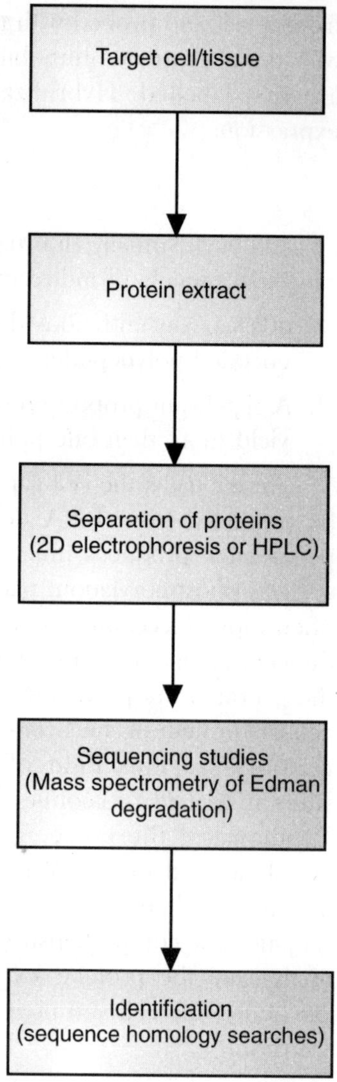

Fig. 7.4. The proteomics approach.

PROTEOMICS

Related to the discipline of proteomics is that of structural genomics. The latter focuses upon the large-scale systematic study of gene product structure. While this embraces rRNA and tRNA, in practice the field focuses upon protein structure. The basic approach to structural genomics entails the cloning and recombinant expression of cellular proteins, followed by their purification and three-dimensional structural analysis. High-resolution determination of a protein's structure is amongst the most challenging of molecular investigations. By the year 2000, protein structure databanks housed in the region of 12000 entries. However, such databanks are highly redundant, often containing multiple entries describing variants of the same molecule. For example, in excess of 50 different structures of 'insulin' have been deposited (e.g. both native and mutated/engineered forms from various species, as well as insulins in various polymeric forms and in the presence of various stabilizers and other chemicals). In reality, by the year 2000, the three-dimensional structure of approximately 2000 truly different proteins had been resolved. Until quite recently, X-ray crystallography was the technique used almost exclusively to resolve the three-dimensional structure of proteins.

As well as itself being technically challenging, a major limitation of X-ray crystallography is the requirement for the target protein to be in crystalline form. It has thus far proven difficult/impossible to induce the majority of proteins to crystallize. NMR is an analytical technique that can also be used to determine the three-dimensional structure of a molecule, and without the necessity for crystallization. For many years, even the most powerful NMR machines could resolve the three-dimensional structure of only relatively small proteins (less than 20-25 kDa). However, recent analytical advances now render it possible to analyse much larger proteins by this technique successfully. The ultimate goal of structural genomics is to provide a complete three-dimensional description of any gene product. Also, as the structures of more and more proteins of known function are elucidated, it should become increasingly possible to link specific functional attributes to specific structural attributes. As such, it may prove ultimately feasible to predict protein function if its structure is known, and vice versa.

SEQUENCE VARIATIONS

Pharmacogenetics relates to the emerging discipline of correlating specific gene DNA sequence information (specifically sequence variations) to drug response. As such, the pursuit will ultimately impinge directly upon the drug development process and should allow doctors to make better-informed decisions regarding what exact drug to prescribe to individual patients. Different people respond differently to any given drug, even if they present with essentially identical disease symptoms. Optimum dose requirements, for example, can vary significantly. Furthermore, not all patients respond positively to a specific drug (e.g. IFN-β is of clinical benefit to only one in three multiple sclerosis patients). The range and severity of adverse effects induced by a drug can also vary significantly within a patient population base. While the basis of such differential responses can sometimes be non-genetic (e.g. general state of health, etc.), genetic variation amongst individuals remains the predominant factor. Although all humans display almost identical genome sequences, some differences are evident.

The most prominent widespread-type variations amongst individuals are known as single nucleotide polymorphisms (SNPs, sometimes pronounced '*snips*'). SNPs occur in the general population at an average incidence of 1 in every 1000 nucleotide bases; hence, the entire human genome harbours 3 million or so. SNPs are not mutations; the latter arise more infrequently, are more diverse and are generally caused by spontaneous/mutagen-induced mistakes in DNA repair/ replication. SNPs occurring in structural genes/gene regulatory sequences can alter amino acid sequence/expression levels of a protein and, hence, affect its functional attributes. SNPs largely account for natural physical variations evident in the human population (e.g. height, colour of eyes, etc.). The presence of an SNP within the regulatory or structural regions of a gene coding for a protein that interacts with a drug could obviously influence the effect of the drug on the body. In this context, the protein product could, for example, be the drug target or perhaps an enzyme involved in metabolizing the drug.

The identification and characterization of SNPs within the human genomes is, therefore, of both academic and applied interest. Several research groups continue to map human SNPs, and over 1.5 million have thus far been identified. By identifying and comparing SNP patterns from a group of patients responsive to a particular drug with patterns displayed by a group of unresponsive patients, it may be possible to identify specific SNP characteristics linked to drug efficacy. In the same way, SNP patterns/characteristics associated with adverse reactions (or even a predisposition to a disease) may be uncovered. This could usher a new era of drug therapy where drug treatment could be tailored to the individual patient. Furthermore, different drugs could be developed with the foreknowledge that each would be efficacious when administered to specific (SNP-determined) patient sub-types. A (distant) futuristic scenario could be visualized where all individuals could carry chips encoded with SNP details relating to their specific genome, allowing medical staff to choose the most appropriate drugs to prescribe in any given circumstance. Linking specific genetic determinants to many diseases, however, is unlikely to be as straightforward as implied thus far. The progress of most diseases, and the relative effectiveness of allied drug treatment, is dependent upon many factors, including the interplay of multiple gene products. 'Environmental' factors such as patient age, sex and general health also play a prominent role. The term '*pharmacogenomics*' is one that has entered the '*genomic*' vocabulary. Although sometimes used almost interchangeably with pharmacogenetics, it more specifically refers to studying the pattern of expression of gene products involved in a drug response.

PHYSICOCHEMICAL PROPERTIES

The physicochemical and other properties of any newly identified drug must be extensively characterized prior to its entry into clinical trials. As the vast bulk of biopharmaceuticals are proteins, a summary overview of the approach taken to initial characterization of these biomolecules is presented. A prerequisite to such characterization is initial purification of the protein. Purification to homogeneity usually requires a combination of three or more high-resolution chromatographic steps. The purification protocol is designed carefully, as it usually forms the basis of subsequent pilot- and process-scale purification systems. The purified product is then subjected to a battery of tests that aim to characterize it fully. Moreover, once these characteristics have been defined, they form the basis of many of the QC identity tests routinely performed on the product during its subsequent commercial manufacture. In addition, stability characteristics of the protein with regard to e.g. temperature, pH and

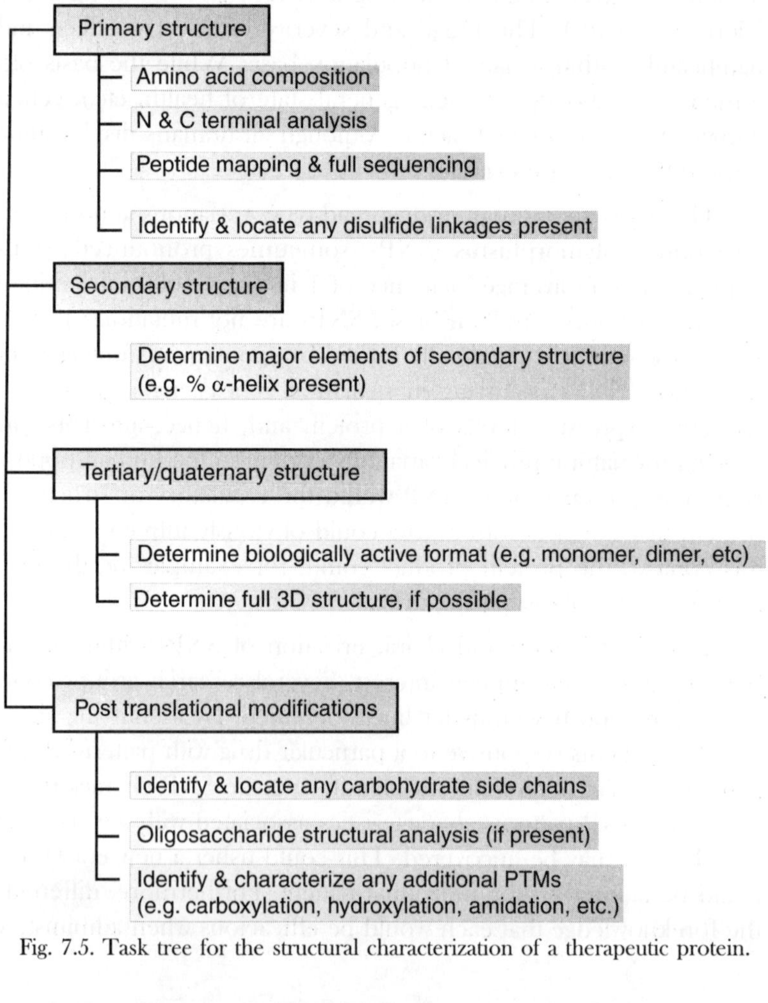

Fig. 7.5. Task tree for the structural characterization of a therapeutic protein.

incubation with various potential excipients are studied. Such information is required in order to identify a suitable final product formulation, and to give an early indication of the likely useful shelf-life of the product.

POTENTIAL PHARMACEUTICAL APPLICATION

The discovery and initial characterization of any substance of potential pharmaceutical application is followed by its patenting. The more detail given relevant to the drug's physicochemical characteristics, a method of synthesis and its biological effects, the better the chances of successfully securing a patent. Thus, patenting may not take place until preclinical trials and phase I clinical trials are completed. Patenting, once successfully completed, does not grant the patent holder an automatic right to utilize/sell the patented product; first, it must be proven safe and effective in subsequent clinical trials, and then be approved for general medical use by the relevant regulatory authorities.

Monopoly of Government

A patent may be described as a monopoly granted by a government to an inventor, such that only the inventor may exploit the invention/innovation for a fixed period of time (up to 20 years). In return, the inventor makes available a detailed technical description of the invention/innovation so that, when the monopoly period has expired, it may be exploited by others without the inventor's permission. A patent, therefore, encourages innovation by promoting research and development. It can also be regarded as a physical asset, which can be sold or licensed to third parties for cash. Patents also represent a unique source of technical information regarding the patented product. The philosophy underlining patent law is fairly similar throughout the world. Thus, although there is no worldwide patenting office, patent practice in different world regions is often quite similar. This is fortuitous, as there is a growing tendency towards world harmonization of patent law, fuelled by multinational trade agreements. In order to be considered patentable, an invention/innovation must satisfy several criteria, the most important four of which are:

1. Novelty
2. Non-obviousness
3. Sufficiency of disclosure
4. Utility.

Patenting of Gene/DNA

Many products of nature (e.g. specific antibiotics, microorganisms, proteins, etc.) have been successfully patented. It might be argued that simply to find any substance naturally occurring on the Earth is categorized as a discovery and would be unpatentable because it lacks true novelty or any inventive step. However, if you enrich, purify or modify a product of nature such that you make available the substance for the first time in an industrially useful format, that product/process is generally patentable. In other words, patenting is possible if the 'hand of man' has played an obvious part in developing the product. In the USA, purity alone often facilitates patenting of a product of nature. The US Patent and Trademark Office (PTO) recognizes purity as a change in form of the natural material. For example, although vitamin B_{12} was a known product of nature for many years, it was only available in the form of a crude liver extract, which was of no use therapeutically.

Development of a suitable production (fermentation) and purification protocol allowed production of pure, crystalline vitamin B_{12} which could be used clinically. On this basis, a product patent was granted in the USA. Using the same logic, the PTO has granted patents, for example, for pure cultures of specific microorganisms and for medically important proteins (e.g. Factor VIII purified from blood and EPO purified from urine). Rapid technological advances in the biological sciences raises complex patenting issues, and patenting law as applied to modern biotechnology is still evolving. In the late 1980s, the PTO confirmed they would consider issuing patents for non-human multicellular organisms, including animals.

The first transgenic animal was patented in 1988 by Harvard University. The '*Harvard mouse*' carried a gene that made it more susceptible to cancer and, hence, more sensitive in detecting possible carcinogens. Another area of biotechnology patent law relates to the patenting of genes and DNA sequences. Thus far, patents have been issued for some human genes, largely on the basis of thc use of their cloned products (e.g. EPO and tPA). A consensus has emerged that patent protection should only be considered for nucleotide sequences that can be used for specific purposes, e.g. for a sequence that can serve as a diagnostic marker or that codes for a protein product of medical value. This appears to be a reasonably approach, as it balances issues of public interest with encouraging innovation in the area. The issue of patenting genetic material or transgenic plants/animals remains a contentious one. The debate is not confined to technical and legal argument: ethical and political issues, including public opinion, also impinge on the decision-making process.

The increasing technical complexity and sophistication of the biological principles/processes upon which biotechnological innovations are based also render resolution of legal patenting issues more difficult. A major step in clarifying European Union (EU) -wide law with regard to patenting in biotechnology stems from the introduction of the 1998 European Patent Directive. This directive (EU law) confirms that biological material (e.g. specific cells, proteins, genes, nucleotide sequences, antibiotics, etc.) that previously existed in nature are potentially patentable. However, in order actually to be patentable, they must (a) be isolated/purified from their natural environment and/or be produced via a technical process (e.g. rDNA technology in the case of recombinant proteins) and (b) they must conform to the general patentability principles regarding novelty, non-obviousness, utility and sufficiency of disclosure. The 'utility' condition, therefore, in effect prevents patenting of gene/genome sequences of unknown function. The directive also prohibits the possibility of patenting inventions if their exploitation would be contrary to public order or morality. Thus, it is not possible to patent:

1. The human body;
2. The cloning of humans;
3. The use of human embryos for commercial purposes;
4. Modifying germ line identity in humans;

5. Modifying the genetic complement of an animal if the modifications cause suffering without resultant substantial medical benefits to the animal/to humans.

PARENTERAL ADMINISTRATION

An important issue that must be addressed during the preclinical phase of the drug development process relates to the route by which the drug will be delivered/administered. To date, the vast majority of biopharmaceuticals approved for general medical use are administered by direct injection (i.e. parenterally) usually by intravenous (i.v.), subcutaneous (s.c., i.e. directly under the skin) or intramuscular (i.m., i.e. into muscle tissue) routes. Administration via the s.c. or i.m. route is generally followed by slow release of the drug from its depot site into the bloodstream. Amongst the few exceptions to this parenteral route are the enzyme DNase, used to treat cystic fibrosis, and platelet-derived growth factor (PDGF), used to treat certain skin ulcers. However, neither of these products is required to reach the bloodstream in order to achieve its therapeutic effect.

In fact, in each case the delivery system delivers the biopharmaceutical directly to its site of action (DNase is delivered directly to the lungs via aerosol inhalation, and PDGF is applied topically, i.e. directly on the ulcer surface, as a gel). Parenteral administration is not perceived as a problem in the context of drugs which are administered infrequently, or as a once-off dose to a patient. However, in the case of products administered frequently/daily (e.g. insulin to diabetics), non-parenteral delivery routes would be preferred. Such routes would be more convenient, less invasive, less painful and generally would achieve better patient compliance. Alternative potential delivery routes include oral, nasal, transmucosal, transdermal or pulmonary routes. Although such routes have proven possible in the context of many drugs, routine administration of biopharmaceuticals by such means has proven to be technically challenging. Obstacles encountered include their high molecular mass, their susceptibility to enzymatic inactivation and their potential to aggregate.

Biopharmaceutical Delivery

Oral delivery is usually the preferred system for drug delivery, owing to its convenience and the high level of associated patient compliance generally attained. Biopharmaceutical delivery via this route has proven problematic for a number of reasons:

1. Inactivation due to stomach acid. Prior to consumption of a meal, stomach pH is usually below 2.0. Although the buffering action of food can increase the pH to neutrality, the associated stimulation of stomach acid secretion subsequently reduces the ambient pH back down to 3.0-3.5. Virtually all biopharmaceuticals are acid labile and are inactivated at low pH values.
2. Inactivation due to digestive proteases. Therapeutic proteins would represent potential targets for digestive proteases such as pepsin, trypsin and chymotrypsin.
3. Their (relatively) large size and hydrophilic nature renders difficult the passage of intact biopharmaceuticals across the intestinal mucosa.
4. Orally absorbed drugs are subjected to first-pass metabolism. Upon entry into the bloodstream, the first organ encountered is the liver, which usually removes a significant proportion of absorbed drugs from circulation.

Given such difficulties, it is not unsurprising that bioavailabilities below 1 per cent are often recorded in the context of oral biopharmaceutical drug delivery. Strategies pursued to improve bioavailability include physically protecting the drug via encapsulation and formulation as microemulsions/microparticulates, as well as inclusion of protease inhibitors and permeability enhancers. Encapsulation within an enteric coat (resistant to low pH values) protects the product during stomach transit. Microcapsules/spheres utilized have been made from various polymeric substances, including cellulose, polyvinyl alcohol, polymethylacrylates and polystyrene.

Delivery systems based upon the use of liposomes and cyclodextrin-protective coats have also been developed. Included in some such systems also are protease inhibitors, such as aprotinin and ovomucoids. Permeation enhancers employed are usually detergent-based substances, which can enhance absorption through the gastrointestinal lining. More recently, increasing research attention has focused upon the use of '*mucoadhesive delivery systems*' in which the biopharmaceutical is formulated with/encapsulated in molecules that interact with the intestinal mucosa membranes. The strategy is obviously to retain the drug at the absorbing surface for a prolonged period. Non-specific (charge-based) interactions can be achieved by the use of polyacrylic acid, whereas more biospecific interactions are achieved by using selected lectins or bacterial adhesion proteins. Despite intensive efforts, however, the successful delivery of biopharmaceuticals via the oral route remains some way off.

Pulmonary Delivery

Pulmonary delivery currently represents the most promising alternative to parenteral delivery systems for biopharmaceuticals. Delivery via the pulmonary route moved from concept to reality in 2006 with the approval of Exubera, an inhalable insulin product. Although the lung is not particularly permeable to solutes of low molecular mass (e.g. sucrose or urea), macromolecules can be absorbed into the blood via the lungs surprisingly well.

In fact, pulmonary macromolecular absorption generally appears to be inversely related to molecular mass, up to a mass of about 500 kDa. Many peptides/proteins delivered to the deep lung are detected in the blood within minutes, and bioavailabilities approaching/exceeding 50 per cent (relative to s.c. injection) have been reported for therapeutic proteins such as colony-stimulating factors and some interferons. Although not completely understood, such high pulmonary bioavailability may stem from:

1. The lung's very large surface area;
2. Their low surface fluid volume;
3. Thin diffusional layer;
4. Relatively slow cell surface clearance;
5. The presence of proteolytic inhibitors.

Additional advantages associated with the pulmonary route include:

1. The avoidance of first-pass metabolism;
2. The availability of reliable, metered nebulizer-based delivery systems capable of accurate dosage delivery, either in powder or liquid form;

3. Levels of absorption achieved without the need to include penetration enhancers which are generally too irritating for long-term use.

Although obviously occurring in practice, macromolecules absorbed via the pulmonary route must cross a number of biological barriers to get into the blood. These are:

1. A protective monolayer of insoluble phospholipid, termed 'lung surfactant', and its underlying surface lining fluid, which lies immediately above the lung epithelial cells;
2. The epithelial cells lining the lung;
3. The interstitium (an extracellular space), and the basement membrane, composed of a layer of interstitial fibrous material;
4. The vascular endothelium, i.e. the monolayer of cells that constitute the walls of the blood vessels.

Passage through the epithelium and endothelial cellular barriers likely represents the greatest challenge to absorption. Although the molecular details remain unclear, this absorption process appears to occur via one of two possible means: transcytosis or paracellular transport.

Nasal Based Biopharmaceutical Delivery

A nasal-based biopharmaceutical delivery route is considered potentially attractive as:

1. It is easily accessible;
2. Nasal cavities are serviced by a high density of blood vessels;
3. Nasal microvilli generate a large potential absorption surface area;
4. Nasal delivery ensures the drug bypasses first-pass metabolism.

However, the route does display some disadvantages, including:

1. Clearance of a proportion of administered drug occurs due to its deposition upon the nasal mucous blanket, which is constantly cleared by ciliary action;
2. The existence of extracellular nasal proteases/peptidases;
3. Low uptake rates for larger peptides/polypeptides.

Peptide/protein uptake rates across the nasal epithelia are dependent upon molecular mass. Relatively small peptides (such as oxytocin, desmopressin and luteinizing hormone-releasing hormone analogues) cross relatively easily, and several such products used medically are routinely delivered nasally. Larger molecules (of molecular mass greater than 10 kDa) generally do not cross the epithelial barrier without the concurrent administration of detergent-like uptake enhancers. Long-term use of enhancers is prohibited due to their damaging cellular effects. Research efforts also continue to explore mucosal delivery of peptides/proteins via the buccal, vaginal and rectal routes. Again, bioavailabilities recorded are low, with modest increases observed upon inclusion of permeation enhancers.

Additional barriers also exist relating, for example, to low surface areas, relatively rapid clearance from the mouth (buccal) cavity and the cyclic changes characteristic of vaginal tissue. Various strategies have been adopted in an attempt to achieve biopharmaceutical delivery across the skin (transdermal systems). Most have met with, at best, modest success thus far. Strategies employed include the use of a jet of helium or the application of a low voltage to accelerate proteins through the skin.

CLINICAL TRIALS

In order to gain approval for general medical use, the quality, safety and efficacy of any product must be demonstrated. Demonstration of conformance to these requirements, particularly safety and efficacy, is largely attained by undertaking clinical trials. However, preliminary data, especially safety data, must be obtained prior to the drug's administration to human volunteers. Regulatory authority approval to commence clinical trials is based largely upon preclinical pharmacological and toxicological assessment of the potential new drug in animals. Such preclinical studies can take up to 3 years to complete, and at a cost of anywhere between US$10 million and US$30 million. On average, approximately 10 per cent of potential new drugs survive preclinical trials. In many instances, there is no strict set of rules governing the range of tests that must be undertaken during preclinical studies. However, guidelines are usually provided by regulatory authorities. Most of these tests are equally applicable to biopharmaceutical products.

CHARACTERISTICS OF DRUGS

Pharmacology may be described as the study of the properties of drugs and how they interact with/affect the body. Within this broad discipline exist (somewhat artificial) subdisciplines, including pharmacokinetics and pharmacodynamics. Pharmacokinetics relates to the fate of a drug in the body, particularly its ADME, i.e. its absorption into the body, its distribution within the body, its metabolism by the body, and its excretion from the body. The results of such studies not only help to identify any toxic effects, but also point to the most appropriate method of drug administration, as well as the most likely effective dosage regime to employ. Generally, ADME studies are undertaken in two species, usually rats and dogs, and studies are repeated at various different dosage levels. All studies are undertaken in both males and females. If initial clinical trials reveal differences in human versus animal model pharmacokinetic profiles, additional pharmacokinetic studies may be necessary using primates.

Pharmacodynamic studies deal more specifically with how the drug brings about its characteristic effects. Emphasis in such studies is often placed upon how a drug interacts with a cell/organ type, the effects and side effects it induces, and observed dose-response curves. Bioavailability and bioequivalence are also usually assessed in animals. Such studies are undertaken as part of pharmacokinetic and/or pharmacodynamic studies. Bioavailability relates to the proportion of a drug that actually reaches its site of action after administration. As most biopharmaceuticals are delivered parenterally (e.g. by injection), their bioavailability is virtually 100 per cent. On the other hand, administration of biopharmaceuticals by mouth would, in most instances, yield a bioavailability at or near 0 per cent.

Bioavailability studies would be rendered more complex if, for example, a therapeutic peptide was being administered intranasally. Bioequivalence studies come into play if any change in product production/delivery systems was being contemplated. These studies would seek to identify whether such modifications still yield a product equivalent to the original one in terms of safety and efficacy. Modifications could include an altered formulation or method of administration, dosage regimes, etc.

Pharmacokinetics and Pharmacodynamic Studies

A prerequisite to pharmacokinetic/pharmacodynamic studies is the availability of a sufficiently selective and sensitive assay. The assay must be capable of detecting and accurately quantifying the therapeutic protein in the presence of a complex soup of 'contaminant' molecules characteristic of tissue extracts/body fluids. Specific proteins are usually detected and quantified either via immunoassay or bioassay. Additional analytical approaches occasionally used include liquid chromatography (e.g. HPLC) or the use of radioactively labelled protein. The macromolecular structure of drugs and the fact that relatively minor structural alterations can potentially have a major influence upon bioactivity are often complicating factors. For example, an immunoassay may be blind to the oxidation of an amino acid residue, or very limited proteolytic processing, although such events can activate or decrease bioactivity. As outlined previously, i.v. or s.c. administration is by far the most common delivery approach in the context of biopharmaceuticals. Whole-body distribution studies are undertaken mainly in order to assess tissue targeting and to identify the major elimination routes.

The large molecular weight of therapeutic proteins, along with additional properties (e.g. charge), generally impairs their passage through biomembranes; hence, their initial distribution is usually limited to the volume of the extracellular space (mainly the plasma volume). Distribution volume usually subsequently increases, as the protein is taken up into tissue during its metabolism/ elimination. The metabolism/elimination of therapeutic proteins occurs via processes identical to those pertaining to native endogenous proteins. Ultimately this entails proteolytic degradation, with amino acid residues released either being incorporated into newly synthesized protein or being further degraded by standard metabolic pathways. Although the therapeutic protein may be subject to limited proteolysis in the blood, extensive and full metabolism occurs intracellularly, subsequent to product cellular uptake.

Clearance of protein drugs from systemic circulation commences with passage across the capillary endothelia. The rate of passage depends upon the protein's physicochemical properties (e.g. mass and charge). Final product excretion is, in the main, either renal and/or hepatic mediated. Many proteins of molecular mass <30 kDa are eliminated by the kidneys via glomerular filtration. In addition to size, filtration is also dependent upon the protein's charge characteristics. Owing to the presence of glycosaminoglycans, the glomerular filter is itself negatively charged, so negatively charged proteins are poorly filtered due to charge repulsion. After initial filtration many proteins are actively reabsorbed (endocytosed) by the proximal tubules and subjected to lysosomal degradation, with subsequent amino acid reabsorption. Thus, very little intact protein actually enters the urine. Uptake of protein by hepatocytes can occur via one of two mechanisms: (i) receptor-mediated endocytosis or (ii) non-selective pinocytosis, again with subsequent protein proteolysis.

Similarly, a proportion of some proteins are likely degraded within the target tissue, as binding to their functional cell surface receptors triggers endocytotic internalization of the receptor ligand complex. Cellular uptake of some glycosylated therapeutic proteins occurs via specific sugar-binding

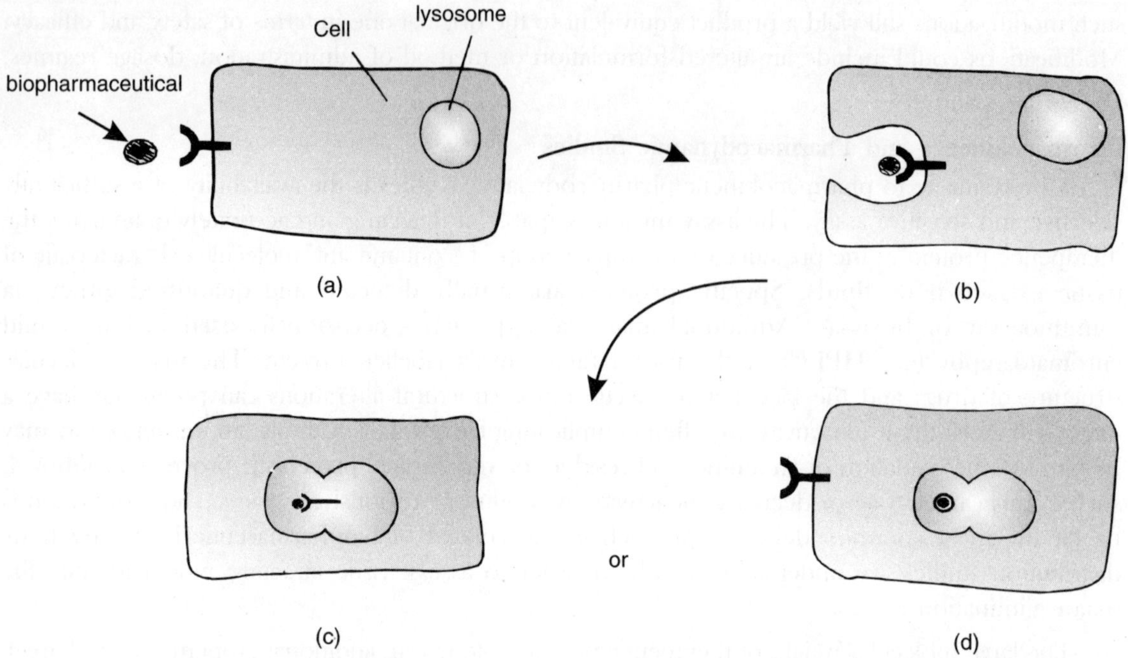

Fig. 7.6. The process of receptor-mediated endocytosis. Binding of ligand, in this case a therapeutic protein, to its cell surface receptor (a) triggers invagination of the immediately surrounding area of plasma-membrane, internalizing the receptor and its ligand in an intracellular vesicle (b). This is usually followed by fusion of the internalized vesicle with a lysosome and, therefore, degradation of both ligand and receptor by lysosomal hydrolases (c). In some cases, however, a variation can occur in which the ligand disassociated from the receptor, with subsequent budding off a small receptor-containing section of the vesicle, which returns the receptor to the cell surface. In this situation, only the ligand is available for degradation upon subsequent vesicular fusion with the lysosome (d).

cell surface receptors. Cell surface mannose receptors, for example, are capable of binding glycoproteins whose sugar side chains terminate in mannose, fucose, *N*-acetyl glucosamine or *N*-acetyl galactosamine. Evidence suggests that a liver-specific form of the mannose receptor mediates clearance of luteinizing hormone (LH). The sugar side chains of many glycoproteins exhibit terminal sialic acid residues (sialic acid caps). The hepatic asialoglycoprotein receptor binds glycoproteins whose sialic acid caps have been removed, likely mediating their removal from general circulation. Pharmacokinetic and indeed pharmacodynamic characteristics of therapeutic proteins can be rendered (even more) complicated by a number of factors, including:

Binding proteins

Some biopharmaceuticals (including insulin-like growth factor (IGF), GH and certain cytokines) are notable in that the blood contains proteins that specifically bind them. Such binding proteins can function naturally as transporters or activators, and binding can affect characteristics such as serum elimination rates.

Therapeutic proteins

Many, if not most, therapeutic proteins are potentially immunogenic when administered to humans. The likelihood that non-human proteins (e.g. murine monoclonal antibodies) are immunogenic in humans is an obvious one. However, human proteins can also be potentially immunogenic. Antibodies raised in this way can bind the therapeutic protein, neutralizing its activity and/or affecting its serum half-life.

Therapeutic glycoproteins

Expression of a therapeutic glycoprotein in different eukaryotic expression systems results in a product displaying differences in exact glycosylation detail. The exact glycosylation pattern can influence protein activity and stability *in vivo*, and some sugar motifs characteristic of yeast-, insect- and plant-based expression systems are immunogenic in man.

Protein Engineering

A number of different approaches may be used in order to alter a protein's pharmacokinetic profile. This can be desirable in order to achieve a predefined therapeutic goal, such as generating a faster- or slower-acting product, lengthening a product's serum half-life or altering a product's tissue distribution profile. The approach taken usually relies upon protein engineering, be it alteration of amino acid sequence, alteration of a native post-translational modification (usually glycosylation) or the attachment of a chemical moiety to the protein's backbone (often the attachment of PEG, i.e. PEGylation).

Receptor Binding

Different protein therapeutics bring about their therapeutic effect in different ways. Hormones and additional regulatory molecules invariably achieve their effect by binding to a specific cell surface receptor, with receptor binding triggering intracellular signal transduction event(s) that ultimately mediate the observed physiological effect(s). Many antibodies, on the other hand, bring about their effect by binding to their specific target molecule, which either inactivates/triggers destruction of the target molecule or (in the case of diagnostic applications) effectively tags the target molecules/ cells. Therapeutic enzymes bring about their effect via a catalytic mechanism. However, it is important to note that the exact molecular detail underpinning the effects of many such drugs has not been fully characterized. A significant element of preclinical studies, therefore, centres upon identification of a drug's mode of action at a molecular level, in addition to investigating the full range of resultant physiological effects. Pharmacodynamic studies will invariably include monitoring effects (and the timing of effects) of the therapeutic protein at different known drug concentrations and drug delivery schedules.

TOXICOLOGICAL APPROACH

Toxicity studies are carried out on all putative new drugs, largely via testing in animals, in order to ascertain whether the product exhibits any short-term or long-term toxicity. Acute toxicity is usually assessed by administration of a single high dose of the test drug to rodents. Both rats and mice (male and female) are usually employed. The test material is administered by two means, one of which should represent the proposed therapeutic method of administration. The animals are then monitored for 7–14 days, with all fatalities undergoing extensive post-mortem analysis. Earlier studies demanded calculation of an LD_{50} value (i.e. the quantity of the drug required to cause death of 50 per cent of the test animals). Such studies required large quantities of animals,

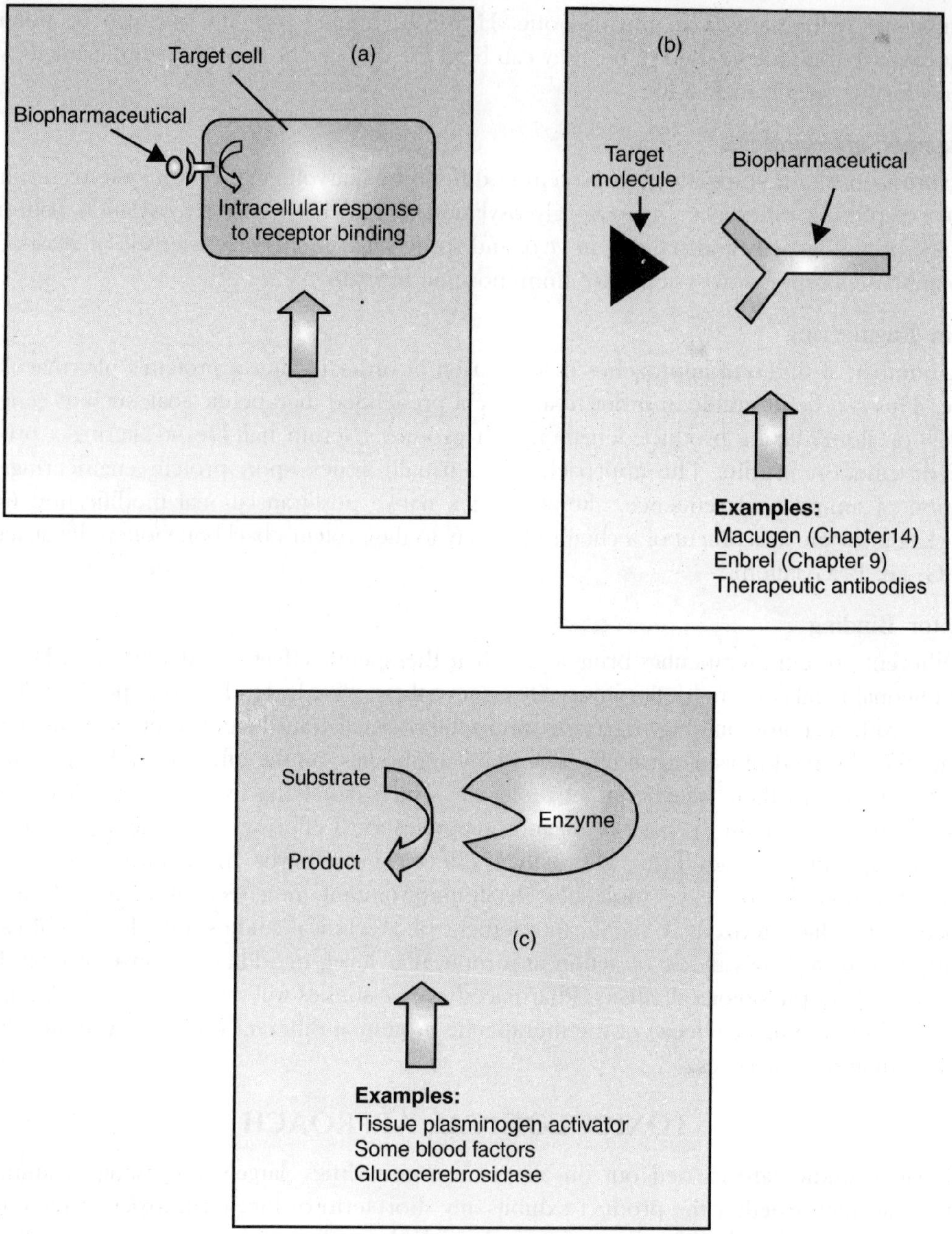

Fig. 7.7. Overview of the mode of action of biopharmaceutical.

were expensive, and attracted much attention from animal welfare groups. Its physiological relevance to humans was often also questioned. Nowadays, in most world regions, calculation of the approximate lethal dose is sufficient. Chronic toxicity studies also require large numbers of animals and, in some instances, can last for up to 2 years.

Most chronic toxicity studies demand daily administration of the test drug (parenterally for most biopharmaceuticals). Studies lasting 1-4 weeks are initially carried out in order to, for example, assess drug levels required to induce an observable toxic effect. The main studies are then initiated and generally involve administration of the drug at three different dosage levels. The highest level should ideally induce a mild but observable toxic effect, whereas the lowest level should not induce any ill effects. The studies are normally carried out in two different species, usually rats and dogs, and using both males and females. All animals are subjected to routine clinical examination, and periodic analyses of, for example, blood and urine are undertaken. Extensive pathological examination of all animals is undertaken at the end of the study. The duration of such toxicity tests varies. In the USA, the FDA usually recommends a period of up to 2 years, whereas in Europe the recommended duration is usually much shorter.

Chronic toxicity studies of biopharmaceuticals can also be complicated by their likely stimulation of an immune response in the recipient animals. In the context of new chemical entities (NCEs, i.e. low molecular weight traditional chemicals), not only can the drug itself exhibit a toxic effect, but so potentially can drug breakdown products. As proteins are degraded to amino acids, any potentially toxicity associated with protein-based drugs is typically associated with the protein itself and not degradation products.

Teratogenical Studies

All reproductive studies entail ongoing administration of the proposed drug at three different dosage levels (ranging from non-toxic to slightly toxic) to different groups of the chosen target species (usually rodents). Fertility studies aim to assess the nature of any effect of the substance on male or female reproductive function. The drug is administered to males for at least 60 days (one full spermatogenesis cycle). Females are dosed for at least 14 days before they are mated. Specific tests carried out include assessment of male spermatogenesis and female follicular development, as well as fertilization, implantation and early foetal development. These reproductive toxicity studies complement teratogenicity studies, which aim to assess whether the drug promotes any developmental abnormalities in the foetus. (A teratogen is any substance/agent that can induce foetal developmental abnormalities. Examples include alcohol, radiation and some viruses.)

Daily doses of the drug are administered to pregnant females of at least two species (usually rats and rabbits). The animals are sacrificed close to term and a full autopsy on the mother and foetus ensues. Post-natal toxicity evaluation often forms an extension of such studies. This entails administration of the drug to females both during and after pregnancy, with assessment of mother and progeny not only during pregnancy, but also during the lactation period. Therapeutic proteins rarely display any signs of reproductive toxicity or teratogenicity.

Mutagenicity Tests

Mutagenicity tests aim to determine whether the proposed drug is capable of inducing DNA damage, either by inducing alterations in chromosomal structure or by promoting changes in nucleotide base sequence. Although mutagenicity tests are prudent and necessary in the case of chemical-based drugs, they are less so for most biopharmaceutical substances. In many cases, biopharmaceutical mutagenicity testing is likely to focus more so on any novel excipients added to the final product, rather than the biopharmaceutical itself. (Excipient refers to any substance other than the active ingredient that is present in the final drug formulation). Mutagenicity tests are usually carried out *in vitro* and *in vivo*, often using both prokaryotic and eukaryotic organisms. A well-known example is the Ames test, which assesses the ability of a drug to induce mutation reversions in *E. coli* and *Salmonella typhimurium*.

Longer-term carcinogenicity tests are undertaken, particularly if (a) the product's likely therapeutic indication will necessitate its administration over prolonged periods (a few weeks or more) or (b) if there is any reason to suspect that the active ingredient or other constituents could be carcinogenic. These tests normally entail ongoing administration of the product to rodents at various dosage levels for periods of up to (or above) 2 years. Some additional animal investigations are also undertaken during preclinical trials. These include immunotoxicity and local toxicity tests. Again, for many biopharmaceuticals, immunotoxicity tests (i.e. the product's ability to induce an allergic or hypersensitive response, or even a clinically relevant antibody response) are often impractical.

The regulatory guidelines suggest that further studies should be carried out if a biotechnology drug is found capable of inducing an immune response. However, many of the most prominent biopharmaceuticals (e.g. cytokines) actually function to modulate immunological activities in the first place. Prediction and preclinical assessment of the immunogenic potential of any biopharmaceutical in humans is by no means straightforward. The use of animal models is inappropriate, as the human protein will be automatically seen as foreign by their immune system, almost certainly stimulating an immune response. Some factors, such as the extent and nature of post-translational modifications, the mode and frequency of administration and whether the protein sequence is of human origin or not, provide pointers but are by no means accurate predictors.

One potential predictive approach entails the development and use of transgenic animals that are made immunotolerant for the human protein under development. Such animals can then provide some basis for the study of breaking immunological tolerance. Many drugs, including many biopharmaceuticals, are administered to localized areas within the body by, for example, s.c. or i.m. injection. Local toxicity tests appraise whether there is any associated toxicity at/surrounding the site of injection. Predictably, these are generally carried out by s.c. or i.m. injection of product to test animals, followed by observation of the site of injection. The exact cause of any adverse response noted (i.e. active ingredient or excipient) is usually determined by their separate subsequent administration. Preclinical pharmacological and toxicological assessment entails the use of thousands of animals. This is both costly and, in many cases, politically contentious. Attempts have been made to develop alternatives to using animals for toxicity tests, and these have mainly centred around animal cell culture systems.

A whole range of animal and human cell types may be cultured, at least transiently, *in vitro*. Large-scale and fairly rapid screening can be undertaken by, for example, microculture of the target animal cells in microtitre plates, followed by addition of the drug and an indicator molecule. The indicator molecule serves to assess the state of health of the cultured cells. The dye neutral red is often used (healthy cells assimilate the dye, dead cells do not). The major drawback to such systems is that they do not reflect the complexities of living animals and, hence, may not accurately reflect likely results of whole-body toxicity studies. Regulatory authorities are (rightly) slow to allow replacement of animal-based test protocols until the replacement system is proven to be reliable and is fully validated. The exact range of preclinical tests that regulatory authorities suggest be undertaken for biopharmaceutical substances remains flexible. (Generally, only a subgroup of the standard tests for chemical-based drugs is appropriate. Biopharmaceuticals pose several particular difficulties, especially in relation to preclinical toxicological assessment. These difficulties stem from several factors (some of which have already been mentioned). These include:

1. The species specificity exhibited by some biopharmaceuticals, e.g. GH and several cytokines, means that the biological activity they induce in man is not mirrored in test animals;
2. For biopharmaceuticals, greater batch-to-batch variability exists compared with equivalent chemical-based products;
3. Induction of an immunological response is likely during long-term toxicological studies;
4. Lack of appropriate analytical methodologies in some cases.

In addition, tests for mutagenicity and carcinogenicity are not likely required for most biopharmaceutical substances. The regulatory guidelines and industrial practices relating to biopharmaceutical preclinical trials thus remain in an evolutionary mode, and each product is taken on a case-by-case basis.

Therapeutic Intervention

Clinical trials serve to assess the safety and efficacy of any potential new therapeutic '*intervention*' in its intended target species. In our context, an intervention represents the use of a new biopharmaceutical. Examples of other interventions could be, for example, a new surgical procedure or a novel medical device. Veterinary clinical trials are based upon the same principles, but this discussion is restricted to investigations in humans. Clinical trials are also prospective rather than retrospective in nature, i.e. participants receiving the intervention are followed forward with time. Clinical trials may be divided into three consecutive phases. During phase I trials, the drug is normally administered to a small group of healthy volunteers. The aims of these studies are largely to establish:

1. The pharmacological properties of the drug in humans (including pharmacokinetic and pharmacodynamic considerations);
2. The toxicological properties of the drug in humans (with establishment of the maximally tolerated dose);
3. The appropriate route and frequency of administration of the drug to humans.

Thus, the emphasis of phase I trials largely remains upon assessing drug safety. If satisfactory results are obtained during phase I studies, the drug then enters phase II trials. These studies aim to assess both the safety and effectiveness of the drug when administered to volunteer patients (i.e. persons suffering from the condition the drug claims to cure/alleviate). The design of phase II trials is influenced by the phase I results. Phase II studies typically last for anything up to 2 years, with anywhere between a few dozen and a hundred or more patients participating, depending upon the trial size. If the drug proves safe and effective, phase III trials are initiated. (In the context of clinical trials, safe and effective are rarely used in the absolute sense. 'Safe' generally refers to a favourable risk:benefit ratio, i.e. the benefits should outweigh any associated risk. A drug is rarely 100 per cent effective in all patients. Thus, an acceptable level of efficacy must be defined, ideally prior to trial commencement.

Depending upon the trial context, 'efficacy' could be defined as prevention of death/prolonging of life by a specific time-frame. It could also be defined as alleviation of disease symptoms or enhancement of the quality of life of sufferers (often difficult parameters to measure objectively). An acceptable incidence of efficacy should also be defined (particularly for phase II and III trials), e.g. the drug should be efficacious in, say, 25 per cent of all patients. If the observed incidence is below the minimal acceptable level, then clinical trials are normally terminated. Phase III clinical trials are designed to assess the safety and efficacy characteristics of a drug in greater detail. Depending upon the trial size, usually hundreds if not thousands of patients are recruited, and the trial may last for up to 3 years.

These trials serve to assess the potential role of the new drug in routine clinical practice; the phase III results will largely dictate whether or not the prospective drug subsequently gains approval for general medical use. Even if a product gains marketing approval (on average, 10-20 per cent of prospective drugs that enter clinical trials are eventually commercialized), the regulatory authorities may demand further post-marketing surveillance studies. These are often termed 'phase IV clinical trials'. They aim to assess the long-term safety of a drug, particularly if the drug is administered to patients for periods of time longer than the phase III clinical trials. The discovery of more long-term unexpected side effects can result in subsequent withdrawal of the product from the market. Both preclinical and clinical trials are underpinned by a necessity to produce sufficient quantities of the prospective drug for its evaluation.

Depending on the biopharmaceutical product, this could require from several hundred grams to over a kilogram of active ingredient. It is important that a suitable production process be designed prior to commencement of preclinical trials, that the process is amenable to scaling up and that, as far as is practicable, it is optimized. The material used for preclinical and clinical trials should be produced using the same process by which it is intended to undertake final-scale commercial manufacture. Extensive early development work is thus essential. Any significant deviation from the production protocol used to generate the trial material could invalidate all the clinical trial results with respect to the proposed commercialized product.

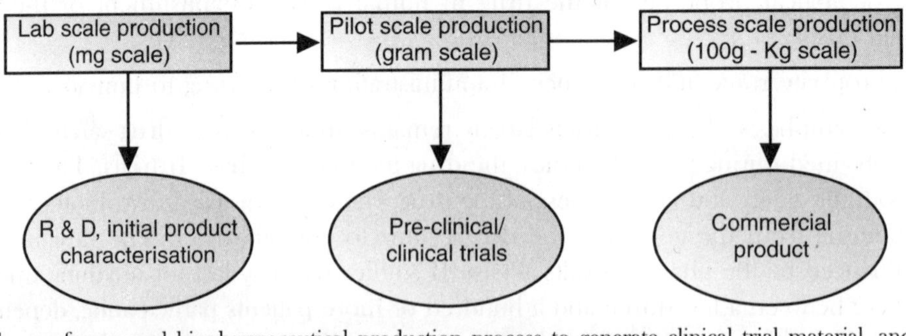

Fig. 7.8. Scale-up of proposed biopharmaceutical production process to generate clinical trial material, and eventually commercial product.

Comprehensive Planning

Proper and comprehensive planning of a clinical trial is essential to the successful development of any drug. The first issue to be considered when developing a trial protocol is to define precisely what questions the trial results should be capable of answering. As discussed previously, the terms safety and efficacy are difficult to define in a therapeutic context. An acceptable meaning of these concepts, however, should be committed to paper prior to planning of the trial.

Clinical Trials

A clinical trial must obviously have a control group, against which the test (intervention) group can be compared. The control group may receive: (a) no intervention at all; (b) a placebo (i.e. a substance such as saline, which will have no pharmacological or other effect); (c) the therapy most commonly used at that time to combat the target disease/condition. The size of the trial will be limited by a number of factors, including:

1. Economic considerations (level of supporting financial resources);
2. Size of population with target condition;
3. Size of population with target condition after additional trial criteria have been imposed (e.g. specific age bracket, lack of complicating medical conditions, etc.);
4. Size of eligible population willing to participate in the trial.

Whereas a comprehensive phase III trial would normally require at least several hundred patients, smaller trials would suffice if, for example:

1. The target disease is very serious/fatal;
2. There are no existing acceptable alternative treatments;
3. The target disease population is quite small;
4. The new drug is clearly effective and exhibits little toxicity.

Choosing the study population is obviously critical to adequate trial design. The specific criteria of patient eligibility should be clearly predefined as part of the primary question the trial strives to answer. A number of trial design types may be used, each having its own unique advantages and disadvantages. The most scientifically pure is a randomized, double-blind trial. However, in many instances, alternative trial designs are chosen based on ethical or other grounds. In most cases, two groups are considered: control and test. However, these designs can be adapted to facilitate more complex subgrouping. Clinical trial design is a subject whose scope is too broad to be undertaken in this text.

PHARMACEUTICAL INDUSTRY

The pharmaceutical industry is one of the most highly regulated industries known. Governments in virtually all world regions continue to pass tough laws to ensure that every aspect of pharmaceutical activity is tightly controlled. All regulations pertaining to the pharmaceutical industry are enforced by government-established regulatory agencies. The role and remit of some of the major world regulatory authorities is outlined below. In the context of this chapter, particular emphasis is placed upon their role with regard to the drug development process.

FDA

The *Food and Drug Administration* (FDA) represents the American regulatory authority. Its mission statement defines its goal simply as being to '*protect public health*'. In fulfilling this role, it regulates many products/consumer items, the total annual value of which is estimated to be US$1 trillion. Its work entails inspecting/regulating almost 100000 establishments in the USA. The agency employs over 9000 people, of whom in the region of 4000 are concerned with enforcing drug law. The FDA's total annual budget is in the region of US$1 billion. The FDA was founded in 1930. An act of Congress officially established it as a governmental agency in 1988, and it now forms a part of the US Department of Health and Human Services. The FDA Commissioner is appointed directly by the President. The FDA derives most of its statutory powers from the Federal Food, Drug and Cosmetic Act. This legislation was originally signed into law in 1930, but has been amended several times since. The agency interprets and enforces these laws. In order to achieve this, it draws up regulations based upon the legislation.

Most of the regulations themselves are worded in general terms, and are supported by various FDA publications that explain/interpret these regulations in far greater detail. The publications include: 'Written Guidelines', 'Letters to Industry' and the 'Points to Consider' series of documents. As technological and other advances are made, the FDA further supplements their support publication list. The core activities of biopharmaceutical drug approval/regulation is undertaken mainly by the Center for Drug Evaluation and Research (CDER) and the Center for Biologics Evaluation and Research (CBER). The major FDA responsibilities with regard to drugs include:

1. Assessing preclinical data to decide whether a potential drug is safe enough to allow commencement of clinical trials;
2. Protecting the interests and rights of patients participating in clinical trials;
3. Assessing preclinical and clinical trial data generated by a drug and deciding whether that drug should be made available for general medical use;

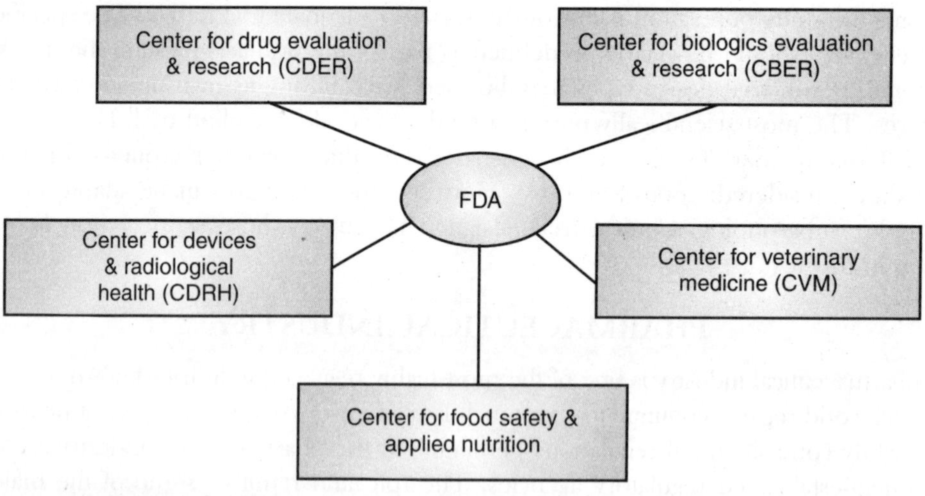

Fig. 7.9. Partial organizational structure of the FDA, displaying the various centres primarily responsible for regulating drugs, devices, and food.

4. Overseeing the manufacture of safe effective drugs (inspecting and approving drug manufacturing facilities on the basis of compliance to the principles of good manufacturing practice as applied to pharmaceuticals);
5. Ensuring the safety of the US blood supply.

In relation to the drug development process, the CDER has traditionally overseen and regulated the development and marketing approval of mainly chemical-based drugs. The CBER is more concerned with biologics. 'Biologic' has traditionally been defined in a narrow sense and has been taken to refer to vaccines and viruses, to blood and blood products, and to antiserum, toxins and antitoxins used for therapeutic purposes. Because of this, many established pharmaceutical products (e.g. microbial metabolites and hormones) have come under appraisal by the CDER, even though one might initially assume they would come under the biologics umbrella.

The CDER has now also been assigned regulatory responsibility for the majority of products of pharmaceutical biotechnology. The criteria used by the CBER and CDER regulators in assessing product performance during the drug development process are similar, i.e. safety, quality and efficacy. However, the administrative details can vary in both name and content. Upon concluding preclinical trials, all the data generated regarding any potential new drug are compiled in a dossier and submitted to the CDER or CBER in the form of an investigational new drug application (IND application). The FDA then assesses the application; if it does not object within a specific time-frame (usually 30 days), then clinical trials can begin. The FDA usually meet with the drug developers at various stages to be updated, and often to give informal guidance/advice. Once clinical trials have been completed, all the data generated during the entire development process are compiled in a multi-volume dossier. The dossier submitted to the CDER is known as a new drug application (NDA), which, if approved, allows the drug to be marketed. If the drug is a CBER-regulated one, then a biologics licence application (BLA) is submitted.

Biochemical Based Substances

An investigational new drug is a new chemical-based, biologic or biopharmaceutical substance for which the FDA has given approval to undergo clinical trials. An IND application should contain information detailing preclinical findings, method of product manufacture and proposed protocol for initial clinical trials. In some instances, the FDA and drug sponsor (company/institution submitting the IND) will agree to hold a pre-IND meeting. This aims to acquaint the FDA officials with the background to/content of the IND application, and to get a feel for whether the IND application will be regarded as adequate by the FDA. An IND application can consist of up to 15 volumes of approximately 400 pages each. Once received by the FDA, it is studied to ensure that:

1. It contains sufficient/complete information required;
2. The information supplied supports the conclusion that clinical trial subjects would not be exposed to an unreasonable risk of illness/injury (the primary FDA role being to protect public health);
3. The clinical investigator named is qualified to conduct the clinical trials;
4. The sponsor's product brochures are not misleading/incomplete.

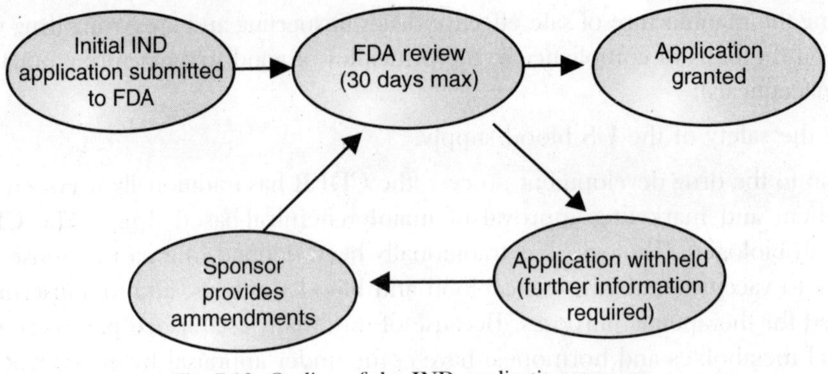

Fig. 7.10. Outline of the IND application process.

Based on their findings, the FDA may grant the application immediately or may require additional information which the sponsor then submits as IND amendments. Once clinical trials begin, the sponsor must provide the FDA with periodic updates, usually in the form of annual reports. Unscheduled reports must also be submitted under a variety of circumstances, including:

1. If any amendments to the trial protocol are being considered;
2. If any new scientific information regarding the product is obtained;
3. If any unexpected safety observations are made.

During the clinical trial phase, the sponsor and FDA will meet on one or more occasions. A particularly important meeting is often the end of phase II meeting. This aims primarily to evaluate and agree upon phase III plans and protocols. This is particularly important, as phase III trials are the most costly and generate the greatest quantity of data, used later to support the drug approval application.

NAD (New Drug Application)

Upon completion of clinical trials, the sponsor will collate all the preclinical, clinical and other pertinent data and submit this to FDA in support of an application to allow the new drug to be placed upon the market. For CDER-related drugs, this submission document is termed an NDA. The NDA must be an integrated document. It often consists of 200-300 volumes, which can represent over 120000 pages. Several copies of the entire document, and sections thereof, are provided to the CDER. The FDA then classify the NDA based upon the chemical type of the drug and its therapeutic potential.

Generally, drugs of high therapeutic potential (e.g. new drugs capable of curing/alleviating serious/terminal medical conditions) are appraised by the CDER in the shortest time. After initial submission of the NDA, the FDA has 45 days in which to undertake a preliminary inspection of the document, to ensure that everything is in order. They then 'file' the NDA; or, if more information/better information management is needed, they refuse to file until such changes are implemented by the sponsor. Once filed, an NDA undergoes several layers of review. A primary review panel generally consists of a chemist, microbiologist, pharmacologist, biostatistician, medical officer and biopharmaceutics scientist. Most hold PhDs in their relevant discipline. The team is organized by a project manager or *consumer safety officer* (CSO). The CSO initially forwards relevant portions of the NDA to the primary review panel member with the appropriate expertise. Each reviewer then prepares a review report. This is forwarded to their supervisory officers, who undertake a second review.

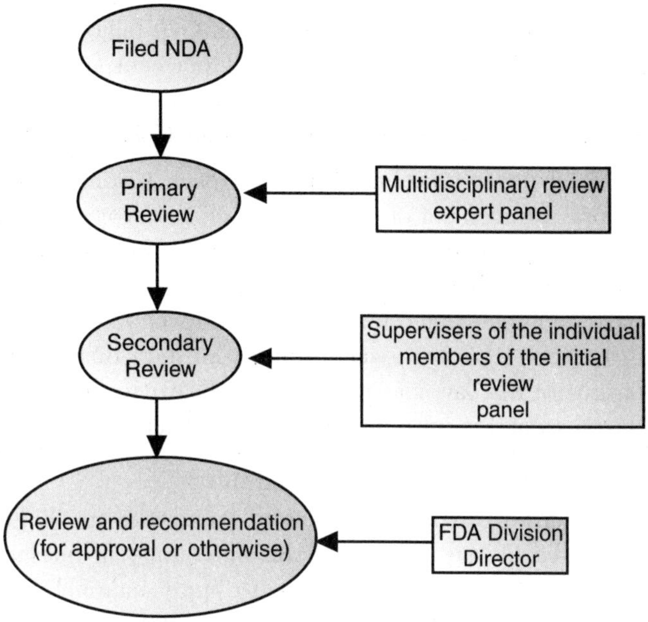

Fig. 7.11. The CDER review process for a typical NDA.

All of the reports are then sent to the division director who, in turn, recommends rejection or approval, or asks the sponsor to provide more information. On average, this entire process takes some 12 months. Even when the NDA is approved and the product goes on sale, the sponsor must provide the FDA with further occasional reports. These can be in the form of scheduled annual reports, but also unscheduled reports are required in instances such as the occurrence of an unexpected adverse response to the drug. A similar general approach is taken by the CBER with regard to drugs being developed under their auspices. The CBER '*licensing process*' for a new drug consists of three phases: the IND phase (already discussed), the pre-marketing approval phase (licensure phase) and the post-marketing surveillance phase.

The pre-marketing approval phase (i.e. clinical trial phase) aims to generate data that prove the potency, purity and safety of the product. Upon completion of clinical trials, the sponsor collates the data generated and submits it to the FDA in the form of a BLA, which must provide a comprehensive description of both the product and product manufacture (including methods of QC analysis, product stability data, labelling data and, of course, safety and efficacy data). A small number of biotechnology products are classified as medical devices and, hence, are regulated by the Center for Devices and Radiological Health (CDRH). The first approved biotech product to come under the auspices of the CDRH was OP-1 implant. Marketed by Stryker Biotech, OP-1 implant is a sterile powder composed of recombinant human oestrogenic protein-1 (OP-1) along with bovine collagen. It is used to treat fractured bones that fail to heal. The product is mixed with sterile saline immediately before application, and entails surgical insertion of the paste into the fracture.

European Pharmaceutical Marketing

The overall philosophy behind granting a marketing authorization for a new drug is broadly similar in the USA and Europe. There are, however, major differences in the systems by which these philosophies are implemented in the two regions. The EU is currently composed of 27 member states and a total population of 480 million (compared with 250 million in the USA and 125 million in Japan, the other two world pharmaceutical markets). The total European pharmaceutical market value stands at € 70 billion, representing almost 33 per cent of world sales (the US and Japanese figures are 31 per cent and 21 per cent respectively). Annual expenditure on European pharmaceutical R&D stands at about €17 billion. Prominent within the EU organizational structure is the European Commission.

The major function of the European Commission is to propose new EU legislation and to ensure enforcement of existing legislation. European drug law, therefore, comes under the auspices of the Commission. Two forms of legal instrument can be issued from the centralized European authorities: a '*regulation*' and a '*directive*'. A regulation is a strong legal instruction that, once passed, must be implemented immediately, and without modification, by national governments. A directive is a looser legal term, and provides an individual member state with 18 months to translate the flavour of that law into national law. Pharmaceutical law within the EU has been shaped by both directives and regulations, as discussed later.

NRA (National Regulatory Authorities)

There are national regulatory authorities in all European countries. These authorities are appointed by the government of the country in question and are usually located in the national Ministry of Health. They serve to apply national and European law with regard to the drug development process. In many countries, different arms of the regulatory authorities are responsible for authorizing and assessing clinical trials, assessing the resultant drug dossier and deciding on that basis whether or not to grant a (national) marketing authorization/product licence. They are also often responsible for issuing manufacturing licences to companies. In the past, a company wishing to gain a marketing licence within Europe usually applied for separate marketing authorizations on a country-by-country basis. This entailed significant duplication of effort as:

1. The drug dossiers needed to be translated into various European languages;
2. National laws often differed and, hence, different expectations/dossier requirements were associated with different countries;
3. The time-scale taken for dossier assessment varied from country to country.

Attempts to harmonize European pharmaceutical laws were accelerated in the 1980s. From 1985 onwards, a substantial number of European pharmaceutical directives have been adopted. This entire legislation has been published in the form of a nine-volume series entitled 'The Rules Governing Medicinal Products in the European Union'. These volumes form the basis of EU-wide regulation of virtually every aspect of pharmaceutical activity.

Drug Marketing

In 1993, a significant advance in simplifying the procedures relating to drug marketing authorization in the EU was made. At that time, the legal basis of a new drug approval system was formed by an EU regulation and three directives. Central to this was the establishment of a European agency for the evaluation of medicinal products, the European Medicines Agency (EMEA), which is based in London and began work in January 1995. Two new marketing authorization (drug registration) procedures for human or veterinary drugs are now in place within the EU:

1. A centralized procedure in which applications for a marketing licence are forwarded directly to the EMEA;
2. A decentralized procedure based upon mutual acceptance or recognition of national authority decisions; disputes arising from this system are arbitrated by the EMEA.

The EMEA is comprised of:

1. A management board, consisting of representatives of each EU member state and representatives of the European Commission and European Parliament. This group functions mainly to coordinate EMEA activities and manage its budget.
2. Pre-authorization evaluation of medicines for human use unit. This unit provides scientific advice relating to quality, safety and efficacy issues, as well as relating to orphan drugs.
3. Post-authorization evaluation of medicines for human use. This regulatory affairs unit is responsible for issues such as post-marketing surveillance of drugs.
4. Veterinary medicines and inspections. This unit is responsible for veterinary marketing authorization procedures and inspections.

Central to the functioning of the EMEA are three committees:

1. The Committee for Medicinal Products for Human use (CHMP). This committee is composed of 35 technical experts drawn from the various EU member countries. It is primarily responsible for formulating the EMEA's opinion on any medicinal product being considered for marketing approval under the centralized procedure.
2. The Committee for Medicinal Products for Veterinary use (CVMP), whose structure and role is similar to the CPMP, except that it is concerned with animal medicines.
3. The Committee for Orphan Medicinal Products (COMP), composed of a representative from each EU member state and EMEA and patient representatives. The COMP assesses applications relating to experimental medicines being granted '*orphan*' status. Orphan medicines are those intended to treat rare diseases, and orphan designation results in a reduction of the fees charged by the EMEA when assessing marketing authorization applications.

Unlike the FDA, the EMEA itself does not directly undertake appraisals of drug dossiers submitted to support marketing authorization applications under the centralized procedure. Instead, they forward the dossier to selected national EU regulatory bodies, who undertake the appraisal, and the EMEA makes a recommendation to approve (or not) the application based upon the national body's report. The overall role of the EMEA is thus to coordinate and manage the new system. The EMEA's annual budget is of the order of €120 million. The key objectives of the EMEA may be summarized as:

1. Protection of public and animal health by ensuring the quality, safety and efficacy of medicinal products for human and veterinary use;
2. To strengthen the ideal of a single European market for human and veterinary pharmaceuticals;
3. To support the European pharmaceutical industry as part of the EU industrial policy.

Applications

Under the centralized procedure, applications are accepted with regard to (a) products of biotechnology; (b) NCEs (drugs in which the active ingredient is new). Biotech products are grouped as 'list A' and NCEs as 'list B'. Marketing approval application for biotech products must be considered under the centralized procedure, whereas NCEs can be considered under centralized or decentralized mechanisms. Upon receipt of a marketing authorization application under centralized procedures, the EMEA staff carry out an initial appraisal to ensure that it is complete and has been compiled in accordance with the appropriate EU guidelines. This appraisal must be completed within 10 days, at which time (if the application is in order), it is given a filing date.

The sponsor also pays an appropriate fee. The EMEA then has 210 days to consider the application. In the case of human drugs, the application immediately comes before the CPMP (which convenes for 2-3 days each month). Two rappoteurs are then appointed (committee members who are responsible for getting the application assessed). The rappoteurs generally arrange to have the application assessed by their respective home national regulatory authorities. Once assessment is complete, the reports are presented via the rappoteur to the CPMP.

After discussion, the CPMP issue an '*opinion*' (i.e. a recommendation that the application be accepted, or not). This opinion is then forwarded to the European Commission, who have another 90 days to consider it. They usually accept the opinion and 'convert' it into a decision (they – and not the EMEA directly – have the authority to issue a marketing authorization). The single marketing authorization covers the entire EU. Once granted it is valid for 5 years, after which it must be renewed. The total time in which a decision must be taken is 300 days (210 + 90). The 300-day deadline only applies in situations where the CPMP do not require any additional information from the sponsor. In many cases, additional information is required; when this occurs, the 300-day 'clock' stops until the additional information is received.

Products of Biotechnology

The mutual recognition procedure is an alternative means by which a marketing authorization may be sought. It is open to all drug types except products of biotechnology. Briefly, if this procedure is adopted by a sponsor, then the sponsor applies for a marketing licence not to the EMEA, but to a specific national regulatory authority (chosen by the sponsor). The national authority then has 210 days to assess the application. If adopted by the national authority in question, the sponsor can seek marketing licences in other countries on that basis. For this bilateral phase, other states in which marketing authorization are sought have 60 days in which to review the application.

The theory, of course, is that no substantive difficulty should arise at this stage, as all countries are working to the same set of standards as laid down in 'The Rules Governing Medicinal Products in the European Union'. A further 30 days is set aside in which any difficulties that arise may be resolved. The total application duration is 300 days. If one or more states refuse to grant the marketing authorization (i.e. mutual recognition breaks down), then the difficulties are referred back to the EMEA. The CPMP will then make a decision ('opinion'), which is sent to the European Commission. The Commission, taking into account the CPMP opinion, will make a final decision that is a binding.

Consumers of Pharmaceutical Products

The Japanese are the greatest consumers of pharmaceutical products per capita in the world. The Ministry of Health and Welfare in Japan has overall responsibility to implement Japanese pharmaceutical law. Within the department is the Pharmaceutical Affairs Bureau (PAB), which exercises this authority. There are three basic steps in the Japanese regulatory process:

1. Approval ('shonin') must be obtained to manufacture or import a drug;
2. A licence ('kyoka') must also be obtained;
3. An official price for the drug must be set.

The PAB undertakes drug dossier evaluations, a process that normally takes 18 months. The approval requirements/process for pharmaceuticals (including biopharmaceuticals) are, in broad terms, quite similar to the USA. The PAB have issued specific requirements for submission of recombinant protein drugs. Applications are initially carefully checked by a single regulatory examiner to ensure conformance to guidelines. Subsequently, the application is reviewed in detail by a subcommittee of specialists. Clear evidence of safety, quality and efficacy are required prior to approval. In addition, the Japanese normally insist that at least some clinical trials be carried out in Japan itself. This position is adopted due to, for example, differences in body size and metabolism of Japanese, compared with US and European citizens. Also, the quantity of active ingredient present in Japanese drugs is lower than in many other world regions. Hence, trials must be undertaken to prove product efficacy under intended Japanese usage conditions.

CONCLUDING REMARKS

There is a growing trend in the global pharmaceutical industry towards internationalization. Increasingly, mergers and other strategic alliances are a feature of world pharmaceutical activity. Many companies are developing drugs that they aim to register in several world regions. Differences in regulatory practices and requirements in these different regions considerably complicate this process. Development of harmonized requirements for world drug registrations would be of considerable benefit to pharmaceutical companies and to patients, for whom many drugs would be available much more quickly. The International Conference on Harmonization (ICH) process (of technical requirements for registration of pharmaceuticals for human use) brings together experts from both the regulatory authorities and pharmaceutical industries based in the EU, USA and Japan.

The aim is to achieve greater harmonization of technical/guidelines relating to product registration in these world regions, and considerable progress has been made in this regard over the last decade. The 'members' comprising the ICH are experts from the European Commission, the European Federation of Pharmaceutical Industries Association, the Ministry of Health and Welfare, Japan, the Japanese Pharmaceutical Manufacturers Association, the FDA and the Pharmaceutical Research and Manufacturers of America. The ICH process is supported by an ICH secretariat, based in Geneva, Switzerland. The drug discovery and development process is a long and expensive one. A wide range of strategies may be adopted in the quest for identifying new therapeutic substances.

Most biopharmaceuticals, however, have been discovered directly as a consequence of an increased understanding of the molecular mechanisms underlining how the body functions, both in health and disease. Before any newly discovered drug is placed on the market, it must undergo extensive testing, in order to assure that it is both safe and effective in achieving its claimed therapeutic effect. The data generated by these tests (i.e. preclinical and clinical trials), are then appraised by independent, government-appointed regulatory agencies, who ultimately decide whether a drug should gain a marketing licence. While the drug development process may seem cumbersome and protracted, the cautious attitude adopted by regulatory authorities has served the public well in ensuring that only drugs of the highest quality finally come on to the market.

8

Onychopathology

Nails have long been recognized as indicators of systemic health problems by dermatologists, and the study of nails in this field is known as *onychopathology*. Disease states as diverse as cardiac failure and leprosy can affect the pathology of nails. Nails will also be affected from the ingestion of a wide variety of exogenous agents including antibiotics, chemotherapeutic drugs, heavy metals, and antimalerial drugs. As a matrix for detecting drug ingestion, nails are best associated with arsenic poisoning and the presence of Mees' lines. Nails have also been specifically analyzed for transition metals in occupational and environmental exposure cases. In addition, nail analysis for therapeutic drug monitoring is practiced in the clinical chemistry arena.

With the advent of hair testing for drugs-of-abuse, nail analysis was the natural progression for forensic toxicologists and was first reported in 1984 for the detection of methamphetamine abuse. To date, the application of nail analysis for drugs-of-abuse has been described in antemortem and postmortem toxicology, and in prenatal drug exposure. This chapter will review the analysis of nails for drugs-of-abuse and their utility in forensic toxicology.

HISTOMORPHOLOGY OF NAILS

Before the utility of nails in forensic toxicology can be evaluated, it is necessary to understand their anatomy. Nails perform two functions: to protect and to enhance the sense of touch of the fingertip. There are six major components of the nail, namely the germinal matrix, the lunula, the nail bed, the hyponychium, the nail plate, and the nail folds.

Nail Root

The germinal matrix of the nail, or the nail root, is the site of the majority of nail production. It lies beneath the skin, forms the floor of the proximal nail groove, and is protected by the proximal nail fold. The germinal matrix produces keratin cells that constitute the nail plate. As new keratin cells are formed, they push the older ones out through the cuticle where they flatten, undergo cytoplasmic condensation, and form the translucent nail plate. The germinal matrix contains few melanocytes, cells responsible for producing melanin, and hence nails lack pigmentation. (The pink color seen in the nail plate is actually the blood supply to the nail bed beneath the nail plate.) The size and shape of the germinal matrix determines the shape and thickness of the nail plate: the longer the matrix, the more cells it can produce and the thicker the nail. The shape of the matrix varies between individuals. A curved matrix will produce a curved nail plate while a flat matrix generates a flat nail.

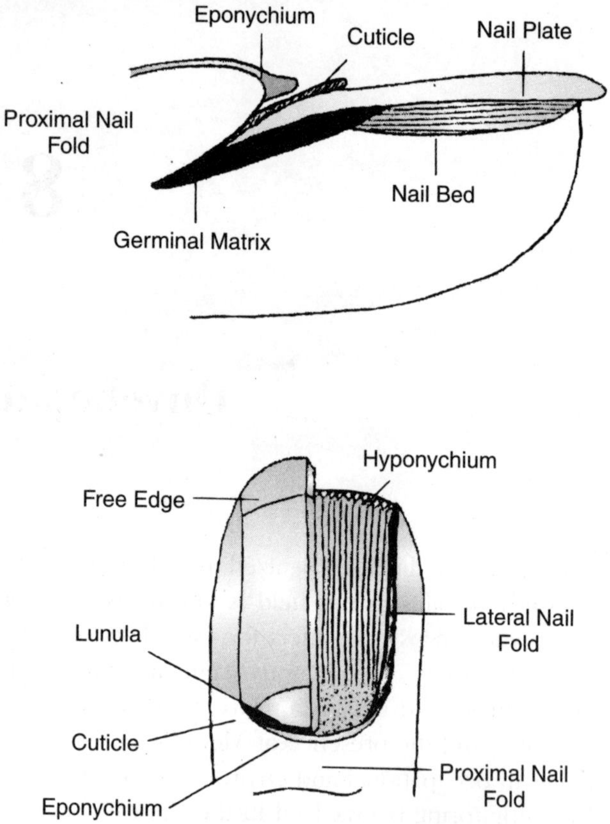

Fig. 8.1. Schematic of anatomy of nail.

Half Moon Shaped Area

The *lunula* is the white, half-moon-shaped area seen at the base of the nail. It is most prominent in the thumbnail and is where the matrix extends beyond the proximal nail fold. It is hypothesized that the white color is a result of light reflection or of a zone of incomplete keratinization.

Nail Bed

The *nail bed* extends from the lunula to the hyponychium. It is the flat surface to which the nail plate adheres, is very vascular, and it consists of longitudinal epidermal ridges and dermal papillae.

Hyponychium

The area under the nail plate, where the nail bed ends and the normal epidermis begins, is referred to as the hyponychium. It is the area between the free edge of the nail plate and the fingertip, and it acts as a waterproof barrier.

Broadest Part of Nail

The *nail plate* is the hard, translucent material known more simply as the nail. It is composed of interlocking, dead, keratinous cells that lack nuclei and organelles. The cells are so tightly knit they form a smooth surface that is not conducive to exfoliation, unlike the skin. The nail plate is approximately 0.5 mm thick, is generally thicker in men than women, and thickens with age. The nail plate at the free edge is thinner than the plate over the nail bed. It is the nail plate that will be discussed as an alternate biological matrix for use in forensic toxicology.

Nail Folds

The nail plate sits on the nail bed in three groves made by three folds of skin: the *proximal nail fold* and *two lateral nail folds*. The nail folds act as protective barriers against bacteria for the matrix and nail bed. The nail plate grows from under the proximal nail fold, which is an extension of the epidermis and dermis and contains sweat glands. The epithelium of the proximal nail fold, called the eponychium, attaches to the nail plate and moves with it as it grows. The cuticle is a build-up of dead skin cells that are sloughed off from the underside of the proximal nail fold as the nail plate grows out from under it and also attaches to the nail plate. Cosmetically, the eponychium and the cuticle are pushed back during a manicure.

Growth Rates

Fingernails grow at an average rate of 0.1 mm per day (3 mm per month; range 1.9-4.4 mm per month). It takes about 2 months for the new nail to grow the 5 mm out from under the proximal nail fold and about 6 months to completely replace a nail that has been removed. Toenails grow at about one-third the rate of fingernails. Consequently, it takes from 12 to 18 months for a new toenail to grow out. The rate of growth varies between individuals and even between digits. Generally, the rate of nail growth is proportional to the length of the digit, the "pinkie" fingernail being the slowest followed by the thumbnail.

The nail on the index finger tends to have the fastest growth rate. Nails grow faster in children, their peak growth rate (up to 0.15 mm per day) occurring between the ages of 10-14 years and then slow down significantly in the second decade. With slower growth rates, the cells in the nail plate are larger resulting in thicker nails. This correlates with an increase in thickness of the nail plate with increased age. Nails grow faster in the dominant hand and in nail biters due to external stresses and in males. Other factors such as the weather, illness, diet, and physiologic changes such as pregnancy will also determine growth rates.

Nail Formation

It was once widely accepted that nail is formed only at the germinal matrix. A second school of thought suggests that about 20% of the total nail mass is continuously being generated as the nail grows out over the nail bed. Studies have shown that the nail plate increases in thickness from the proximal to distal end at a rate of 0.02 mm/mm (0.22% per each 1% increment in length). This latter theory would explain how a moving nail can remain attached to the nail bed and the kinetics of drug incorporation into the distal nail when the presence of an antifungal drug is detected in the distal nail plate long before it should if the drug was only incorporated during nail formation in the germinal matrix.

ACCUMULATION OF DRUGS IN NAIL

As there is debate as to where and precisely how the nail plate is grown, the mechanism of drug incorporation in the matrix is still unclear. It is probable that there are at least two mechanisms of drug incorporation corresponding to the two mechanisms of nail formation. Thus, drugs become embedded in the nail plate from the blood supplying the germinal matrix during keratinization and from the vasculature in the nail bed that also contributes to nail formation. Other routes of drug incorporation must also be considered, such as from the sweat generated in the proximal nail fold that bathes the nail plate as it passes through it during growth and longitudinal diffusion from the proximal to distal end. Because of the difficulty of administering known amounts of drugs-of-abuse to subjects for the purposes of determining pharmacokinetics, much of the information regarding drug incorporation into nail comes from the therapeutic drug monitoring literature. There does appear to be a direct correlation between the concentration of an antifungal drug that is incorporated into nail and the dose, which also correlates with an improved cure rate.

The dose-concentration relationship is also evident with cocaine and haloperidol. The correlation is less evident, however, with codeine and methadone although the latter information relied upon self-report, a notably inaccurate measurement. In addition, incorporation of drugs appears to reach steady state with chronic use. After a therapeutic oral dosing study of the antifungal drug, itraconazole, the drug was detected in the distal nail clippings of the toenail up to 9 months and in the fingernail clippings up to 6 months after the end of dosing. A similar finding was observed with terbinafine. This may exemplify the permanent incorporation of drugs into the nail matrix and their stability once present. In paired hair and nail studies, nail contained a smaller mean peak concentration of drug for terbinafine, haloperidol, cocaine, and codeine per gram compared to hair although amphetamines tend to be less consistent in this observation. For example, Suzuki et al. discovered that methamphetamine was present at higher concentrations in nail than in hair approximately 50% of the time, Cirimele et al. found that amphetamines were present in approximately equal concentrations in fingernail scrapings and a paired hair sample, and Lin et al. demonstrated hair specimens contain higher methamphetamine and amphetamine concentrations than fingernail clippings.

The difference in the concentration of drugs in nail compared to hair could be due to differences in the mechanism of incorporation of drugs between hair and nail, the different growth rates of the matrices, sampling the correct sections of paired nail and hair that corresponded to drug use, or differences in the structure of the matrices; nails contain no melanin, which has been shown to influence the concentration of drugs detected in hair: the more melanin present, the more drug is detected. A study of methamphetamine disposition in nails compared to the corresponding hair segments of drug abusers undergoing treatment revealed evidence that drugs are deposited along the entire nail plate and not exclusively at the root. Other factors such as the physiochemical nature of the drug, its molecular weight, lipophilicity, dissociation constant, amount of protein binding, bioavailability, and route of administration are important in its ability to be incorporated into nail.

Influence of Environment

Exposure to external sources of drug such as from the environment, sweat, sebum, and saliva cannot be dismissed as viable mechanisms of drug incorporation into nail. Pounds et al. suggested that the small concentration of arsenic observed in the fingernail clippings of volunteers given therapeutic amounts of arsenic approximately 1 month after ingestion was due to contamination of the ventral nail surface from sweat. This suggestion was made upon detecting the majority of the arsenic in the portion of nail that corresponded to nail formation at the root at the time of arsenic ingestion. Access through sweat was also the mechanism postulated during the oral administration of fluconazole for the treatment of onychomycosis (fungal infection of the nail). The antifungal drug was detected in the nail plate after only 8 h postdose and because of the high water solubility of fluconazole, incorporation through sweat appeared to be a reasonable hypothesis.

The results from one of the few controlled studies involving the disposition of drugs-of-abuse, opiates and cocaine, in hair and nails also support the hypothesis that drugs are transferred from bodily secretions, such as sweat, into these matrices. Incoroporation of drugs into nail from environmental contamination such as handling drugs, however, has been suggested to be minimal.

In a study to evaluate this, cocaine analytes and opiates were analyzed in the finger and toenails, and the corresponding nail washes, of decedents known to have used these drugs. As toenails are less likely to be subjected to environmental contamination than fingernails, one would expect to see significantly less drug present in the wash of the toenails compared to that found in the washes of the fingernails. This was not observed. The wash/nail ratios of drug present were similar for both finger and toenails for all analytes except benzoylecgonine. The higher benzoylecgonine finger/toenail wash ratio was explained by the authors as possible spontaneous hydrolysis of cocaine, stuck under the nails, to benzoylecgonine. The authors concluded that drug incorporation into nail from environmental contamination is probably not a significant mechanism for three reasons:

1. Toenails contain drugs despite their exposure to this type of contamination being less likely than fingernails;
2. The similar wash/nail ratios for finger and toenails suggest that environmental contamination is minimal; and
3. The ratio of norcocaine to cocaine in nails is higher than that observed in illicit cocaine, an observation that would not exist if cocaine had not been metabolized enzymatically to norcocaine.

Extrapolating from therapeutic drug monitoring studies, drug penetration from external sources may depend not so much on the lipophilicity or the state of ionization of the drug (although drugs were demonstrated to be absorbed to a lesser extent when ionized than unionized) as much as molecular size. Drugs with increasing molecular weight were observed to penetrate nail samples to a lesser extent than drugs with smaller molecular weights. This may correlate with the observation that the nail plate contains intercellular channels of finite size that leads to a certain permeability, especially to polar compounds that would otherwise be unable to pass the membranes.

In addition, as the dorsal surface of the nail plate is permeable, the medium that it is exposed to blood, sweat, sebum, saliva, water, urine, and the relative proportion of drugs contained therein will determine how much and which drugs-of-abuse and their metabolites will become incorporated in the nail through this mechanism. The extent of drug incorporation through any of the mechanisms suggested is unknown and probably varies considerably depending on the many confounding factors involved. These include the type of drug, the physical state of the individual, and environmental conditions.

DRUGS OF ABUSE IN NAIL

The first reported instance of nails being utilized for the purpose of detecting illicit drug abuse was in 1984 with the detection of methamphetamine and amphetamine in the nails of methamphetamine users. This was followed 5 years later by a second report of detecting methamphetamine and amphetamine in nails. This study also compared these findings with the hair, sweat, and saliva of habitual methamphetamine users. The authors concluded that the concentration of methamphetamine in nails predominated its metabolite, amphetamine, and was detectable for longer periods than in the saliva and sweat of the same individuals. It was not until the mid-1990s that nail analysis for drugs-of-abuse became more prevalent.

In 1994, Tiess et al. presented a report on the analysis for cocaine in the fingernail clippings and the entire nail plate of the toe in an acute cocaine intoxication victim. Tiess et al. found comparable amounts of cocaine in the fingernail clippings of both left and right hands, which were considerably greater than in the toenail. In addition, more cocaine was detected in the root of the toenail than the rest of the nail plate. From these observations and the results of the corresponding biological fluids and hair analysis, Tiess et al. concluded that the deceased had a chronic cocaine habit and had used the drug shortly before his death. In the same year, Miller et al. extracted cocaine and its primary metabolite, benzoylecgonine, from hair, and finger and toenails.

The findings demonstrated that cocaine and benzoylecgonine were present in nails at approximately equal concentrations but at lower concentrations than the corresponding hair. In addition to amphetamine, 3,4-methylenedioxymethamphetamine (MDMA) and 3,4-methylenedioxyamphetamine (MDA) were detected in the scrapings of a single fingernail and in the head hair of a drug abuser. Cirimele et al. found the concentration of each analyte to be approximately equal in both matrices, but the amount of MDMA detected was much higher than either amphetamine or MDA. MDMA was not present in the blood or urine from this individual. From these analyses, the authors were able to conclude that the individual tested was a chronic MDMA user, a fact not evident from the more traditional blood and urine testing. Nails have been proposed as an optimal specimen for the detection of intrauterine drug exposure. Unless authorities suspect the mother of drug abuse before the birth, meconium may not be collected in such cases. As the window of detection in blood and urine is very small, prenatal drug use may go undetected.

Nail formation begins *in utero* at the beginning of the second trimester. Any drugs absorbed by the mother during pregnancy should, therefore, be incorporated in the nail matrix through systemic exposure and through being bathed in amniotic fluid. While this is also true of hair, prenatal hair may be sparse and is lost in the perinatal period. Nails grown during gestation survive for a longer postnatal period than hair. Unless the baby is breast-fed, the presence of drugs in nail is strong evidence of prenatal exposure. The possibility of prenatal cocaine use was investigated by Skopp and Potsch, utilizing nail analysis in a suspected case of *sudden infant death syndrome* (SIDS). The case involved a 3-month-old baby where no anatomical or toxicological cause of death could be found using conventional methods. A slight heart abnormality in the infant was an indicator of possible maternal drug use.

Nail clippings from the fingers and toes were analyzed for cocaine and opiates in the absence of hair. Parent cocaine was detected. While the presence of drugs in the nails of a neonate may confirm intrauterine exposure, if one can exclude environmental contamination, it cannot be used on its own to determine the cause of death. This information may be useful, however, in identifying and monitoring high-risk children and in epidemiological research for SIDS. Toward the end of the 1990s, two studies on postmortem nail analysis were published. The first involved the study of toenails as a viable matrix for forensic toxicologists in postmortem cases. In this study, by Engelhart et al., cocaine and cocaine metabolites were analyzed simultaneously with opiates, and the results were compared to those of more conventional biological specimens.

The majority of the nail specimens that tested positive for cocaine and/or opiates had negative toxicology results in the corresponding biological fluids. These results showed that toenails could be utilized by forensic toxicologists to gather information about a deceased person and that while recent drug use may not be reflected, a past or chronic use history could be deduced. Garside et al. published the second of these studies, which investigated the presence of cocaine and its metabolites in the finger and toenails of decedents suspected of cocaine use. The results revealed that 82% of the subjects tested were positive for cocaine analytes utilizing nails for the analysis. Only 30% of these cases were positive for cocaine analytes when blood or urine were the specimens of choice. Anhydroecgonine methyl ester, a marker of cocaine smoking, was one of the analytes detected in nail. Finally, although the authors also tested for *m*-hydroxybenzoylecgonine during this study, none was detected. In 1999, the first report for the detection of cannabinoids in nails was published by Lemos et al. Fingernail clippings from volunteers who consumed marijuana and were attending a drug clinic were analyzed.

During the study, it was determined that the extensive pre-wash procedure of the nails removed possible external contamination since at least one of the last three methanolic washes were negative for cannabinoids. Both RIA and GC/MS were used successfully for the detection of cannabinoids. It was discovered, however, that the resulting basic pH of the nail digest, which was ideal for the subsequent liquid/liquid extraction of Δ9-tetrahydrocannabinol (THC), was ineffective for the recovery of the carboxylic acid metabolite, 11-nor-Δ9-tetrahydrocannabinoinl-9-carboxylic acid (THC-COOH). To recover THC-COOH, the nail digest had to be made acidic before the solvent extraction of the metabolite. There was no recovery of the parent THC under the acidic conditions. In early 2000, the same authors who reported a method for the detection of cannabinoids in nail reported their findings of the usefulness of nail as a matrix in the detection of morphine in known heroin users.

The conclusions were such that nail may be utilized to detect morphine and can, therefore, be an alternate matrix to hair in detecting past heroin use. Also in 2000, with a method for cannabinoids and morphine successfully developed, Lemos et al. applied their techniques to the analysis of nail for methadone. Both EIA and GC/MS were effective in detecting methadone in the extracts of fingernail clippings from individuals participating in a methadone maintenance program. Nails may be an appropriate matrix, therefore, for monitoring individual compliance in methadone maintenance programs. The first paper to describe disposition patterns of drugs in nail was reported in 2000. They investigated dose-response relationships of cocaine and codeine in paired hair and nail specimens in a controlled dosing study. The study enrolled eight black males, as inpatients, who were administered codeine, and cocaine orally and by subcutaneous injection, respectively. The study was designed to simulate a short, multi-drug use session typical of regular users.

The design eliminated the possibility of environmental contamination. The authors reported that decontamination washes remove more drug out of nail than hair, a dose-response relationship does exist in both hair and nail for cocaine and to a lesser degree for codeine, the concentrations of codeine and cocaine were higher in hair than in nails, and drugs appear in hair and nail within 3 days of administration. The authors acknowledged that the method of nail collection (scraping the dorsal surface) may have influenced the results. In a study to evaluate the extent of drug incorporation into nail due to external contamination, Engelhart and Jenkins analyzed the content of finger and toenails for drugs-of-abuse and the corresponding nail washes.

Nail clippings, blood, and urine were collected from decedents who had a history of drug abuse and were analyzed for cocaine analytes and opiates. It was found that with both cocaine analytes and opiates, the concentration of drug in fingernail was significantly 7-20 and 30 times, respectively, higher than in toenails and the concentration of analytes in the washes of both finger and toenails was less than detected in the nails themselves. Through their observations of nail/ wash ratios of finger and toenails, the authors concluded that although external contamination cannot be excluded as a route of drug incorporation into nail, it is probably not a significant mechanism of drug incorporation into nail. In addition, it was observed that there was no correlation between nail and blood concentrations. The simultaneous detection of morphine, 6-acetylmorphine, and cocaine in toenails was reported in 2004 by Cingolani et al.

The concentrations of the drugs found in the proximal section of the big toenail were compared to those found in the corresponding proximal hair samples. Cocaine and morphine were present in toenails at approximately twice the concentration found in hair while 6-acetylmorphine was present at approximately equal amounts. Ragoucy-Sengler et al. were able to detect cocaine and cocaine analytes, including anhydroecgonine methyl ester, in the nails of two subjects known to smoke cocaine. Another recent study evaluated the disposition of drugs into nails. Lin et al. compared the concentration of methamphetamine in corresponding segments of nail and hair, estimated by the rate of growth of each matrix.

The analyses showed that if drugs are deposited into nail exclusively at the matrix, then the proportion of drug seen in the distal clippings when compared to the corresponding portion of hair would have a similar distribution pattern. Instead, after abstinence, the concentration of methamphetamine in the distal portion of the nail decreased similarly to the new growth of hair. It was suggested that drugs must be deposited along the length of the nail bed, whereas in hair they are deposited at the scalp (root).

Additional observations in this study included the concentration ratio of amphetamine to methamphetamine was the same for hair and nail, and the concentration of these analytes were greater in hair than in fingernails. Recently, the detection of phencyclidine (PCP) in nails from four homicide victims was reported by Jenkins and Engelhart. The corresponding blood and urine samples were also positive for PCP. Of PCP in the fingernails was generally greater than in the toenails.

ANALYZING NAILS

Analysis of nails may be described as discrete processes, namely, decontamination, preparation and extraction, clean-up, and detection.

Removal of Contamination

The first step in the analysis of nails involves washing the specimen to remove potential external contamination. Many different wash protocols have been designed including rinsing with hot or cold water, soap solution, buffer solution, and organic solvents such as methanol, isopropanol or methylene chloride, and occasionally utilizing sonication. The length of time for each wash step ranges from 15 s to 30 min. Controversy exists as to whether this step can remove all external contamination, and at what point it extracts internally incorporated drug.

Extraction of Analytes

Following decontamination, the analytes of interest must be extracted from the nail matrix similar to hair analysis. This may be performed by solvent extraction, or the matrix may be digested with an acid, a base, or enzymatically. Typically, the nail specimens are pulverized first or cut into small pieces to aid in the extraction process. The analytes to be extracted have to be taken into consideration during this step. For example, if cocaine is the analyte of interest, a basic digestion would be ill advised as it would result in hydrolysis to benzoylecgonine.

Cleaning

Once the analytes have been extracted from the matrix, the extract has to be cleaned to remove the remains of the matrix and other unwanted compounds. This is achieved by solid phase or liquid/liquid extraction. If required, derivatization of the isolated analytes also takes place for optimal chromatography.

Detection Devices

A variety of detection devices have been utilized in the analysis of nails. The most common technique is gas chromatography followed by mass spectrometry. *High-performance liquid chromatography* (HPLC) has also been employed. HPLC coupled with an electrochemical detector was used in the final step of the analysis for morphine in nail with a basic mobile phase comprised of phosphate buffer/isopropanol/tetrahydrofuran. HPLC is likely to be used more frequently in the future.

INTERPRETING RESULTS

Nail analysis has proved a useful tool for detecting drug exposure in both antemortem and postmortem specimens, especially in the absence of more conventional samples and even hair. The answers to the obvious questions of how much drug and when did exposure occur are elusive, however. The major obstacle is the relative paucity of disposition studies and lack of complete understanding of the mechanisms of drug incorporation. Of the studies that have been conducted, it is difficult to compare the findings because of the lack of standardized testing methods and sampling techniques. In addition, the quality of the results in the studies that have been conducted are hampered by unreliable self-reports of drug use, limited paired specimens, and lack of controlled dosing. It appears that drugs are incorporated along the nail bed from the lunula to the free edge in addition to the nail matrix and that external factors such as sweat and environmental contamination must be considered. This makes interpretation of timelines seemingly impossible. It is intuitative that toenails would be less exposed to environmental contamination than fingernails although they are probably bathed in more sweat than fingernails. Fingernails, although more likely to come into contact with drugs from external sources, are washed more frequently than toenails.

Growth rates will also vary, as discussed above, with age, sex, health, digit, and environment. One study demonstrated a large difference in drug disposition between individuals of the same sex and race, with controlled dosing, utilizing the same analytical method and minimizing environmental contamination. Although there appears to be a dose-response relationship for many drugs, there does not appear to be any correlation between blood concentrations and the concentration found in nail. Caution must be utilized in interpreting results of nail analysis to account for the possibility of external contamination or incorporation versus systemic exposure through ingestion or parenteral use. The presence of some drug metabolites is essential for distinguishing between these two scenarios. Cocaine, benzoylecgonine, and anhydroecgonine methyl ester could all be present from environmental contamination, but norcocaine and norbenzoylecgonine are indicative of internal metabolism.

APPLICATIONS

The most obvious advantage and probably the most utilitarian application of nail analysis is the slow growth rate and, therefore, long retrospective analysis that is possible. This is especially true of the big toenail that may potentially represent 12 months of exposure. Other advantages are the ease of and non-invasive collection of the sample (clippings), stability of the drug once incorporated in the matrix, which makes storage easy at room temperature, and the longevity of nail as a sample. Nail is preserved for thousands of years and is often the only viable specimen remaining on skeletonized remains. Not only can this be useful in contemperory forensic cases but also in historic investigations such as studying the lifestyle of ancient people through analysis of nails in mummies.

The small sample size required for nail analysis, typically 10–50 mg, is also attractive, and unlike more conventional samples, nails are difficult to adulterate. The exact mechanism(s) of incorporation of drugs into nails are unknown, which hinders interpretation much beyond qualitative answers and even then, without metabolites, systemic (ingestion or parenteral) exposure is ambiguous. 6-Acetylmorphine can be detected in nail, however, which allows for the determination of heroin exposure compared with morphine or codeine and provides an alternate matrix if urine is unavailable to make this definitive interpretation. The presence of cocaethylene and anhydroecgoninemethyl ester also provides additional information. The absence of more information and a proficiency testing program is a distinct disadvantage to the reliability of the nail analysis test results.

9

Clinical Study of Oral Fluid

Saliva is the term used to describe the liquid excreted from more than 14 saliva glands of the head and mouth. When mixed saliva from the various glands is collected by expectoration or by placing absorbent collectors in the mouth, the resulting specimen is termed "*oral fluid*" and may contain blood, lymph, crevicular fluid, and cells from the gums and cheeks in addition to mixed saliva. The use of oral fluid as a diagnostic tool has increased in recent years due to significant improvements in instrumental sensitivity and through a greater under-standing of the mechanisms of drug transfer into the oral cavity. Oral fluid is recognized as a robust alternative testing matrix with numerous applications including monitoring for infectious diseases, monitoring compliance in drug treatment programs, and employee testing.

WORK DONE

Saliva or oral fluid has long been of interest as a non-invasive specimen that would mirror blood levels of drugs and hormones. Saliva was collected as a specimen matrix for therapeutic drug monitoring of antiepileptic drugs in children to avoid repeated blood draws in these young patients. It is a preferred specimen for studies that require repeated specimens over an extended period, such as pharmacokinetic studies, to minimize trauma to the patient. However, the drugs of interest in monitoring are mostly acidic (*antiepileptic drugs*) and/or strongly protein bound, or have very low saliva/plasma ratios.

Therefore, resultant saliva concentrations were below the limit of detection of analytical methods available at the time. In 1984, Peel et al. of the Royal Canadian Mounted Police published a study of oral fluid testing for drugs involved in driving under the influence offenses. As many illegal drugs have saliva/plasma ratios greater than 1, the oral fluid concentrations are greater than those in blood and provide the advantage of enhanced concentrations as well as the potential for correlation with impaired driving. Saliva is recognized as an advantageous specimen for roadside DUI testing and criminal justice testing for drugs-of-abuse.

CHEMICAL MAKE-UP OF ORAL FLUID

The largest glands and main source of saliva are the parotid, submandibular (submaximal), and sublingual glands. The parotid gland produces approximately 25% of resting saliva and 50% of stimulated saliva and produces only serous saliva. The submandibular gland (70% of resting saliva) secretes both serous and mucus saliva. The sublingual, labial, and palatal glands secrete 70% of the mucins. Saliva is composed of 99% water, 0.7% protein (largely amylase), and 0.26% mucins. Human saliva contains two mucins: oligomeric mucous glycoprotein MG1 and monomeric mucous glycoprotein MG2. As was first recorded by Pavlov, the mucin composition of saliva depends on the nature of the stimulus for saliva secretion and which glands respond.

Mechanical stimuli produce serous saliva; food stimuli produce thicker, mucous-containing saliva. In general, parasympathetic stimulation of the parotid gland leads to high rates of fluid output while sympathetic stimulation of the sublingual gland leads to high levels of protein secretion. Salivary glands are composed of two regions: the acinar region that contains the water-permeable cells that are capable of secretion and the ductal region lined with water-impermeable cells that carry the secretions to the outlets in the mouth. Sodium and exocrine proteins are secreted by the secretory cells in the acinar region, and water passes into the lumen of the gland due to the osmotic pressure. The fluid that collects in the acinar lumen is isotonic with plasma.

As the fluid travels down the saliva ducts, sodium and chloride are reabsorbed while potassium, lithium, and bicarbonate are secreted. When the fluid reaches the mouth, it is hypotonic to plasma. Resting saliva has greater acidity than plasma. When saliva moves rapidly through the ducts, less time is available for sodium reabsorption and carbonate secretion, therefore the pH of the saliva is higher. After secretion, saliva becomes more alkaline as the dissolved carbon dioxide is lost. Saliva pH ranges from 5.6 to 7.9. In healthy donors, gingival crevicular fluid from the tooth/gum margin may constitute up to 0.5% of oral fluid collected and is of similar composition to that of plasma.

SAMPLING TECHNIQUE

Saliva can be collected using small flexible plastic cups that are applied over the opening of the saliva gland into the mouth. Collection in this manner can demonstrate the differences, if any, between drugs in saliva and drugs in oral fluid. By definition, oral fluid is collected by absorbents placed into the mouth, by draining saliva from the mouth, or by expectoration (spitting). Oral fluid collected by draining or spitting is collected in a cup, vial, or test tube. A number of commercial devices employing absorbent materials are available for oral fluid collection.

Absorbent collectors may be a dental cotton roll (Salivette) or plastic material placed in the mouth that is then removed and centrifuged or compressed to remove the oral fluid. More acceptable to donors are collectors consisting of an absorbent pad attached to a plastic straw or rod that the donor can place in their own mouth and then remove themselves to hand to the collector. The pad is usually placed in a preservative buffer after collection and oral fluids and/or drugs eluted by mixing with the buffer. Some devices have used an absorbent material to wipe the tongue to collect oral fluid and cells.

Deposition of Drug in Oral Fluid

Drugs such as digoxin, steroids, and antibiotics (penicillin and tetracycline) are actively secreted into saliva. Most drugs enter saliva by passive diffusion across the cell membranes. Saliva concentrations of drugs that cross the cells by passive diffusion are related to blood or plasma concentrations of the unbound, unionized parent drug or its lipophilic metabolites. The theoretical saliva plasma ratio (*S/P* ratio) can be calculated from the following equation derived from the Henderson-Hasselbach equation:

For basic drugs:

$$S/P = \frac{[1 + 10(pKb - pHs)]}{[1 + 10(pKb - pHp)]} + \frac{fp}{fs}$$

For Acidic Drugs:

$$S/P = \frac{[1 + 10(pHs - pKa)]}{[1 + 10(pHp - pKa)]} + \frac{fp}{fs}$$

S/P is the saliva to plasma ratio: *S* is the drug concentration in saliva and *P* is the drug concentration in plasma, pKb is the log of the ionization constant for basic drugs, pKa is the log ionization constant for acidic drugs, pHs is the pH of saliva, and pHp is the plasma pH. fp is the fraction of drug protein bound in plasma and fs is the fraction protein bound in saliva. Drugs have been classified according to their S/P ratio.

Acidic drugs (pKa less than 5.5) and/or those that are highly protein bound generally have a S/P ratio of less than 1.0. Examples are benzodiazepines, barbiturates, and cannabinoids. Neutral drugs (pKa less than 8.5 but greater than 5.5) and alcohol have an S/P ratio of approximately 1.0 that does not vary with rate of saliva formation or conditions of collection. The third group includes drugs with a S/P ratio greater than 1. These drugs are found in higher concentrations in oral fluid than in plasma. This group includes drugs that are actively transported into oral fluid such as digoxin and penicillin as well as basic drugs (pKa greater than 8.5) and drugs with low protein binding that are ionized at saliva pH and ion trapped in the oral fluid. Drugs-of-abuse can be classified into those that enter oral fluid by passive diffusion and those that enter oral fluid from depots in mouth tissues.

In general, drugs that enter the oral fluid from depots in the oral tissues are found in higher concentrations than expected from their theoretical S/P ratios calculated from the Henderson-Hasselbach equation. Drugs that are abused are often smoked, creating substantial oral tissue depots. S/P ratios elevated more than a 100-fold after smoking have been reported for cocaine, heroin, cannabis, methamphetamine and phencyclidine (PCP). Drugs that are administered by sublingual absorption, such as fentanyl or buprenorphine, and those given as liquid preparations (codeine, morphine, and methadone) would be expected to form oral tissue depots and therefore have elevated S/P ratios after oral dosing.

Drugs normally taken in tablet form can also have elevated S/P ratios if the pill material is held in the mouth or if pill fragments remain in the mouth. This has been reported for "*ecstasy*" and for phenytoin. Insufflated (snorted) drugs, for example, cocaine and heroin, also form tissue depots bathed by oral fluid.

Influence of Drug on Oral Fluid

Oral fluid has been collected from donors by expectoration, draining, absorption, and suction. During collection, saliva production can be stimulated by citrate salts or acidic sour candy or by mechanical means such as chewing paraffin or rubber bands. The presence of an absorptive collector in the mouth will stimulate saliva flow. The stimulation method will influence the mucin content and pH of the saliva. By stimulating saliva flow and increasing the saliva pH, the drug content of oral fluid is expected to be reduced for basic drugs as per the Henderson-Hasselbach equation. Collection devices can also affect recovery of drugs from oral fluid due to retention of oral fluid in the absorbent, adsorption of the drug on the device components, and the effectiveness of device buffers to release drugs from the collector.

After controlled administration of codeine, oral fluid codeine concentrations were higher in specimens obtained by expectoration or draining of unstimulated oral fluid than those obtained by absorptive devices and were higher in specimens collected without stimulation than in those with stimulation of saliva flow. Kato et al. compared cocaine oral fluid concentrations in unstimulated to stimulated saliva. Unstimulated saliva collected by expectoration contained on average 5.2 (range 3.0-9.5) times as much cocaine compared with saliva collected under stimulated conditions (citric acid sour candy). The mean ratios of unstimulated to stimulated area under the concentration-time curve for benzoylegconine and ecgonine methyl ester were 6.0 (range 3.3-9.0) and 5.5 (range 2.5-8.8), respectively.

EXPERIMENTAL PROCEDURE

There is a lack of data published on the subject of drug stability in biological matrices although it is widely recognized that drug levels can decrease over time. The degradation of drugs can be attributed to a number of factors including storage conditions, microbial action, and the adherence of the drug to the surfaces of the sample container. Current practices in oral fluid testing involve transport of the sample after collection for several days at varying temperatures and a significant length of time may pass between an initial presumptive screening test and a confirmation test required prior to the case going to court. Manufacturers of oral fluid collection devices have attempted to address this issue with the addition of buffers, to release the sample and drugs from the collection device, and preservatives, to stabilize the drugs in oral fluid.

Handling Samples

Oral fluid samples diluted with buffer do not require any special pretreatment for immunoassay screening and can also be analyzed directly by *liquid chromatography-mass spectrometry* (LC-MS). *Solid-phase extraction* (SPE) or *liquid-liquid extraction* (LLE) is recommended to reduce matrix interference. Subjecting the collected oral fluid samples to several freeze-thaw cycles followed by centrifugation can further reduce interference effects from mucins.

Screening Tests

Most screening tests for drugs-of-abuse are immunoassays. Antibodies used in immunoassays for detection of drugs in saliva must cross-react with the parent drug and lipophilic metabolites. For example, heroin and 6-acetylmorphine (6-AM), cocaine and ecgonine methyl ester, and Δ-9-

tetrahydrocannabinol (Δ-9-THC) predominate in saliva. When drugs are leached into saliva from buccal depots such as is the case for smoked drugs, such as marijuana, parent drug and pyrolysis products will predominate in saliva.

Immunoassays with cross-reactivity to free morphine and to 6-AM are most useful for the detection of opiates in oral fluid. Immunoassays that have been developed to detect the hydrophilic metabolites of drugs in urine will not be appropriate for saliva screening. For effective evaluation of commercially available immunoassays for oral fluid, samples should be collected from both field and clinical trials and compared with a recognized reference method, for example, gas chromatography (GC)-MS. Controlled administration studies are immensely important for establishing a greater under-standing of pharmacokinetics while the analysis of samples from known drug users provides important information on expected drug profiles. With advancements in technology, a number of researchers have investigated the use of LC-MS for multiple-drug screening as an alternative to immunoassays. The main advantage of this technique is the use of a small sample volume coupled with the separation and identification power of the LC-MS. Immunoassays provide greater flexibility with respect to reduced sample handling and reduced costs for high-throughput testing laboratories.

Immunoassays

A number of on-site or point-of-care testing (POCT) devices have been introduced for screening drugs in oral fluid. The technology used in the majority of these devices is based on lateral flow of the oral fluid sample across a membrane impregnated with lines of labeled immobilized drug. These disposable cartridges are available for a range of drug groups including amphetamines, cocaine, opiates, PCP, and cannabinoids. The collection of oral fluid for on-site testing has provided a non-invasive alternative to blood or urine samples with application at the roadside, in clinical settings, and custody suits. An early study conducted in 1999 to assess on-site tests by the roadside (ROSITA - Roadside) evaluated three devices: the Securetec Drugwipe, the Avitar OralScreen, and the Cozart RapiScan. The authors concluded that in general the sensitivity of the devices was not sufficient for low-drug concentrations seen for benzodiazepines and cannabinoids and the average total time for sampling and analysis was considered too long at 20 min (range 13-33).

A follow-up to this study, ROSITA 2, was started in 2003 to evaluate on-site devices by the roadside as before but to now include new and improved devices that had not been available for the initial study. Evaluation of the devices was extended to include use by police officers. The findings of the latest study were that six of the nine devices evaluated had an unacceptable number of failures (greater than 25%). In addition, the sensitivity and specificity of the devices did not meet the minimum specifications and no single device could be recommended for use at the roadside. The final report is available on the ROSITA website. The Cozart RapiScan (CRS) oral fluid drug testing system was the first fully integrated hand-held system for the detection of drugs. The CRS has been used successfully in police custody for testing individuals for use of opiates and cocaine who have been charged with a "*trigger offence*," for example, theft and drug offences. The CRS has also been used successfully for monitoring heroin addicts and detecting codeine and cocaine in controlled administration studies.

Chromatography with Mass Spectrometry

Chromatography coupled with MS is the technique of choice for confirming the presence of drugs in biological matrices. GC-MS and LC-MS are routinely used in analytical laboratories as they provide definitive identification of analytes of interest. The confirmation of drugs in oral fluid has proven a challenge to toxicologists due to limited sample volumes available for analysis and low cutoff concentrations. In response, a number of papers have been published recently employing tandem-MS to achieve greater sensitivity with small sample volumes.

LC-MS-MS publications have been far greater with methods for amphetamines, benzodiazepines, and cannabinoids while GC-MS-MS methodologies predominantly involved cannabinoids and opiates. The use of two-dimensional GC coupled to MS (GC-GC-MS) was reported for the sensitive analysis of 11-nor-delta-9-tetrahydrocannabinol-9-carboxylic acid (THC-COOH) in hair. The application of this method to the analysis of drugs in oral fluid could provide an alternative approach for achieving the required sensitivity without the cost implications associated with tandem MS.

CLINICALLY REPORTED DRUGS IN ORAL FLUID

Amphetamines

Amphetamine, methamphetamine, 3,4-methylenedioxymethamphetamine (MDMA), 3,4-methlyenedioxyamphetamine (MDA), and other amphetamine class drugs have saliva to plasma ratios (S/P ratios) of greater than one, ranging from 2 to 20. Saliva/plasma ratios of 2-4 during the elimination phase have been reported for amphetamine, and for methamphetamine and 3-20 for MDMA. When 10 mg oral doses of amphetamine were given to volunteers, oral fluid concentrations of 10-60 ng/mL, peaking about 5 h after dosing (range 2-10 h depending on pH) were reported. When amphetamine was administered to subjects as a racemic mixture, both D and L isomers were found in saliva.

After sustained release oral doses of methamphetamine, peak concentrations in oral fluid ranged from 14.5 to 33.9 ng/mL (10 mg) and 26.2 to 44.3 ng/mL (20 mg) and occurred within 2-12 h. Kintz reported *N*-methyl-1-(3,4-methylenedioxyphenyl)-2-butanamine (MBDB) and its metabolite (BDB) in oral fluid for 17 h after oral administration of 100 mg MBDB (concentration range 14-1045 ng/mL). The window of detection for methamphetamine in oral fluid has been reported from a minimum of 24 h after use to as long as 50 h.

Cannabinoids

Concentrations of the active ingredient of marijuana, Δ-9-THC, in oral fluid correlate well with Δ-9-THC concentrations in blood. As little THC, and very little THCA, is secreted in saliva this is most likely because of oral mucosa depots of cannabinoids as the source of Δ-9-THC in both oral fluid and serum. Oral fluid concentrations of Δ-9-THC range from 1 to 400 ng/mL. Teixeira et al. reported that concentrations of THC in specimens obtained by direct spitting ranged from 50 to 6552 ng/mL while oral fluid retrieved from Salivette collectors by centrifugation from the same subjects at the same time contained from 0 to 134 ng/mL Δ-9-THC. Oral fluid specimens were positive at a cutoff of 1 ng/mL for an average of 5.7 h after smoking a 1.75% THC cigarette and for 8.8 h after a 3.55% Δ-9-THC cigarette. Others report oral fluid to be positive after marijuana smoking for 2–10 h. Niedbala et al. reported that oral fluid specimens could be positive after passive exposure to marijuana smoke for up to 30 min after exposure.

A follow-up study investigated the role of environmental contamination from the exposure of the collector to the cannabis smoke during the study. Sample collection was carried out in a smoke-free environment in contrast to the initial study. All passive subjects were negative at screening/confirmation cut-off concentrations throughout the study. The authors concluded that the risk of a positive test for THC was virtually eliminated when specimens were collected in the absence of THC smoke.

Cocaine

Cocaine and cocaine metabolites are found in oral fluid. The S/P ratio for cocaine ranges from 0.5 to 10. The S/P ratio for benzoylecgonine ranges from 0.5 to 2.5, for ecgonine methyl ester from 2.3 to 5.1, for norcocaine 5.6 to 13.6, and for *para*-hydroxycocaine 2.4 to 10.8. Cocaethylene is found in saliva when ethanol is ingested concurrently with cocaine (S/P ratio ~1). After smoking cocaine, oral fluid concentrations of cocaine are elevated and the pyrolysis product of cocaine, anhydroecgonine methyl ester (AEME), is detected in oral fluid for up to 6 h but not in plasma. However, the elevated saliva/plasma ratios of cocaine due to contamination of oral fluids clears within 2 h after smoking cocaine. Terminal half-lives in oral fluid are 7.9 h for cocaine, 9.2 h for benzoylecgonine, and 10 h for ecgonine methyl ester. The detection time for cocaine in oral fluid ranges from 4 to 8 h after intravenous injection, 5 to 12 h after insufflation, and 12 h to 16 days after smoking cocaine.

Opioids

Although all opioids may be subject to abuse, the principle illegal drug is heroin. The S/P ratios after intravenous administration of heroin are 0.1-1.9 for heroin, 0.7-7.2 for 6-AM, and 0.5-1.8 for morphine. The presence of opiates in saliva after intravenous administration of heroin and intramuscular administration of morphine has been reported for up to 24 h. After 10 and 20 mg of heroin is administered intravenously, diacetylmorphine (6-20 ng/mL) is found in oral fluid for the first hour [terminal half life 8.5-582 min] and then the primary metabolite of heroin, 6-AM (18-131 ng/mL) and morphine (10-16 ng/mL) are found for 4-8 h. A longer window of detection would be expected where tolerant individuals use larger amounts of heroin. Heroin may be smoked. Elevated S/P ratios (heroin 20-400, 6-AM 8-350, morphine 2-30) and higher drug concentrations (10-20,000 ng/mL heroin, 10-4000 ng/mL 6-AM, and 2-150 ng/mL morphine) may be found in oral fluid after smoking heroin. After smoking, heroin was found in oral fluid for up to 24 h and 6-AM and morphine for up to 8 h.

Phencyclidine

PCP has a theoretical S/P ratio of greater than 1 (pKa 9.4, plasma protein binding <10%). An *S/P* ratio of 1.5-3.0 was found in expectorated oral fluid after oral (1 mg) and intravenous (0.1 or 1 mg) administration of radiolabeled PCP to healthy male volunteers. PCP and metabolites were detectable in saliva for 75 h. McCarron et al. reported PCP concentrations ranging from 2 to 600 ng/mL in saliva samples from 100 patients suspected of PCP intoxication.

EFFECTIVE INTERPRETATION OF DRUGS

There are a number of factors that must be considered for effective interpretation of drugs in oral fluid. These include sample collection, dilution, and storage; the relationship between saliva and plasma for determining impairment; and the contribution of buccal drug depots and passive exposure. Sample collection is the most important step in the process of oral fluid testing and can have a significant impact on the drugs detected. Collectors that cause excessive stimulation of saliva will reduce drug levels, and many sample collection devices have significant variability in the volume collected. Crouch reported percentage volumes recovered between 18 and 83%, while Dickson et al. reported excellent sample recoveries (CV < 10%) but highlighted the potential increase in error associated with incorporating a dilution step with the addition of preservative buffers.

The use of a sample adequacy indicator will reduce the impact of the use of sample buffers. A recent paper by Moore et al. described losses for the Quantisal oral fluid collection device between 10 and 20% for Δ-9-THC when stored in the refrigerator and at room temperature, respectively. Crouch reported losses of THC of greater than 40% with the Intercept when stored under similar conditions. THC remained relatively stable in samples stored at –20 °C. Dickson et al. investigated the effects of storage conditions (5°C and room temperature) on drug stability and recovery using collection devices supplied by three different manufacturers. The study was restricted to investigating the recovery of THC, benzodiazepines, methamphetamine, and morphine. They concluded that storage at 5°C or room temperature had no significant effect on drug recoveries. The variability of materials and preservatives used by different manufacturers will have a significant impact on the stability of drugs in oral fluid as demonstrated by the difference in THC stability for the two collection devices discussed previously.

The impact of the collection device on drug stability must be assessed when interpreting an analytical result. As it is the free lipophilic drug and drug metabolites that cross cell membranes such as the blood/brain barrier, and cause physiological effects, free drug in plasma, and related levels in saliva, may be correlated with drug effects. Through the use of the S/P ratio, there is the potential to relate drug levels in oral fluid to impairment, but this cannot be achieved with our current understanding of the mechanisms effecting drug levels in oral fluid. Saliva drug concentrations cannot be extrapolated to provide related levels in blood without knowing the saliva pH at the time of collection. The route of administration must also be considered as drug concentrations can be very high for several hours after smoking, snorting, or swallowing. The levels detected in oral fluid will not reflect those detected in plasma until the oral contamination has cleared. The detection of marijuana in oral fluid is thought to be as a result of the deposition of cannabinoids in the oral mucosa following smoking of the drug. The measurement of Δ-9-THC, however, appears to follow the on-set and subsequent decline in physiological and pharmacological effects of marijuana.

APPLICATIONS

The main advantage afforded by oral fluid is that the collection process is non-invasive and can be conducted under direct observation, therefore, minimizing the potential for adulteration of the sample. In addition, many collection devices have incorporated sample volume adequacy indicators ensuring that the required volume has been collected. It is extremely difficult to hold substances in the mouth in an attempt to adulterate or substitute the oral fluid sample, and this can be easily avoided by incorporating an oral cavity check as part of the collection procedure and/or a 10-min observation period prior to collection.

The collection can be carried out without the need for special facilities, so maintaining the dignity of the donor, and does not require same-sex collectors. The simplicity of the collection procedure has established oral fluid as a suitable alternative for near-patient testing and for investigating the role of drugs in impaired driving by sample collection and testing at the roadside. The combination of low-drug concentrations and small sample volume presents a challenge to the analyst. Laboratories offering oral fluid tests have had to invest heavily in state-of-the-art instrumentation to achieve the required sensitivities.

Another disadvantage associated with oral fluid testing is the short window of detection with many drugs being undetectable in oral fluid within 24 h. In the case of marijuana, A-9-tetrahydrocannabinol may be detected for only a few hours following smoking. The advantage of a short window of detection is the association with recent use and potential links to assessment of impairment. The inability to provide a sample on demand (comparable with a "*shy bladder*" for urine testing) is also a feature of oral fluid testing. Dry mouth is a common symptom associated with stress and some drugs will inhibit saliva secretions. Offering the donor a drink of water and waiting for 10 min before attempting another collection is recommended, but if the problem persists an alternate biological specimen should be collected.

CONCLUDING REMARKS

The full potential of oral fluid has yet to be realized, and this presents a number of challenges to those already working with oral fluid and for those interested in incorporating oral fluid testing into their laboratory. At present, there are no recognized guidelines or quality standards associated with the analysis and interpretation of drugs in oral fluid. In the workplace testing field, draft guidelines are in circulation both in the USA and the United Kingdom. The formalization of these guidelines will provide much needed guidance on best practice including cut-off concentrations and recommended collection procedures. Testing laboratories worldwide recognize the importance of quality assurance in maintaining standards and work to both national and international quality standards.

Including oral fluid into laboratory accreditation schemes can be difficult as the auditors do not have access to industry standards, and effective assessment of laboratory performance requires external quality assessment (EQA). Only recently have commercially quality control materials and proficiency testing schemes for oral fluid become available. One laboratory took the decision to commission an agency to design and instigate an in-house EQA scheme. Oral fluid testing has the potential to allow testing of drivers at the roadside and ultimately will provide an indication of an individuals' impairment due to drugs. Further research is required to optimize the benefits of using oral fluid in both clinical and criminal justice settings.

10

Discovery of Antibacterial Compounds

The current crisis in medicine represented by antibacterial drug resistance is poised to send us back to the preantibiotic era if novel, potent, and safe drugs to treat infections are not developed soon. One approach that has been discussed as having some promise for success is that of identifying bacterial functions with specific roles in human infection and developing drugs to inhibit them. This strategy would represent a significant shift in antibacterial drug development, which historically relied on either whole-cell screens or targeted assays aimed at inhibiting a small number of targets involved in essential bacterial functions. The most successful and widely used antibiotics interfere with either bacterial cell wall synthesis, DNA replication, or protein synthesis; other drugs target the cell membrane, RNA polymerase, or a metabolic pathway. Drugs against proven targets such as these have been exploited to yield third- and even fourth-generation antibiotics in some classes, often with limited utility due to resistance issues.

The need to develop novel drugs is undisputed; however, there does not appear to be a novel target class or pathway that has the appeal or promise of any of the proven targets. More widespread use of animal infection models and clever strategies to understand and identify virulence determinants begs our consideration of such functions as targets for antibacterial drug discovery. There are several advantages to such an approach, making pathogenesis genes attractive targets:

1. Virulence genes are likely to be specific to bacteria, with little chance of having close homologs in the human genome;
2. Drugs targeting pathogenesis determinants will not necessarily interfere with bacterial growth, thereby perhaps removing the selective pressure for survival and subsequent resistance problems that plague traditional antibiotics;
3. Antivirulence drugs may be used in conjunction with conventional therapies with possible synergistic effects or to *"customize"* more broad-spectrum antibiotics; and
4. Pathogenesis targets are likely to be novel since they may not be the same functions as those required for laboratory growth and thus are unlikely to already have drugs developed against them or be cross-resistant with current drugs.

There are however, several significant drawbacks to using these types of targets as the basis for antibacterial therapy. The major obstacle involves the difficulty of setting up screens to inhibit them *in vitro*; traditional *minimal inhibitory concentration* (MIC) screening would be ineffective in this scenario. The challenge will be to formulate assays in the absence of knowledge of function, *in vitro* activity, or phenotypic effects. In addition, it is unclear whether inhibition of these targets *in vivo* will have a significant enough effect to clear the infection. Given the distinctive nature of infections caused by different pathogens even within the same host site, it may be that an agent developed against a virulence factor from one organism will have no effect on a different organism. In other words, such targets may not be broad spectrum enough to be useful. This chapter focuses on current strategies to identify pathogenesis targets and how this formidable challenge for drug discovery might be overcome.

WORK DONE

It has long been observed that passage of virulent bacteria on laboratory media resulted in a decreased ability of those organisms to cause infection upon reintroduction into animals. This observation emphasizes the notion that bacteria are compact, efficient factories, generating only those products that are required for their immediate growth and/or survival. Such conservation of energy at times can result in genetic alteration, rendering the bacteria completely attenuated for virulence while their laboratory growth is unaffected. Indeed, genetic manipulations of bacteria have also generated mutants that show little to no growth defect *in vitro* but that are unable to cause infection in an animal model.

Numerous examples of this phenomenon exist; one obvious explanation is that requirements for growth *in vivo* are more stringent than those *in vitro* such that mutations that are silent in rich laboratory media can have significant effects *in vivo*. Likewise it is possible that a mutation can have a more pleiotropic effect *in vivo* due to the complex environment of the host. Investigators have exploited this phenomenon by using gene banks to complement defects and restore virulence, thus identifying the gene(s) responsible for that trait. Complementation of mutant strains helped to identify several genes required for full virulence in a number of organisms.

Surrogates

Early studies of pathogenesis relied on cell culture to dissect the multiple aspects of virulence into its individual components. Cell lines and assays were developed to study bacterial adhesion; uptake/internalization/invasion; evasion of host defenses such as phagocytosis, complement binding, and phagosome-lysosome fusion; survival and growth within cells; and host cell killing in a variety of systems. These methods, combined with the power of bacterial genetics, enabled researchers to identify bacterial factors important for these functions.

The use of cell culture models as surrogates for animal infection allowed the development of assays designed to study individual aspects of pathogenesis, and bacterial proteins required at the different stages were identified for a number of pathogens. It was not always the case, however, that bacterial factors identified via such screens were found to be required for *in vivo* pathogenesis. Investigators were forced to deal with the fact that virulence is multifactorial and dynamic, and organisms have evolved redundant and/or compensatory functions to be able to adjust accordingly. By way of example, the invasin protein of *Yersinia pseudotuberculosis* was identified in cell culture assays as being critical for cell adhesion. Invasin was the first demonstration that a single protein could, when expressed in noninvasive *Escherichia coli,* render that organism invasive.

The expectation was that an invasin mutant would be severely attenuated *in vivo.* In fact, further studies showed that an invasin loss-of-function mutant was still virulent in murine infection models and that *Y. pseudotuberculosis* has evolved alternate pathways to infect its host. It soon became clear that cell culture screens do not take into account the temporal aspect of pathogenesis, such that a protein involved at an early stage in infection may not be required at later time points or for survival in another host site.

Multiple Facets of Infection

In order to fully understand the multiple facets of the infection process it was necessary to develop infection models in animals that more closely resemble human disease. Such models should ideally take into account the route of infection, the bacterial load, the tissue tropism observed, the course of the infection through the host, and the eventual outcome as compared to those in humans. There exists a wide range of animal infection models covering the major bacterial pathogens affecting humans, the use of which has facilitated antibacterial discovery but which have been used only recently to study pathogenesis at the molecular level.

More widespread use of animal models has resulted in a pool of information on molecular pathogenesis with the potential to identify new bacterial targets. The most significant limitations to animal models compared to cell culture models are due to those of cost, space, and ethics. For a genetics approach to be successful, a large collection of mutants must be screened for loss of a given phenotype, making an animal infection screen impractical. For these reasons, researchers have developed clever, high-throughput strategies to identify bacterial factors important for the establishment, maintenance, and outcome of infection, ostensibly to understand how the infection process occurs for a given organism, but also to potentially develop inhibitors of that process for use as antibacterial therapy. Two general approaches have been taken: one involves a mutagenesis protocol and the other involves strategies for studying conditional (*in vivo*) gene expression.

In the first scenario, *signature-tagged mutagenesis* (STM), a pool of uniquely tagged mutants is used to infect animals, and, via comparison of input and output populations, attenuated mutants may be identified. In the second scenario, either a library of promoter fusions is screened for expression via cell sorting (*differential fluorescence induction,* or DFI) or selection (*in vivo* expression technology, or IVET), or a global analysis of mRNA is performed following infection (microarrays or transcript profiling).

Beneficial Features

Beneficial features common to all these approaches include the feasibility of tapping into multiple host sites at different times following infection, the adaptability to high-throughput format, and the ability to work in a wide range of genetically tractable gram-negative and gram-positive bacteria.

Mutant selection

This method relies on the power of a mutant selection to identify bacterial genes required during infection of an animal. The perennial limitation to using animals to screen mutants has been that mutants must be screened individually; STM avoids this by employing pools of uniquely tagged mutants in an *in vivo* negative selection scheme to identify those mutants that are unable to cause infection. This enables the investigator to infect mice with up to 100 different mutants each. As a direct consequence of this, each infection is in effect a competition experiment, whereby small differences in virulence between mutant strains are amplified in the context of a pool of pathogenically wild-type clones. In addition, the infection model chosen can influence the outcome. In the case of a model in which the organisms must disseminate to deeper tissues, or traverse some barrier in order to access their preferred niche, it is likely that some clones in the diverse infection pool would be killed or otherwise lost prior to spreading to the site that is sampled.

As a result, such clones may be scored as being attenuated and yet when screened individually are found to be virulent. Technical problems due to the complexity of the pools of tags have arisen, requiring researchers to either purify 96 unrelated tags or to infect with a lower number of tagged mutants. Last, because this is a mutant screen, all mutants involving *in vitro* essential genes would be lost prior to infection. However, despite these caveats, STM has been applied to numerous organisms, with the result that known virulence determinants as well as proteins involved in every cellular function have been identified. Most of these latter proteins had never before been implicated in virulence. Some of the more interesting findings came about when multiple infection sites were polled [9]; these sorts of experiments allow one to determine whether specific proteins are required to infect certain niches, whereas others are involved more broadly in survival, dissemination, and replication functions regardless of the particular host site.

Selection strategy

IVET is a selection strategy based on complementation of a defined auxotrophy. The auxotrophy [either *thyA* or *puraA* have been used successfully] renders the organism avirulent, and the selectivity comes from a promoter-trap library cloned upstream of the complementing gene lacking its own promoter. When auxotrophic organisms carrying this library are introduced into an animal, the majority of the bacteria will be unable to survive; however, those clones carrying a fragment of DNA containing a promoter that is up-regulated in the host will express the complementing gene and thus survive and multiply. Several permutations of IVET have been reported, including those that rely on antibiotic resistance or recombinase [RIVET] as a reporter, depending on the organism of interest. Each permutation is a subtle variation on the original theme, yet each can add value by overcoming some of the limitations of the original. For example, RIVET (recombinase-based *in vivo* expression technology) removes the selective pressure and allows the identification of transiently expressed genes.

In a twist on a traditional application, a recent report used RIVET in *Vibrio cholerae* to identify regulators of *toxT* and *ctxA* that function *in vivo* to direct virulence. Yet another variation of this is *in vivo* induced antigen technology (IVIAT), whereby sera from infected patients are pooled and used to screen an inducible expression library to identify clones carrying host-induced genes. Unlike classical IVET, IVIAT is not a selection. This method allows the screening of potentially thousands of clones in a high-throughput fashion following a human infection, one of the rare methods to date to do so.

Promoter-trap screen

DFI is a promoter-trap screen for genes induced under various *in vitro* or *in vivo* conditions. The screen relies on a promoterless *gfp* (green fluorescent protein) gene cloned downstream of random chromosomal fragments from the organism of interest. If a specific DNA fragment contains a promoter active under a given growth condition, *gfp* will be expressed and such clones can be sorted on the basis of their fluorescence using flow cytometry. Unlike the IVET system, DFI is not a selection, and so a threshold level of expression is not required for cell viability at any stage of the screen. This allows investigators to isolate promoter clones with even modest inducing effects.

In addition, this screen does not necessarily depend on a strict "off" reading in the noninducing condition to be successful. On the other hand, because it is not a selection, it is necessary to individually screen hundreds of sorted clones to identify particular clones of interest. In addition, the stability of Gfp makes it difficult to use DFI to determine whether promoters are only transiently expressed, and it has been observed that *gfp* expression can vary in some bacteria depending on the growth phase of the cells.

Expression profiling

The vast amount of information that will be obtained through genomics and microarray technology (expression profiling) promises to have an enormous impact on antimicrobial drug discovery. With nearly 40% of genes from sequenced organisms having no assigned function a wealth of novel targets is waiting to be identified. Microarrays can point toward functions for unknown genes; likewise, by bundling gene expression profiles we may gain understanding of potential regulons. In addition, as we learn more about bacterial requirements for and during infection we can identify genes and pathways needed for pathogens in a given niche or host environment.

One recent study compared gene expression profiles of *Pseudomonas aeruginosa* grown in broth culture with that grown in a biofilm. It is known that many bacterial infections involve biofilm formation, and the assumption is that significant differential gene expression is required for the bacteria to convert from a planktonic state to a biofilm. However, the authors found that only about 1% of genes were differentially expressed, with half being induced and half repressed in the biofilm condition. That microarray technology may be applied to human infections opens up a huge opportunity to study aspects of disease never possible before.

Bacteria (or bacterial mRNA) can be recovered from infected human tissue and gene expression analyzed, organisms identified, and antibiotic susceptibilities determined. These studies are really an extension of IVET and DFI, but on a more global scale, potentially in a human host, and with the addition of gene annotation to glean maximal information. A potential drawback to wide use of this technology is the small numbers of bacteria that can routinely be recovered from infection sites. However, as technology progresses and more sensitive tools are developed, this should no longer be a hindrance.

Virulence determinants

One interesting approach makes use of the fact that a single pathogenic bacterium can cause disease in a wide range of genera, with the hope that identifying common virulence determinants will provide greater understanding of the infection pathway for this organism as well as provide surrogate models in which to study infection. It has been shown that a clinical isolate of *Pseudomonas aeruginosa* (PA14) can infect mice in a burn model, cause lesions in an*Arabidopsis thaliana* leaf infiltration model, and kill *Caenorhabditis elegans* when it is used as a food source. The authors demonstrated that several nonessential genes act as virulence determinants in all three systems. More recent work has shown that *C. elegans* is also susceptible to killing by gram-positive bacterial pathogens, implicating specific proteins as being important for this pathogenesis.

It is still too early to tell whether these types of surrogate infection models will have utility as screens for antibacterial therapies or indeed whether this approach can be applied to a more broad range of bacterial pathogens. Such studies can lead to the identification of highly conserved virulence mechanisms and, through the use of a genetic system such as *C. elegans*, a means for screening potential compounds for *in vivo* efficacy and drug resistance mechanisms. Though the *C. elegans* and *A. thaliana* models are far removed from human disease, their usefulness as *in vitro* assays for high-throughput screening of compounds may lie in the ability to observe bacterial killing or growth inhibition in the absence of an MIC screen and outside of an animal. A number of groups are extending these studies by attempting to identify genes whose expression is induced during human infection. For obvious reasons, the only reagents available for such studies are patient sera and infected tissue samples. One group described a surface polysaccharide [poly-*N*-succinyl-β-1-6 glucosamine, (PNSG)] of *Staphylococcus aureus* that is preferentially expressed during mouse and human infections. This antigen was identified in lung tissue samples taken from *S. aureus*-infected patients using specific immune rabbit serum.

When the same antiserum was used to probe *S. aureus* isolates grown under *in vitro* conditions, antigen expression was found to be decreased, indicating specific expression of the polysaccharide in the host. Subsequent studies indicated that this antigen would be a good vaccine target, and indeed, immunization with PNSG protected mice from challenge with *S. aureus*. All *S. aureus* strains examined contained the genes involved in PNSG biosynthesis, *icaADBC*. A surprising finding in light of these results is that *S. epidermidis* and *S. aureus* strains carrying deletions of the *ica* locus are still capable of causing infection. This would indicate that either the expression of this antigen is model specific or that an inhibitory antibody perhaps can block more than PNSG, thus leading to attenuation of virulence.

Genomic Progresses

As the era of genomics progresses and more information is gained on the genetic structure of a multitude of bacterial pathogens, it is likely that a target's attractiveness will be determined as much by its importance in the disease process as its spectrum. Several common themes in the pathogenic arsenal are emerging–namely two-component signal transduction systems (TCSTS's), quorum sensing systems, and type III secretion systems. These may not have direct sequence homologies among pathogens but could easily have structural and functional homologies that may be exploited and to which small molecule inhibitors may be directed. TCSTS's are of great interest due to their capacity to regulate bacterial gene expression and the finding that several have been shown to be essential. It is known that in a wide range of gram-positive and gram- negative pathogens, quorum sensing plays a role in virulence gene regulation by coordinating gene expression of a population of cells.

The AgrC/AgrA system of *S. aureus* and the LasI/LasR system of *P. aeruginosa* are two notable examples. In the case of type III secretion systems, used by bacteria to inject potentially toxic molecules into host cells, it is often the case that the secretory apparatuses are highly conserved, whereas the injected molecules vary in structure and function. The recent identification of a type III-like function in a gram-positive pathogen may serve to broaden the spectrum of this highly appealing target. Vaccines represent a sort of proof-of-concept that virulence factor inhibition may be a successful strategy against bacterial pathogens. The majority of current vaccines against bacterial virulence factors are aimed at either preventing colonization or blocking the function of a toxin. Such vaccines have been highly effective in preventing incidence and spread of disease, thereby lending support to the idea of inhibiting virulence factor production or function to fight infection.

Comprehensive Analysis

The most comprehensive analyses to search for *in vivo* targets have been performed with STM, IVET, and DFI. These methods have been applied to a wide variety of gram-negative and gram-positive bacterial pathogens, many using different animal infection models, with some overlap in results in terms of identified genes. Using different approaches or the same approach in a different animal model is likely to lead to a greater understanding of the genetic requirements for virulence in a given organism. What is striking from these studies is that they do not point toward a particular gene, or class of genes, that would be an ideal target(s) for antibacterial therapy. A number of genes were identified either in more than one application of a given screen or in different screens entirely. No single category of genes stands out as being isolated a disproportionate number of times from these screens, which is most likely indicative of the somewhat subtle requirements imposed by each.

The disparate genes identified in the different screens also reflect the strain and model differences utilized by the different groups. A gene must be mutated in order for its role in virulence to be assessed. As expected, because it is a mutant screen, most of the genes identified by STM are important for virulence, since mutants in those genes show attenuation. In contrast, the two other strategies assume that genes expressed at a high level *in vivo* are important for infection. At any rate, no gene was identified in all three screens, which would require that a gene is not only induced but is essential for virulence in all the models examined, which include systemic, kidney abscess, subcutaneous abscess, wound, and rabbit endocarditis.

In combination these approaches have yielded targets and methods for understanding pathogenesis on a much larger scale than was possible even as late as the early 1990s. There has been a significant overlap in identified genes but also it is clear that these screens have the potential to identify genes that are model and/or method specific. Perhaps the complementarity achieved by these approaches will impact the future of the field and enable researchers to look at pathogenesis with a fresh eye.

Molecular Pathogenesis

Since the early 1990s the rise of innovative, creative approaches to *molecular pathogenesis*, with great potential for identifying novel targets for antibacterial therapy. A number of intriguing targets have been isolated by these *in vivo* strategies; however, none has yet led to a successful drug candidate. That is not to say that this is an impossible task or an unfeasible approach. It is far too early at this stage to be able to predict the success of these strategies, but no other current or past approach is showing more promise. With the molecular tools and infection models that have recently been developed, we are poised to reap the benefits. Perhaps more specific models would help move this area forward—allowing us to consider bacterial adaptations *in vivo*, as well as niche-specific, host-specific, and strain-specific responses.

A genetic switch that would allow one to turn gene expression on or off *in vivo* would be a tool of enormous value. Surrogate models such as those discussed above may be helpful in target identification and high-throughput screening of antibacterial compounds. Probably the main reason for avoiding this approach is the inability to generate MIC's and demonstrate *in vitro* activity of compounds directed against virulence targets. In some cases it may be possible to identify *in vitro* conditions under which inhibition of a virulence target would be lethal. It may be that the traditional screening paradigms need to be adapted in order to accommodate this new range of targets. This issue should be given serious consideration in the near future. Perhaps the best bet for compounds against these targets would be in adjunct therapy in order to add specificity to current drugs.

Virulence targets can have an impact on antibacterial drug discovery and all avenues should be explored. This field is still relatively new and further research will be needed to validate both the strategies and the targets they have generated, but the more significant lesson here is the necessity of developing tools for studying pathogenesis in the human host or appropriate surrogates to do so. As we gain a better understanding of molecular pathogenesis, we can only be in a better position to combat the pathogens.

ANTIBACTERIAL COMPOUNDS

Natural products are important sources of drugs, including antibacterials such as β-lactams, aminoglycosides, tetracyclines, macrolides, and streptogramins. Of the 64 antibacterials approved for clinical use between 1983 and 1994, over 78% were unmodified natural products or compounds derived from natural products. For more information on natural products and natural product drugs, several comprehensive reviews should be consulted. Almost all of the natural product drugs were discovered so far through the following steps: collection and identification of natural material, fermentation and extraction, *in vitro* assays for biological activities, isolation and structural elucidation of the active components from the extracts, and finally characterization of the *in vivo* biological efficacy and toxicity profiles. This empirical approach of natural products screening has been and will continue to be effective in the discovery of pharmaceuticals, including anti-infectives.

Nevertheless, the recent developments in the genetics of microbial secondary metabolites biosynthesis, genomics, and metabolic engineering are playing an ever-increasing role in facilitating the natural product drug discovery process. A number of extensive, focused reviews on molecular approaches in natural product drug discovery have been published in recent years. This section attempts to provide readers with some general concepts and recent progress of applications of genetic engineering and genomics in natural products drug discovery. Due to space limitations, we shall focus on the following two areas: (i) tapping the chemistry potential of unculturable and slow-growing organisms by expressing secondary metabolite-encoding genes or gene clusters in heterologous hosts and (ii) generating novel unnatural secondary metabolites by genetically modifying secondary metabolic biosynthesis pathways.

HETEROLOGOUS HOSTS

Efforts to search for bioactive natural products have mainly focused on organisms that can be easily isolated, cultured, and maintained using laboratory media. Approximately 11,900 antibiotic secondary metabolites have been discovered from microorganisms to date and 66% of these were derived from the order actinomycetes, mostly from *Streptomyces* species. The number of prokaryotes on Earth is estimated to be 4-6 × 10^{30} cells. This large population provides enormous potential for prokaryotes to acquire genetic diversity. Recent estimates for the number of prokaryotic species range from 10^5 to 10^7. Similarly, only 70,000 species of the approximately 1.5 × 10^6 species of fungi have been described to date.

It is expected that discoveries of new bioactive secondary metabolites from easily culturable organisms will continue, but expansion of the natural products sources to unculturable and difficult-to-grow microorganisms may lead to new structural classes. Improved cultivation techniques can help to take advantage of some of the less manageable organisms, but the task is likely to become increasingly intractable, and many microorganisms may never be cultivated in the laboratory.

Genetic Diversity

One way to capture the genetic diversity among unculturable or slow-growing microorganisms for secondary metabolite discovery is to introduce large segments of genomic DNA from these unculturable microorganisms into genetically amenable host organisms. In this procedure, the host strain is asked to express genes from the donor organism and to produce secondary metabolites that are not being produced by the hosts. These host strains are typically fast growing and easily fermented on a large scale, and genes can be manipulated genetically. For example, *Escherchia coli* has been used as a host to maintain and express environmental DNA isolated from soil.

Transgenic Approach

The transgenic approach requires that genes or clusters of genes from unculturable organisms be expressed in laboratory strains. This is possible because transcription and translation control sequences from one organism often function in closely related organisms and sometimes even in distantly related organisms. For example, the entire epothilone gene cluster isolated from the Gram-negative myxobacterium *Sorangium cellulosum* was expressed in *Streptomyces coelicolor*, a Gram-positive actinomycete. The erythromycin-encoding gene cluster of *Saccharopolyspora erythraea* was engineered into *E. coli* and *S. coelicolor* for heterologous expression and erythromycin production. It needs to be emphasized that in these cases, promoters from the host strains were used in the heterologous gene expression since the gene donor organisms were very different from the heterologous hosts. Nevertheless, it is remarkable that such large and complex gene clusters can

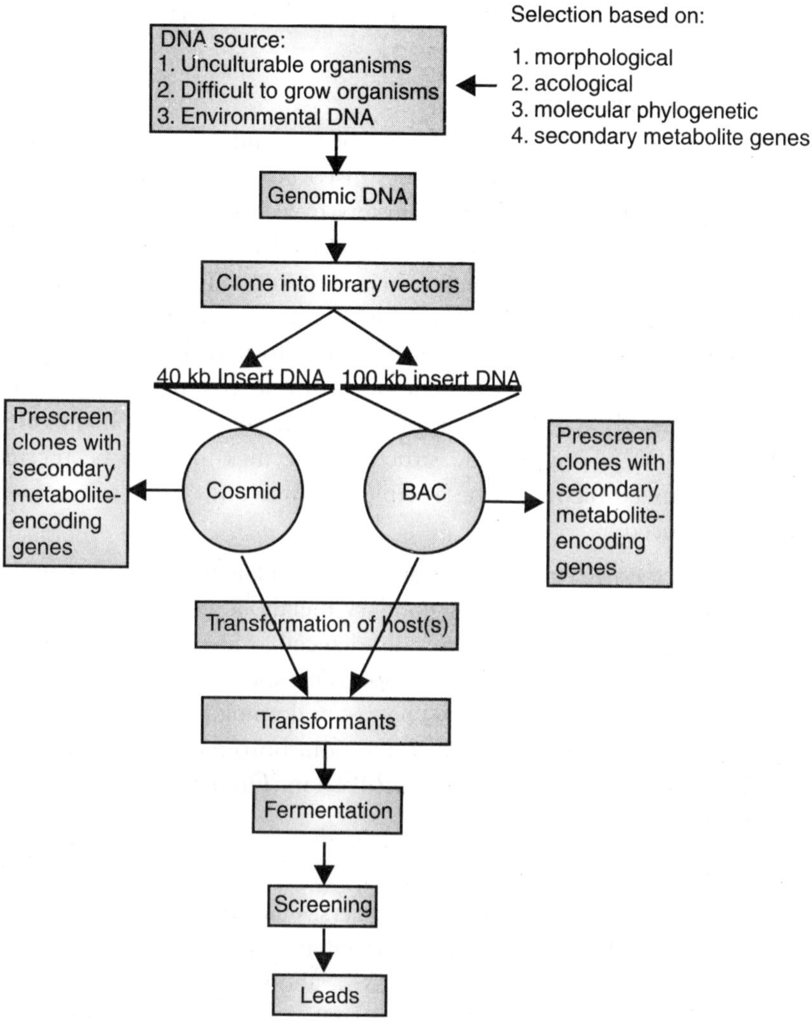

Fig. 10.1. A flow diagram illustrates the concept and steps of capturing the chemistry diversity of untapped microbes by heterologous gene expression.

be functionally expressed in heterologous hosts. Similarly, the fungal transformation vector pAN7-1, which has the *E. coli* hygromycin phosphotransferase gene (*hph*) under the control of the *Aspergillus nidulans* glyceraldehyde-3-phosphate dehydrogenase (*gpd*) gene promoter and *trpC* terminator, efficiently confers hygromycin B resistance to a variety of fungal species.

More than 40 fungal species have been transformed with this vector, ranging from the basidiomycetes *Schizophyllum commune* and *Laccaria laccata* to various ascomycetous species. The *A. niger* glucoamylase gene promoter functioning in *Ustilago maydis* is another example of an ascomycete promoter that functions in a basidiomycete. Some *Aspergillus* genes have been isolated by complementation of *Saccharomyces cerevisiae* mutant strains. These experiments demonstrated that genes of foreign origin can be expressed in heterologous laboratory strains. However, one must still be cautious when selecting a donor-recipient combination or combinations, as many genes have been shown not to be expressed in heterologous strains using their native promoters and codon biases.

Biosynthesis of Microbial Metabolites

Genes involved in the biosynthesis of major microbial secondary metabolites are often clustered. In prokaryotes, essentially all natural products genes discovered to date are found in single clusters. More than 115 bacterial natural products gene clusters have been cloned and analyzed. Relatively fewer natural products biosynthesis genes have been cloned from fungi, but it is clear that genes encoding for natural products in fungi also are often clustered. This tendency for genes to cluster allows us to gather most, if not all, of the genes for a particular secondary metabolite within one large fragment of DNA which can then be introduced into a recipient laboratory strain. Large pieces of DNA may be cloned from unculturable organisms as cosmid clones (35-45 kb), bacterial artificial chromosome clones (BAC) (>100 kb), or yeast artificial chromosomal clones (YAC) (>100 kb). Alternatively, donor DNA may encode structural enzymes that can be recruited to existing pathways of secondary metabolism in the recipient strain and, as a result, novel secondary metabolites may be produced.

Finally, genes encoding transcription and translation regulatory factors that may be introduced to the host genome to either activate silent secondary metabolic pathways in the host or to incite low-activity pathways to biologically and chemically detectable levels. For example, production of the polyketides sterigmatocystin and aflatoxin in *Aspergillus* species is negatively regulated by FadA, the α-subunit of a heterotrimeric G-protein. In contrast, a dominant activating *fadA* allele (*fadA*^{G42R}) stimulates the transcription of a gene involved in the nonribosomal peptide penicillin biosynthesis in *A. nidulans.* Heterologous expression of *fadA*^{G42R} in *Fusarium sporotrichioides* increases trichothecene production.

Prescreening of Secondary Metabolites

The transgenic approach to explore genetic diversity among unculturable and slow-growing organisms for secondary metabolites discovery will generate a large number of transgenic strains. If a typical fungal genome is between 30 and 40 Mb, about 1000 cosmid clones are needed to have a 1X coverage of a given genome. This translates into about 1000 transgenic strains. It is difficult to estimate the number of secondary metabolic pathways for a given fungus, but a search in Chapman & Hall's *Dictionary of Natural Products* revealed over 35 secondary metabolites produced by *A. nidulans.* These metabolites represent at least 12 distinct pathways. Assuming that the average length of a gene cluster for a fungal secondary metabolic pathway is 40 kb and the *A. nidulans* genome is 30 Mb, 12-35 biosynthetic gene clusters constitute about 1.6-4.6% of the genome. Since only a small fraction of a given fungal genome is related to secondary metabolites biosynthesis, the majority of transformants will carry DNA fragments that are not related to secondary metabolite biosynthesis.

Bacterial genomes are smaller than fungal genomes, generally ranging from 1 to 5 Mb. Some bacterial genomes, however, have much larger genomes. For example, the actinomycete *S. coelicolor* genome is about 8 Mb and at least 12 gene clusters that have been identified or predicted to encode various natural products. These gene clusters constitute about 4% of the entire genome.

Different approaches can be used to eliminate nonproductive transformants. One is to screen for biological activities. Biological prescreens have to be highly sensitive and high-throughput in order to process a large numbers of extracts. For example, a macrodroplet high-throughput screening technique has been used to identify transgenic strains that produce antibacterial activities. The disadvantage of using biological screening is that an assay must be developed for each target. Another approach is to select transformants that contain known secondary metabolite biosynthetic genes before they are sent to screening programs.

As discussed earlier, many secondary metabolite-encoding genes have been cloned, sequenced, and characterized. As of 1999, 156 microbial secondary metabolite gene clusters had been reported and sequenced and a considerable number has been discovered since then. By using these genes as probes, one can preselect the clones that most likely contain secondary metabolite-encoding gene(s). Similarly, this approach is effective only in a high-throughput mode. Significant progress has been made in the DNA array technology during the past 5 years and it is now practical to apply these technologies to search for secondary metabolite-encoding gene(s) from a large collection of cloned DNA. For example, one method involves creating a secondary metabolite-encoding gene library with consensus DNA fragments obtained from genes encoding various secondary metabolites, such as polyketides and nonribosomal peptides. These DNA fragments can be used individually as probes to screen libraries for secondary metabolite gene clones. They can also be placed on membranes or silicon to create a secondary metabolite-encoding DNA array.

One of the limitations to this approach is that the molecular genetics are known for only a few classes of microbial secondary metabolites, mainly polyketides and nonribosomal peptides. Of the 156 microbial secondary metabolites gene clusters reported up to 1999, about 40% were polyketides and 34% were nonribosomal peptides. It is well known that in addition to polyketides and nonribosomal peptides, bacteria and fungi also produce shikimates, saccharides, terpenoids, nucleoside, alkaloids, and glycolipids. More secondary metabolite-encoding genes will become available with the continued determination of microbial genomes to come. As of February 2001, there were 40 completed bacterial genomes and an additional 125 bacterial genome projects are in progress.

Fungal genome projects are relatively few and *Saccharomyces cerevisiae* is still the only complete fungal genome in the public database, but there are about 19 eukaryotic microbe genome projects currently under way. Comparison of chemical profiles between the parent strain and the transformants can also be used to separate productive strains from nonproductive strains. Over the past 5 years, there has been an explosion in the use of mass spectral data for high-throughput characterization of organic compounds from small pharmaceuticals to large proteins. The union of a liquid separation technique (LC-MS) with an ionization technique such as electrospray (ESI) has allowed rapid identification of molecular weight and UV characteristics of semipure natural products and other organic molecules (36–38). This and other chemometric tools will continue to be developed to provide higher throughput, increased accuracy, and better reproducibility. One of the most important tasks in using the transgenic approach to capture natural chemical diversity from the unculturable microorganisms is to rescue and analyze the genes that are responsible for the novel biological activities in a heterologous host.

Understanding these new activities at the genetic and molecular levels will enable us to further manipulate the activity. Another major scientific contribution resulting from this approach is the better understanding of heterologous gene expression. Despite recent advances in heterologous gene expression in various microbial systems, our knowledge in the area remains limited. Information on heterologous gene expression will guide scientists to make improved decisions on the selection of unculturable organisms and recipient hosts to maximize the chance of gene expression for the production of novel bioactive compounds. In summary, by exploring genetic diversity from unculturable and slow-growing microbes and by creating "*biocombinatorial*" diversity, we improve the likelihood of discovering novel secondary metabolites. Because the recipient strains are fast-growing, "industrial" organisms, once novel drug-producing transformants are identified, scale-up fermentation for commercial production can be implemented using largely traditional approaches.

BIOCOMBINATORIAL CHEMISTRY

Advances in molecular genetics of secondary metabolite biosynthesis have led to a new approach to drug design known as *biocombinatorial chemistry.* Many academic and industrial laboratories now explore the potential of this genetic engineering/chemistry hybrid approach for developing novel chemicals. Most of the progress so far is in the biocombinatorial synthesis of bacterial polyketides and, to a less extent, nonribosomal peptides.

Polyketide Biosynthetic Pathways

Polyketides constitute perhaps the largest family of microbial secondary metabolites, and the genes for over 60 polyketide biosynthetic pathways from both bacteria and fungi have been cloned and characterized (GenBank database). The controlled variation in chain length, choice of chain-building blocks, and reductive cycle orchestrated by genetically programmed polyketide synthases (PKSs) all lead to the structural diversity of polyketides. In addition, many PKS products undergo further modification by regiospecific glycosylases, methyltransferases, and oxidative enzymes to produce still greater diversity. The tremendous structural diversity of polyketides reflects the wide variety of their biological properties. Many are effective therapeutic agents, ranging from antibiotics (tetracycline and erythromycin) and anticancer agents (doxorubicin) to immunosupressants (FK506 and rapamycin). Some polyketides have multiple pharmacological effects; for example, the immunosuppressant FK506 shows antifungal activity, and the antifungal compound thysanone also inhibits rhinovirus protease. The mode of action of different polyketide drugs varies greatly despite their structural similarities. For example, the immunosuppressants FK506 and rapamycin act very differently on T cells: FK506 inhibits the production of interleukin 2, and rapamycin prevents the proliferative response to interleukin 2 bound at the interleukin 2 receptor. Given the diversity of naturally occurring polyketides, plus their clinical and commercial success, it is expected that more effective polyketide drugs will be discovered and developed in many therapeutic areas.

Genetic organization

Bacterial PKSs are classified into two broad types based on genetic organization and biosynthetic mechanisms. Modular PKSs (type I) have discrete multifunctional enzymes that control the sequential addition of thioester units, and their subsequent modification, to produce macrocyclic compounds (or complex polyketides). Type I PKSs are exemplified by 6-deoxyerythronolide B synthase (DEBS), which catalyzes the formation of the macrolactone portion of erythromycin A, an antibiotic produced by *Saccharopolyspora erythraea.*

There is a total of seven active-site modules in DEBS, with each given module containing three to six active sites. Three domains constitute a minimal module: acyl carrier protein (ACP), acyltransferase (AT), and β-ketoacyl-ACP synthase (KS). Some PKSs contain additional modules for reduction of β-carbons, such as β-ketoacyl-ACP reductase (KR), dehydratase (DH), and/or enoyl reductase (ER). The thioesterase cyclase (TE) protein is present only at the end of module 6.

Monofunctional proteins

Bacterial aromatic PKSs (type II) are composed of several separate, mostly monofunctional proteins, the active sites of which are used iteratively for assembly and functional-group manipulation of the polyketide chain. At least 25 sets of aromatic polyketide PKSs have been cloned from *Streptomyces* and *Saccharopolyspora* species (GenBank database). Best studied among type II PKSs is the PKS for the biosynthesis of the benzoisochromanequinone actinorhodin. Genes for actinorhodin biosynthesis are designated "actI–VII." The actI locus has three genes: KS, AT, and ACP. ActIII encodes KR. ActI and actIII constitute the minimal PKS. The actII locus is responsible for transcriptional regulation of the act genes and for actinorhodin export. The actIV–VII loci encode several postsynthetic modifying functions, e.g., cyclization (VII), aromatization, and subsequent chemical tailoring. Other aromatic PKSs share the same basic architecture with minor structural differences. Type II aromatic PKSs contain a single set of iteratively used active sites, although the basic organization of the two types of PKS is similar.

These characteristics make it difficult to predict the structure of the polyketide produced after modification of a type II PKS gene. Significant advances in research on the combinatorial biosynthesis of novel aromatic polyketides have been made in laboratories of Hopwood, Khosla, and their collaborators in recent years. Hybrid aromatic polyketides have been generated by transferring partial or complete biosynthetic gene clusters to different polyketide producers, followed by screening for new structures. For example, mederrhodin A, a novel aromatic polyketide, was produced by a medermycin-producing *Streptomyces* strain transformed with the actVA gene. This approach of generating novel polyketides was largely empirical, and it required the activity of enzymes for late tailoring steps. During the past few years, tremendous progress has been made in our understanding of genetic programming of aromatic PKSs; consequently, the rational design of novel aromatic polyketides is also advancing. Several novel compounds have been generated by constructing and expressing recombinant PKSs in *Streptomyces* sp. Based on their work and the work of others, McDaniel *et al.* summarized six rules for rational design of novel aromatic polyketides based on chain length, ketoreduction, cyclization of the first ring, first-ring aromatization, second-ring cyclization, and additional cyclization.

Bacterial polyketides

Pioneered by the Hopwood laboratory, molecular genetic analysis of PKS genes has confirmed earlier biochemical and chemical findings that the structural diversity of polyketides is a result of the quantity and types of acyl units involved. This body of research also indicates that novel polyketides can be produced by manipulating the sequence and specificity of enzyme-mediated reduction, dehydration, cyclization, and aromatization through genetic engineering. One of the first examples of the generation of novel polyketides through genetic manipulation of PKS genes was reported in 1990 by the Strohl group. They demonstrated the biosynthesis of anthraquinones aloesaponarin II and desoxyerythrolaccin through the cloning of actinorhodin polyketide biosynthesis genes in the aclarubicin-producing strain *S. galilaeus.* The era of rational design for novel antibiotic structures was ushered in by early successes in synthesis of complex polyketides.

The modular type I PKSs producing complex polyketides contain a unique active site for each enzyme- catalyzed reaction in the pathway, giving rise to final structures that are determined by the numbers and types of active sites. Donadio *et al.* demonstrated in the early 1990s that analogs of the erythromycin polyketide backbone could be generated by eliminating active sites within the PKS. By repositioning a chain-terminating thioesterase domain from the C-terminus of module 6 of DEBS3 to the C-terminus of module 2 of DEBS1, Cortes *et al.* were able to construct a multienzyme unit that catalyzed only the first two rounds of polyketide chain extension. The mutant produced a triketide lactone structure without any trace of the wild-type polyketide erythromycin, indicating premature chain termination and cyclization. By expressing the entire DEBS gene cluster in a heterologous host, substantial quantities of 6-deoxyerythronolide B (the aglycone of the macrolide antibiotic erythromycin) was produced.

Efficient expression

Combinatorial biosynthesis of novel metabolites requires an efficient expression host and sometimes multiple hosts. Several actinomycetes, mostly *Streptomyces* species, have been used as expression host for combinatorial biosynthesis such as *Streptomyces coelicolor*, *Streptomyces lividans*, and *Streptomyces venezuelae.* Each host-vector system has advantages and disadvantages regarding transformation efficiency, growth rate, and genetic understanding of the host. More recently, erythromycin was produced in *E. coli.* This was achieved by multiple genetic engineering steps in the *E. coli* strain, including introduction of the three DEBS genes from *Streptomyces erythraea,* introduction of the *sfp* gene encoding phosphopantetheinyl transferase, deletion of the endogenous *prpRBCD* genes, and overexpression of the endogenous *prpE* and *birA* genes. The feasibility of engineering *E. coli* to produce complex natural products provides an alternative host for combinatorial biosynthesis, because *E. coli* can be easily manipulated at the molecular genetic level. The existence of multiple host-vector systems offers the opportunity to select the optimal system for individual projects.

Manipulating PKS genes

In addition to manipulating the PKS genes themselves, one can also modify various modifying enzymes such as glycosyltransferaseases, methyltransferases, and oxidative enzymes. For example, manipulation of the desosamine biosynthetic genes of the pikromycin biosynthesis pathway in *S. venezuelae* has led to the isolation of several methymycin/pikromycin analogs. In a more elaborate experiment, the combination of the pikromycin pathway in *S. venezuelae* and a sugar biosynthetic gene, CalH, from the calicheamicin pathway in *Micromonospora echinospora* yielded two novel macrolides in a *des*I-deleted *S. venezuelae* strain.

Polykeides of filamentous fungi

Although polyketides are commonly produced by filamentous fungi, nucleotide sequences have been reported for only 14 fungal PKS genes, all of which encode iterative, type I PKSs. Even though combinatorial manipulation of fungal polyketide biosynthetic genes is not as advanced for purposes of genetic engineering as that of bacterial systems, the interative, yet modular fungal PKS systems exhibit several advantages over bacterial systems as follows: (i) fungal polyketides range in chain lengths from 6 to more than 40 carbons; (ii) fungal polyketides can be methylated by a methyltransferase domain that is part of the PKS itself (making it amenable to genetic manipulation), rather than by incorporating propionate as an extender unit as observed in bacteria; (iii) fungal polyketides can be partially or completely reduced, and they undergo unusual cyclizations; both features contribute to their potential as starting materials for production of novel polyketides; (iv) a considerable amount of biochemical information in a relatively short stretch of DNA because fungal PKSs are both structurally modular and functionally iterative; and (v) fungal PKSs described to date are single proteins (bacterial PKSs often form multiple protein complexes; this characteristic explains why only fungal PKSs are sufficiently stable to be completely purified).

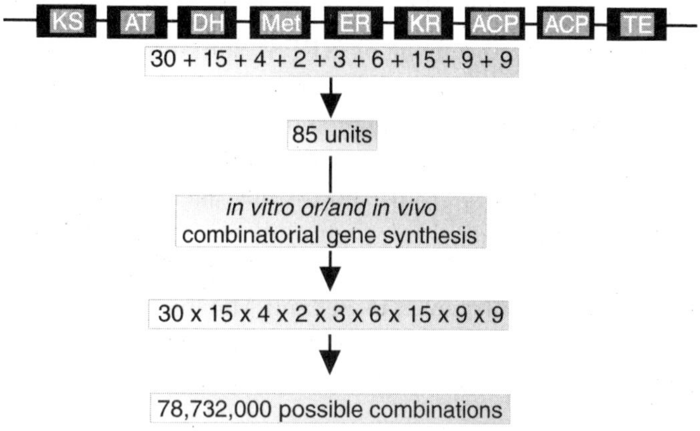

Fig. 10.2. A diagram illustrates the concept of assembling large numbers of chimeric PKS genes by functional domain shuffling.

In addition, availability of single genes encoding single proteins greatly facilitates heterologous expression of PKSs. Similar principles and roles for genetically engineering bacterial polyketides could be applied to combinatorial synthesis of fungal polyketides, but specific protocols need to be developed for fungi because fungal gene regulation and structure are distinct from those of bacteria.

Biosynthesis of Peptides

A wide variety of peptides is synthesized by nonribosomal peptide synthetases in microbes that exhibit a broad range of biological activities. Such activities may act as antibiotics (penicillins, vancomycin, bacitracin, and gramicidin), siderophores (ferrichrome, mycobactin, and enterobactin), toxins (HC toxin and syringomycin), and immunosuppressive agents (cyclosporin and rapamycin). The growing number of nonribosomal peptide synthetase (NRPS) genes isolated and characterized (at least 53 NRPS genes in the GenBank database) is accumulating a wealth of information regarding enzymatic activities and structural features of this family of proteins. One of the most prominent features is modularity.

A module, in the context of NRPSs, is defined as a DNA/protein segment that containing all of the functions required for catalysis of peptide elongation and associated substrate modifications such as acylation, epimerization, heterocyclization, and N-methylation of the amide nitrogen. NRPSs can incorporate hundreds of known amino acids into peptide products. The order and number of modules on an NRPS determine the sequence and number of amino acids in the final peptide product. The increasing number of NRPS genes cloned and characterized allows us to predict in some cases the substrate specificity of adenylation domains of unknown function.

Modular arrangement of NRPSs and substrate specificity of each module in a given NRPS strongly suggests the possibility of creating novel peptide products by genetic and combinatorial manipulations. Although there still are many obstacles in successfully engineering these proteins, recent results have been promising. Working with the three-module surfactin synthetase 1 of *Bacillus subtilis*, Marahiel' s group successfully altered the substrate specificity of multimodular peptide synthetases by targeted substitution of minimal-module encoding regions within NRPS genes. The same group also constructed functional hybrid peptide synthetases by module and domain fusions. In a more recent review, Doekel and Marahiel proposed a strategy to synthesize antipain by genetic engineering and the steps outlined in their strategy summarize the latest development in the combinatorial synthesis of nonribosmal peptides.

11

Analyzing First Fecal Matter

Fetal exposure to drugs, alcohol, or other xenobiotics results in numerous adverse effects for the newborn. Maternal use of cocaine, methamphetamine (MA), and/or phencyclidine (PCP) has been consistently reported as a co-factor in births involving respiratory problems, intracranial bleeding, placental abruption, premature labor, low birth weight, and small head size babies, as well as fetal death. Behavioral consequences later in childhood have also been studied. Beeghly et al. recently showed that newborns with *prenatal cocaine exposure* (PCE) had lower receptive language than children not exposed to cocaine at 6 years, but that difference had modified by age 9. Age, birth weight, and gender seemed to moderate the relationship between PCE and the language ability of school age children. Maternal opiate abuse can result in newborns displaying irritability, tremors, and seizures, which are often consistent with opiate withdrawal symptoms. Although *cannabis* was considered a fairly harmless drug, Hurd et al. recently reported that maternal marijuana use impaired growth in mid-gestation fetuses.

Other researchers have reported adverse effects of marijuana use on the newborn and on the adolescent offspring of mothers smoking marijuana. Porath and Fried recently reported that "maternal cigarette smoking and marijuana use during pregnancy are risk factors for later smoking and marijuana use among adolescent offspring, and add to the weight of evidence that can be used in support of programs aimed at drug use prevention and cessation among women during pregnancy". As these major drug classes cause negative effects in the offspring, it is imperative that a reliable diagnosis of fetal drug exposure be made as soon as possible, in order that the appropriate care and treatment be given to both the newborn and the mother.

BASIS OF ANALYSIS OF MECONIUM

The first reports of the use of meconium to determine fetal drug exposure were published in 1989, and various patents have been awarded based on methods of analysis since that time. Meconium analysis has become routine in many hospitals, as it is a depository for drugs to which the fetus has been exposed during the latter half of pregnancy and provides a much longer history of fetal drug exposure than urine. Various review articles on the analysis of drugs-of-abuse in meconium have been published. Several large-scale studies involving the use of meconium analysis have been performed over recent years. For example, meconium specimens from 8527 newborns were analyzed by immunoassay with gas chromatography/mass spectrometry (GC/MS) confirmation for metabolites of cocaine, opiates, cannabinoids, amphetamines (APs), and PCP as part of The National Maternal Lifestyle Study.

The prevalence of cocaine/opiate exposure was determined to be 10.7% with the majority of neonates (9.5%) exposed to cocaine. However, exposure status varied by site and was higher in low-birth-weight infants (18.6% for very low birth weight and 21.1% for low birth weight). In the cocaine/opiate-exposed group, 38% were cases in which the mother denied use, but the meconium was positive. The report concluded that accurate identification of prenatal drug exposure was improved with GC/MS confirmation of the meconium assay, and maternal interview was taken into account.

Several other researchers have shown increased detection rate of fetal drug exposure when using meconium compared to urine. In 2003, Bar-Oz et al. tested paired samples of neonatal hair and meconium for cocaine, benzoylecgonine (BZE), opiates, cannabis, benzodiazepines, methadone, and barbiturates. They reported that meconium was marginally more sensitive than neonatal hair for the detection of cocaine and marijuana. Both meconium and hair were more effective than urine for the detection of drug exposure.

BIOCHEMISTRY OF MECONIUM

Meconium is the first fecal material passed by a newborn and is normally excreted 1-5 days after birth. It begins to form between the 12th and 16th week of gestation and is a cumulative deposit thereafter. Generally, it represents the intestinal contents of the fetus, providing a history of fetal swallowing and bile excretion, and can provide the physician guidance on gastrointestinal function. The color is generally dark-green because of the presence of bile pigmentation. Meconium consists predominantly of water, but other major components include

1. Mucopolysaccharides
2. Epithelial cells
3. Lipids and proteins
4. Cholesterol and sterol precursors
5. Blood group substances
6. Squamous cells
7. Enzymes
8. Bile acids and salts
9. Residual amniotic fluid.

Meconium is occasionally passed in utero when the fetus has reached gastrointestinal maturity late in gestation or the baby is in distress, producing meconium-stained amniotic fluid. However, in 12-25% of deliveries involving the meconium passage in utero, the cause is not known.

DRUG DEPOSITION

Because of the obvious ethical limitation of providing known amounts of drug to pregnant women, much of the research in the area of drug deposition has been conducted in animals. Studies using pregnant sheep and guinea pigs have demonstrated the presence of both parent drug and metabolites in offspring. However, it still remains difficult to predict the metabolic fate of a drug in the maternal-fetal unit as the various breakdown pathways (oxidation, conjugation, hydrolysis reduction) can be promoted or retarded by fetal or placental enzymes. In general terms, reaction rates increase with gestational age, as maturation of the metabolic pathways occurs, but presence of a specific metabolite in the fetus may not indicate the ability of the fetus to metabolize the drug, as passive diffusion through the placenta may have occurred.

Placental microsomes containing cholinesterase may be responsible for some enzymatic conversion of drugs to their metabolites as they attempt to cross the placental barrier. Placental transfer of drugs is affected by blood flow, drug ionization, and degree of protein binding of the drug to either maternal or fetal plasma. The kinetics of drug transfer have been reviewed, and it has been suggested that drugs reach the fetus by various complex pathways:

1. Passive diffusion of small molecule lipid-soluble drugs across the placental barrier, which may be further metabolized by the fetus depending on gestational age
2. Binding of drugs and/or metabolites to proteins in the amniotic fluid, which is then swallowed by the fetus. Sustained swallowing of amniotic fluid may prolong fetal exposure to drugs.

Therefore, drugs enter the fetal circulation through various pathways that depend upon many parameters. The meconium is the final depository for drugs to which the fetus was exposed, because drugs in the bile are deposited into meconium and drugs in the urine are deposited into the amniotic fluid, which is then swallowed by the fetus. Consequently, meconium is not a homogenous specimen, because it is produced in "layers" as excreted products are stored. This is the main disadvantage to testing meconium for drugs-of-abuse, because frequently not all the specimen is collected. Because of the numerous variables involved over a period of 20 weeks (latter half of pregnancy), such as frequency and nature of drug abuse, nutrition, smoking, other diseases, and general maternal health, the metabolic profile of a drug appearing in the newborn is not the same as in a normal adult. This observation has also contributed to the difficulty of meconium analysis to avoid false-negative results.

SAMPLING TECHNIQUE

Meconium is a complex matrix containing waste products and pigments. The amount of drugs and metabolites generally detected in meconium are much lower than concentrations present in urine samples. There are numerous publications regarding the screening of meconium specimens. In the original patents, meconium (0.5 g) was taken directly from the diaper of the newborn. The sample was mixed with distilled water (10 mL) and concentrated hydrochloric acid (1 mL). This homogenate was filtered through glass wool, the filtrate centrifuged, and the supernatant tested for morphine and BZE using *Abuscreen Radioimmunoassay* (RIA). For cannabinoids, methanol (0.4 mL) was added to the sample (0.1 g). The specimen was mixed, allowed to stand at room temperature, and then centrifuged. An aliquot of the supernatant was tested by RIA. The cut-off values reported were 15, 25, and 50 ng/mL for cocaine, morphine, and cannabinoids, respectively. Since this original work, many extraction procedures have been reported. In general, the meconium is homogenized in an organic solvent, such as methanol or acetonitrile and/or buffer, to help the drugs distribute throughout the matrix.

The concept is to solubilize the drugs in the sample so they can be tested using standard screening formats. The specimen is centrifuged, and the supernatant is analyzed using RIA, *enzyme multiplied immunoassay technique* (EMIT), *enzyme-linked immunosorbent assay* (ELISA), or *fluorescence polarization immunoassay* (FPIA). However, it is difficult to achieve the detection levels required for valid analysis without some form of extraction of drugs from the matrix because of the interference of the pigmentation and turbidity of the extract, especially when enzyme or fluorescent detection is used. Chen and Raisys analyzed the drugs in meconium by homogenization of the sample in methanol, followed by solid-phase extraction (SPE), before screening with EMIT.

Moriya et al. extracted drugs from meconium with chloroform-isopropanol (3:1) and screened by EMIT. The reported detection limits for BZE, D-MA, morphine, and PCP were 250, 730, 110, and 100 ng/g, respectively, which even with the extraction are relatively high. ElSohly et al. described the development and validation of methods for meconium sample preparation for both screening by EMIT and TDx and confirmation by gas chromatography-mass spectrometry (GC-MS) of meconium extracts for cannabinoids, cocaine, opiates, APs, and PCP. The procedures necessitated specimen clean-up prior to screening to achieve the required detection levels. Cut-off levels were administratively set at 20 ng/g for 11-nor-Δ9-tetrahydrocannabinol-9-COOH (THC-COOH) and PCP and at 200 ng/g for BZE, morphine, and APs.

Screening Assays

Several research groups have warned against relying on positive meconium results using immunoassay only, without confirmation. It is necessary for valid, relevant, and reliable detection levels to be observed within the laboratory. Possible reasons for false-negative results include:

1. Insufficient sensitivity or incorrect selection of the target drug
2. Drug metabolites or structurally related compounds contributing to immunoassay response, which are then not part of the confirmatory profile
3. Inadequate sample clean-up prior to analysis
4. Lack of cleanliness of the extract resulting in interference in the assay.

Conversely, false-positive results can be observed due to

1. Increased sensitivity of the immunoassay systems
2. Unrelated compounds contributing to immunoassay response
3. Reliance on screen only results.

ElSohly et al. reported the analysis of 30 meconium specimens by GC-MS analysis for all analytes regardless of the screening results to determine the false-negative rate, if any, of the immunoassay. They determined that there were no false negatives detected using their immunoassay procedures, but the confirmation rate for the positive specimens was generally low. Surprisingly, no specimens were positive for APs. The lowest rate of confirmed positives was found with the cannabinoids. The authors suggested that THC metabolites other than free THC-COOH may be major contributors to the immunoassay response in meconium.

Positive Immunoassay Tests

As previously mentioned, presumptively positive immunoassay tests require that a second assay based on a scientifically separate principle be used to ensure the identification of the drug. For several years, GC-MS was the preferred confirmatory procedure for many drugs-of-abuse. High-pressure liquid chromatography (LC) with ultraviolet, fluorescence, or diode array detection procedures have been reported, but advancing technology has allowed the use of LC coupled to MS to be applied to the analysis of meconium.

Effect of cocaine

The majority of research has focused on cocaine, as it is considered to be a widely abused drug and there are many reports of neonatal consequences of maternal use. At least fifteen metabolites or adducts of cocaine plus alcohol, as well as cocaine itself, are reportedly present in meconium. These include BZE, benzoylnorecgonine (BN), norcocaine (NC), ecgonine (ECG), ecgonine methyl ester (EME), ecgonine ethyl ester (EEE), cocaethylene (CE), norcocaethylene, *meta*-hydroxybenzoylecgonine (*m*-OH-BZE), *para*-hydroxybenzoylecgonine (*p*-OH-BZE), *para*-hydroxycocaine (*p*-OH-COC), *m*-hydroxycocaine (*m*-OH-COC), anhydroecgonine methyl ester (AEME), anhydroecgonine, and cocaine-*N*-oxide.

Detection of metabolites

In urine, the major metabolite detected following cocaine ingestion is BZE. Therefore, most immunoassay platforms and, in earlier research confirmatory assays were targeted to the detection of BZE. The first confirmatory procedure was reported by Clark et al. in 1992 and employed a methanolic extraction, followed by the addition of phosphate buffer, SPE, and derivatization of BZE using trimethylsilyl derivatives. The assay was linear to 10 ng/g for both cocaine and BZE. However, in 1993, Steele et al. first identified m-OH-BZE as a major contributor to immunoassay-positive results in samples collected from cocaine-exposed newborns. Using the standard confirmation profile of cocaine and BZE, these specimens, which had screened positively, did not confirm. This observation caused the development of new procedures for a wide range of cocaine metabolites.

In 1994, Lewis et al. subsequently confirmed this observation and reported that up to 23% of newborns screening positively for cocaine metabolites were not being correctly diagnosed as the confirmation profile did not include m-OH-BZE. Oyler et al. conducted an extensive study into the metabolism of cocaine in the newborn following fetal exposure by analyzing meconium and urine for all the metabolites named above, except cocaine-*N*-oxide. They were the first to identify *p*-OH-BZE in meconium and suggested that this newly identified metabolite, like *m*-OH-BZE, might serve as a valuable marker of fetal cocaine exposure during pregnancy. The presence of cocaine and AEME in meconium was attributed to transfer across the placenta from the mother. However, the origin of the hydrolytic and oxidative metabolites of cocaine could not be established because they were also identified in urine specimens of adult female cocaine users and could have arisen in meconium from either fetal or maternal metabolism.

Later, ElSohly et al. reported the presence of *m*-OH-BZE and also identified *p*-OH-BZE as a cocaine metabolite in meconium. Abusada et al. reported the presence of cocaine, BZE, EME, and CE in meconium. The presence of CE in meconium was the first identification of this important adduct, as its presence indicated maternal intake of alcohol as well as cocaine. Subsequently, CE has been included in many confirmation profiles for cocaine and its metabolites. CE was identified in 31.6% of samples containing cocaine and/or BZE. The authors also reported that "cocaethylene accumulates in greater concentrations in meconium than urine, and is a useful analyte for identifying fetal alcohol exposure."

High performance liquid chromatography (HPLC)

In 1994, Browne et al. used *high-performance liquid chromatography* (HPLC) and GC/MS to analyze meconium for cocaine, NC, and CE in the meconium of premature infants. The report was the first identification of NC in meconium samples, and the protocol was adopted by Dusick et al., who tested the meconium of 323 very low-birth-weight newborns. In 1993, Murphey et al. were the first to report the presence of benzoylenorecgonine (BN) in meconium using HPLC. Following SPE of the drugs from the sample, separation of cocaine, BZE, NC, and BN was achieved using a Microsorb C18 column (100 × 4.6 mm; 3 µm particle size) with a mobile phase of 0.01 M NaH_2PO_4 at pH 2.0 with 58 µm of tetrabutylammonium hydroxide and 13% acetonitrile pumped at a flow rate of 1 mL/min. The assay was linear from 50 to 5000 ng/g. BN, a previously unreported metabolite, was detected in 7 of 11 meconium samples from neonates born to cocaine using women. However, in an interesting article published in 2000, Xia et al. suggested that ECG was in fact the best marker for the determination of cocaine exposure although the authors recommended an eight-metabolite profile for confirmation. Their procedure employed SPE of the drugs followed by analysis using LC with tandem mass spectrometric detection.

The HP 1100 LC system was interfaced to a Micromass Quattro II triple quadrupole MS. The authors reported the presence of ECG, *p*-OH-cocaine, and CE in 21 of 22 samples (one sample was negative for all metabolites). The presence of CE in every sample is surprising, as it is only formed when there is concurrent intake of cocaine and ethanol. This indicated that all the women in the study also exposed their newborns to alcohol. The authors attempted to explain this, stating CE was only present above the level of 10 ng/g in 15 of the 21 samples (71.4%). The detection of ECG in all the specimens is particularly interesting, as the authors note that specific washes of the solid-phase cartridge (hydrochloric acid, followed by extensive methanol washing) as recommended in the manufacturer's procedure would in fact remove all the ECG prior to analysis. Along with the absence of derivatization needed for GC/MS determination, this may account for the reason ECG has previously not been considered a major metabolite in meconium.

ECG was measured at the highest median concentration in all the samples, whereas the median concentration of the *meta*-hydroxylated cocaine and BZE were higher than the corresponding *para*-hydroxylated metabolites. However, the *para*-hydroxylated metabolites were present in more specimens than the *meta*-hydroxy cocaine and BZE. The authors detected all the metabolites in many of the samples. They recommended a profile of eight analytes: cocaine, CE, ECG, *m*-OH-

BZE, *p*-OH-BZE, *p*-OH-cocaine, NC, and BZE be monitored in meconium analysis to provide the greatest utility in the detection of fetal exposure to cocaine using meconium specimens. Cocaine-*N*-oxide, a metabolite that had previously only been identified in a single case study, was detected in 12 of the 21 samples (57%). In 2005, Pichini et al. applied a previously reported LC/MS procedure to the determination of *m*- and *p*-OH-BZE in meconium using nalorphine as internal standard. The method included a methanol extraction, evaporation of the solvent, and sample preparation using solid-phase columns.

The drug recovery was 60-65%. Reversed-phase chromatography with a C8 analytical column (150 mm × 4.6 mm) was used, with a gradient of 1% acetic acid-acetonitrile as the mobile phase. The system was coupled to atmospheric pressure ionization electrospray-MS single-ion monitoring mode. This method was valid from 5 to 1000 ng/g of meconium and when applied to authentic specimens produced quantitative values between 7 and 338 ng/g in meconium for *m*-OH-BZE and values between 7 and 319 ng/g for *p*-OH-BZE. In all positive specimens, cocaine and/or BZE were also present in contrast to other reports. Even though *m*-OH-BZE is considered to be glucuronide bound in meconium, there is sufficient free drug to be detected in a standard confirmation profile using GC/MS without the need for hydrolysis.

Effects of heroin and codeine

Gas chromatography-mass spectrometry. To date, there are no reports of heroin detected in meconium. In 2001, Salem et al. reported a comparison of extraction procedures for the determination of 6-acetylmorphine (6-AM), a primary metabolite of heroin, in specimens from neonates. SPE cartridges were evaluated for their effectiveness in sample preparation. Four different types of commercially available extraction cartridges were used. 6-AM, morphine, and codeine were extracted from meconium samples using these SPE cartridges and analyzed by GC-MS. The limits of quantitation (LOQ) were 20 ng/g for codeine, 10 ng/g for morphine, and 5 ng/g for 6-AM, and the assays were linear over wide concentration ranges. Using nine previously screened opiate-positive meconium specimens, the authors reported the presence of 6-AM in eight of the samples at concentrations ranging from 3.19 to 70.2 ng/g. All of the specimens also contained codeine (31.25-692.71 ng/g) and morphine (264.1-1617.09 ng/g). The detection of 6-AM, however, is not widespread and is rarely reported.

In the Maternal Lifestyle Study previously described, only one sample out of 8527 meconium samples collected was positive for 6-AM using this procedure from 182 positive opiate screens (0.54%). In this study, a screening cut off of 50 ng/g was used, and this may be too high for consistent confirmation. Morphine accounted for the majority of the positive results, detected in 101 specimens (55.5%), codeine in 83 (45.6%), hydrocodone in one sample (0.54%), and hydromorphone in three samples (1.7%). Morphine, a metabolite of both heroin and codeine, is much more commonly encountered in meconium analysis and has been reported in the meconium from stillborn babies.

Liquid chromatography and liquid chromatography-mass spectrometry. The application of HPLC, LC/MS, and LC/MS/MS to opiates in meconium has not been widely reported; however, the ability to analyze glucuronide-bound drugs without hydrolysis is a distinct advantage to these techniques. Pichini et al. were the first to report a procedure based on LC-MS for the simultaneous determination of 6-AM, morphine, morphine-6-glucuronide, morphine-3-glucuronide, codeine, cocaine, benzoylecgonine, and CE in meconium using nalorphine as internal standard. The analytes were extracted from meconium using methanol for 6-AM, morphine, codeine, cocaine, BZE, and CE. Following solubilization, the supernatant was evaporated and part of the specimen reconstituted in phoshate buffer before being subjected to further sample clean-up using mixed mode SPE columns for 6-AM, morphine, codeine, cocaine, BZE, and CE. For the glucuronides, a portion of the methanol extract was reconstituted in 0.01 M ammonium hydrogen carbonate buffer (pH 9.3) and subjected to SPE using ethyl columns.

The analytical column was a C8 reversed-phase column using a gradient of acetic acid 1%-acetonitrile as a mobile phase. Analytes were determined in LC-MS single ion monitoring mode with atmospheric pressure-electrospray (ESI) interface. The linearity of the procedure was 5-1000 ng/g. When applied to patient samples, the presence of 6-AM was confirmed, an observation that had previously only been reported by Salem et al., at concentrations of 5, 6, and 142 ng/g in meconium. The concentrations of 6-AM measured in meconium appear to be low and require sensitive technology to identify fetal exposure to heroin. Pichini et al. were also the first to identify morphine-3- and morphine-6-glucuronide in meconium, with one specimen containing 120 ng/g of morphine-3-glucuronide and 91 ng/g of morphine-6-glucuronide. An important practical advantage of using LC for the chromatographic separation of opiates is that hydrolysis of the specimen is not required.

Effects of hydrocodone and hydromorphone

Gas chromatography-mass spectrometry. In 1995, Moore et al. reported the detection of hydrocodone and hydromorphone, for the first time, as well as codeine and morphine in meconium. Following hydrolysis of the synthetic opiates in 2.4 M hydrochloric acid (110°C/h), 11.8 M potassium hydroxide and buffer salts were added, and the drugs were extracted using tert-butyl methyl ether. Following reacidification and back extraction, the final solvent was evaporated to dryness and the drugs derivatized with BSTFA for analysis using GC/MS. Hydrolysis of the specimens resulted in a substantial increase in the amount of codeine, hydrocodone, and hydromorphone detected but no significant increase in the morphine concentration.

With the exception of morphine, the other opiates appear to be significantly glucuronide bound, and in routine procedures, hydrolysis of meconium specimens is not always conducted. Habrdova et al. reported the presence of hydrocodone, norhydrocodone, and dihydrocodeine in the urine and meconium of a newborn in 2001, but few methodological details were provided. This report is the first identifying dihydrocodeine and/or norhydrocodone in meconium.

Effect of oxycodone

Gas chromatography-mass spectrometry. The detection of oxycodone in meconium was reported for the first time in 2005. The availability of a sustained release formulation in recent years and the potential for use during pregnancy resulting in withdrawal symptoms in the newborn prompted the development of procedures for its analysis in meconium samples. To date, there are no reports of kinetic disposition of oxycodone in meconium. For analysis, the specimens were screened with an oxycodone Direct ELISA kit, then confirmed by GC/MS. The specimen was homogenized in methanol, centrifuged, the supernatant was evaporated to dryness, and refrigerated overnight. The following day, 0.1 M hydrochloric acid was added with methoxyamine hydrochloride. The mix was incubated at room temperature, phosphate buffer added, and the specimen subjected to mixed mode solid phase extraction. The final extract was derivatized with BSTFA +1% TMCS. The authors reported positive results for three specimens, with concentrations of 62, 224, and 490 ng/g oxycodone. No metabolites were reported.

Effects of methadone

Gas chromatography-mass spectrometry. In 2001, ElSohly et al. described an EMIT-ETS d.a.u. immunoassay screening method for methadone in meconium followed by GC-MS confirmation for methadone and metabolites 2-ethylidene-1,5-dimethyl-3,3-diphenylpyrrolidine (EDDP) and 2-ethyl-5-methyl-3,3-diphenylpyrroline (EMDP). The GC-MS method was linear between 25 and 2000 ng/g with reported LOQ of 25 ng/g for all drugs. Fifty meconium samples were screened using a cut off of 200 ng/g, and all samples screened negative. GC/MS analysis showed that four samples contained methadone (35.2-79.9 ng/g), EDDP (28.5-557.2 ng/g), or both, with no detectable amount of EMDP. The authors stated that "the negative results on the four specimens at the cut off used may be explained by the fact that EMIT-ETS d.a.u. antibody for methadone was specific to the parent drug. The results point to the fact that immunoassays should be directed to EDDP for detection of prenatal exposure of methadone through analysis of meconium specimens".

Liquid chromatography and liquid chromatography-mass spectrometry. In 1997, Stolk et al. first developed methods for the analysis of methadone and its principal metabolite, EDDP, in meconium based on FPIA and HPLC with diode array detection. Meconium and urine samples of 16 neonates from 15 methadone-using mothers were analyzed. Methadone, EDDP, or both were detected in both urine and meconium samples from 15 of the newborns. However, the authors reported that the amount of EDDP in meconium was approximately 10 times higher than the amount of methadone. In 2005, Choo et al. reported a validated LC atmospheric pressure chemical ionization tandem mass spectrometry (LC-APCI-MS/MS) method for the quantification of methadone, EDDP, EMDP, and methadol in meconium. The procedure utilized sample clean-up with solid phase extraction before analysis.

The LOQ was 5 ng/g for methadone, EDDP, and EMDP; 25 ng/g for methadol, and the assay was linear from 5 to 500 ng/g. This was the first report of the presence of methadol in meconium and the first LC/MS procedure for all the analytes. The authors reported the analysis of a meconium specimen collected from an infant whose mother was maintained on methadone for 19 weeks during gestation. The meconium contained 2492 ng/g of methadone, 13,188 ng/g of EDDP, and 27 ng/g of EMDP. These results agreed with other publications, indicating the amount of EDDP in meconium is significantly higher than the level of methadone.

Effects of amphetamines

There are few publications regarding the analysis of APs in meconium. In the large-scale Maternal Lifestyle Study (8527 meconium samples), no APs were reported to be present, even from geographic areas of high MA abuse. The lack of positive specimens in this population may have been methodologically related, as APs are notoriously volatile and may be lost during analysis.

Another possibility is that, similar to cocaine and THC, there were metabolites of the APs contributing to the immunoreactive response that were not included in the confirmatory assay. Habrdova et al. reported the presence of MA and ephedrine in the urine and meconium of a newborn in 2001, but extraction recovery for the drugs was low.

Determination of AP

In 1994, Franssen et al. included the determination of AP but not MA in the HPLC assay previously described in the opiate section of this chapter. The detection level was high (500 ng/g) and no positives were reported. More recently, Pichini et al. published a procedure based on LC-MS for the determination of AP, MA, 3,4-methylenedioxy-methamphetamine (MDMA), 3,4-methylenedioxyamphetamine (MDA), 4-hydroxy-3-methoxymethamphetamine (HMMA), 3,4-methylenedioxyethylamphetamine (MDEA), and *N*-methyl-1-(3,4-methylenedioxyphenyl)-2-butamine (MBDB) in meconium, using 3,4- methylendioxypropylamphetamine (MDPA) as internal standard. The analytes were initially extracted from the matrix using a methanol-hydrochloric acid mix, followed by SPE. Chromatography was performed on a C1 8 reversed-phase column using a linear gradient of 10 mM ammonium bicarbonate, pH 9.0-methanol as a mobile phase. Analytes were determined in LC-MS single-ion monitoring mode with an atmospheric pressure ESI interface. The method was validated in the range 5-1000 ng/g using 1 g of meconium per assay. Mean recoveries ranged between 80 and 90% for different analytes, with only AP demonstrating a lower extraction recovery of 60%.

The quantification limits were 5 ng/g meconium for AP, MA, and HMMA; 4 ng/g for MDA, MDMA, MDEA, and MBDB. The method was applied to analysis of meconium in newborns to assess fetal exposure to AP derivatives. Over 600 specimens were analyzed, and none were positive for AP, MA, HMMA, or MDA. Only one was positive for MDMA, at a concentration of 12 ng/g, the first report of such a finding. When the procedure was applied to previously analyzed samples that had been in storage at -20°C for at least 1 year, the concentrations of both AP and MA showed excellent correlation, indicating the stability of both AP and MA in meconium. While not specifically reporting the presence of MDMA in meconium, Ho et al. discussed the characteristics of pregnant women using Ecstasy. They concluded that pregnant women using MDMA tended to be younger, single, had more abortions, and reported higher levels of psychological problems than those not using the drug. Higher rates of unplanned pregnancy, heavy alcohol, cigarette, and illicit drug use were also reported, which may cause more problems in the newborn.

Effects of phencyclidine

With regard to reports of drugs-of-abuse and meconium, the least number of publications have addressed the analysis of PCP and/or its metabolites although some reports are available. PCP remains a widely used illicit drug, especially among adolescents and young adults. The effects in the newborn of PCP are similar to those of cocaine. A study by Tabor et al. compared infants exposed to PCP in utero with those exposed to cocaine. Both groups had a high incidence of intrauterine growth retardation, precipitate labor, and symptoms of neonatal drug withdrawal. However, interestingly, PCP-exposed newborns were more likely to have meconium-stained amniotic fluid, and less likely to be born prematurely as cocaine-exposed infants. Based on this data, it is possible that PCP-exposed babies are not tested as often as those exposed to other drugs, as prematurity is often a risk factor for testing and meconium is less likely to be available.

Confirmatory procedure

In 1996, Moore et al. reported a confirmatory procedure for the determination of PCP in meconium using the *selected ion storage* (SIS) functions of an ion trap mass spectrometer. The method was particularly sensitive, with the authors reporting linearity up to 250 ng/g and a detection level of 5 ng/g. The procedure was useful for meconium analysis, as sample size is often limited and sensitivity is an important factor.

Effect of cannabinoids

Of the drugs discussed, marijuana (THC) and its metabolites are the most difficult to analyze in meconium. Wingert et al. reported difficulty in confirming screen positive results, probably due to the low concentration present in the sample. Additionally, the major urinary metabolite of THC, 11-nor-Δ9-THC-9-carboxylic acid (THC-COOH), is glucuronide bound in meconium, requiring hydrolysis of the sample to release the drug to permit significant drug levels to be detected. Surprisingly, there are no literature reports of the detection of THC-COOH in the urine of newborn babies, so the analysis of meconium may offer a distinct advantage for the confirmation of marijuana exposure in the neonate.

Detection through gas chromatography-mass spectrometry

In 1996, Moore et al. first published a confirmation procedure for the determination of 11-nor-Δ9-tetra-hydrocannabinol-9-carboxylic acid (THC-COOH), a major metabolite of Δ9-THC in meconium. As THC-COOH is significantly glucuronide bound in meconium, the specimen was hydrolyzed with potassium hydroxide and extracted with hexane-ethyl acetate (9:1 v,v). Following back extraction, the final extract was derivatized with MTBSTFA for GC/MS analysis. The confirmation rate at 25 ng/g for a screening concentration was 80%, using 2 ng/g as the limit of detection. The authors suggested that other metabolites of THC may be contributing to the immunoreactive response and were not part of the confirmation profile. This observation was subsequently investigated by other research groups. Feng et al. analyzed the presence of other metabolites in meconium by developing the use of an immunoaffinity extraction column for application to various biological matrices and applying the method to meconium extraction. Using the affinity resin prepared by immobilization of THC antibody onto cyanogen bromide-activated Sepharose 4B, Δ9-THC and its major metabolites including THC-COOH, 11-hydroxy-Δ9- THC (11-OH-THC), and 8β,11-dihydroxy-Δ9-THC were extracted simultaneously from plasma or urine after enzyme hydrolysis.

The modified procedure for meconium produced lower extraction efficiencies, ranging from 52 to 72% at the 10 ng/g level. The assay was linear to 100 ng/g, and the limit of detection for the metabolites ranged from 1 to 2.5 ng/g of meconium. Analysis of 24 meconium specimens showed that 11-OH-THC was in fact an important metabolite in meconium. Having also identified a low confirmation rate for cannabinoids when only THC-COOH was considered, ElSohly et al. investigated the possible contribution of other metabolites of THC, including glucuronides, to the overall response of the immunoassay. Δ9-THC-glucuronide was synthesized, and procedures were developed for the extraction and GC- MS analysis of THC, 11-OH-THC, 8α- and 8β-OH-Δ9-THC, 8 β,11-diOH-Δ9-THC, and THC-COOH after enzymatic hydrolysis of meconium extracts.

The authors concluded that enzymatic hydrolysis of meconium extracts was necessary for efficient recovery of THC metabolites. No significant amounts of THC or its 8-OH metabolite were detected in meconium; however, 11-OH-THC and 8 β,11-diOH-Δ9-THC did show significant contribution to immunoassay response. The authors reported several specimens for which THC-COOH was not present, but 11-OH-THC and/or 8β,11-diOH-Δ9-THC were detected. Only one report of the identification of THC-COOH in meconium using LC with diode array detection is available. The report is a letter to the Editor published in 1995 which provided little analytical information.

CONCLUDING REMARKS

Positive Findings

Because of the considerable number of variables during the latter half of pregnancy (~20 weeks) and the ethical limitations of providing drugs to pregnant women, all attempts at interpreting drug levels in meconium are generally based on anecdotal evidence. The presence of drugs and metabolites in neonatal meconium is scientifically defensible as an indication of drug use by the mother at some point in the latter half of pregnancy. This is provided the testing has been performed with validated laboratory testing methods, and any screening results have been confirmed by a second method based on a scientifically different principle. However, there is insufficient evidence at this time to support the interpretation of quantitative values detected in meconium. Lester et al. in the Maternal Lifestyle Study cautioned against the use of quantitative values to determine the degree of drug exposure.

Negative Findings

The presence of no drugs or metabolites in meconium in cases where drug use is either admitted by the mother or suspected by health professionals should be interpreted with caution. First, an insufficient specimen quantity may be the problem with premature babies or from newborns suffering from meconium-stained amniotic fluid. Second, the detection levels employed by the laboratory should be evaluated, as high cut-off concentrations will cause low-level positive specimens to be missed. Lastly, the metabolites being tested for should be assessed, because as explained earlier in the chapter, neonatal drug profiles are different than those of adults, and specific metabolites must be included in the testing profile to ensure reliability of results.

Advantages of Meconium Analysis

There are several advantages to the use of meconium as a sample matrix:

1. The major advantage of using meconium for the determination of fetal drug exposure is its ability to indicate a longer history of potential exposure than urine. Meconium can provide up to a 20 week "window" of detection assuming that meconium starts to form around the 16th week of gestation. Maternal and/or fetal urine will only indicate the presence of drugs if they were ingested within a few days of sample collection.

2. Meconium, in comparison to urine, is easy to collect. For newborn urine collection, bags are taped to the babies, and these often become detached or urine is spilled. Meconium is simply scraped from a diaper and placed in the collection vial.
3. Meconium is not an invasive sample to collect in contrast to blood or newborn saliva. As it is a normal biological waste product, it is collected rather than discarded.
4. Drugs are stable in meconium at room temperature for up to 2 weeks, which is an important advantage for routine collections at large hospitals. Immediate refrigeration after sample collection is not always possible, and if sending to a testing laboratory, shipping on dry ice increases costs.

Disadvantages of Meconium Analysis

There are several disadvantages associated with meconium analysis:

1. As meconium is not a homogenous sample, it is sometimes difficult to collect all neonatal discharges, which may be days apart. The entire specimen collected should be mixed before testing to allow the drugs to distribute throughout the matrix.
2. Meconium analysis is more labor intensive and requires more time to analyze than urine. Many hospitals are not equipped to analyze meconium, so specimens are sent out to other testing facilities, resulting in increased costs and turnaround time.
3. There are differences in metabolic profiles between meconium and urine. Therefore, standard immunoassay procedures targeted toward urinalysis may cause false-negative results. Similarly, even if samples screen positively, confirmation methods may not include the appropriate metabolites.
4. Often the sample size obtained is small, and the concentration of drugs found in meconium is relatively low. However, advances in technology have improved the sensitivity of testing procedures, so this is increasingly of less importance.
5. There are no standardized procedures, testing profiles, proficiency programs, or control specimens available to ensure laboratory quality regarding meconium analysis.

The detection of drugs, metabolites, or markers of drug intake in the meconium of newborns has been widely reported. This chapter has discussed the major drug classes associated with maternal abuse. However, two other major xenobiotics that have been investigated regarding effects on the newborn are nicotine and alcohol. The detection markers in meconium for the identification of nicotine, alcohol, or both have been reported. The development of faster, more sensitive procedures for the determination of drugs-of-abuse in meconium is an on-going process. The science of meconium analysis is constantly improving as advances in analytical instrumentation and technology are incorporated into hospitals and testing laboratories.

12

Pharmaceutical Studies of Hair

In the 1960s and 1970s, hair analysis was initially used to evaluate exposure to heavy metals, such as arsenic, lead, or mercury. Researchers hypothesized that hair was a preferable specimen to blood and urine for evaluating environmental hazards, because hair could store substances for an extended time period. Therefore, analyses of whole or segmented hairs might provide objective data on the extent of an individual's exposure to metals and the potential for harm. Examination of hair for organic substances, especially pharmaceuticals and drugs-of-abuse, occurred several years later because of the lack of sensitivity of analytical methods for these compounds. In 1979, Baumgartner and colleagues published the first report using *radiommunoassay* (RIA) to detect morphine in the hair of heroin abusers. This paper was followed by many studies that mostly utilized RIA and/or *gas chromatography/mass spectrometry* (GC/MS).

Today, chromatographic procedures, especially those coupled to MS, represent the gold standard for the identification and quantification of drugs in hair because of their separation ability and sensitivity. Technically, testing of hair for drugs is not more difficult or challenging than testing in many other matrices. In fact, the application of analytical methods and instrumental approaches is in most cases quite similar, regardless of the initial sample preparation. Today, hair analysis is routinely used as a tool for detection of xenobiotics (drugs-of-abuse, pharmaceuticals, environmental contaminants, hormones, etc.) in forensic science, traffic medicine, occupational medicine, and clinical toxicology. After a discussion of drug incorporation into this matrix, this view aims to summarize and discuss the various applications of hair analysis.

BIOCHEMISTRY

Hair, a product of differentiated organs in the skin of mammals, is composed of protein (65-95%, keratin essentially), water (15-35%), lipids (1-9%), and minerals (<1%). The hair shaft consists of an outer cuticle that surrounds a cortex, and the cortex surrounding a central medulla in some types of hair. The hair shaft begins in a follicle closely associated with glands (sebaceous and apocrine) and grows in cycles, alternating between periods of growth (*anagen* phase) and periods of quiescence (*catagen* and *telogen* phases). Of the approximately 1 million hair follicles of the adult human scalp, approximately 85% of the hair is in the growing phase and the remaining 15% is in a quiescent stage at any time. Head hair is produced for 4-8 years at a rate of approximately 0.22-0.52 mm/day or 0.6-1.42 cm/month. Hair from other locations typically grows for 6 months. The growth rate depends on hair type (racial difference), age, gender and anatomical location. It is considered that the vertex posterior region of the head provides least variability in growth rate.

MECHANISM OF DRUG DEPOSITION

It is generally proposed that drugs may enter into hair by two processes: adsorption from the external environment and incorporation into the growing hair shaft from blood supplying the hair follicle. Drugs can enter the hair from exposure to chemicals in aerosols, smoke, or secretions from sweat and sebaceous glands. Sweat is known to contain drugs present in blood. Because hair is very porous and can increase its weight up to 18% by absorbing liquids, drugs may be transferred easily to hair through sweat. Finally, chemicals present in air (smoke, vapors, etc.) can be deposited onto hair.

Drugs appear to be incorporated into the hair by at least three methods: from the blood during hair formation from sweat and sebum and from external environment. This model is more able than a passive model (transfer from blood into the growing cells of the hair follicle) to explain several experimental findings such as the following:

1. Drug and metabolite(s) ratios in blood are quite different than those found in hair; and
2. Drug and metabolite(s) concentrations in hair differ markedly in individuals receiving the same dose.

Evidence for the transfer of the drug through sweat and sebum can be suggested as drugs and metabolites are present in sweat and sebum at high concentrations and persist in these secretions longer than in blood. The exact mechanism by which chemicals are bound into hair is not known. It has been suggested that passive diffusion may be augmented by drug binding to intracellular components of the hair cells such as the hair pigment, melanin. For example, the codeine concentration in hair after oral administration is dependent on melanin content. However, this is probably not the only mechanism as drugs are trapped into the hair of albino animals, which lack melanin. Another mechanism proposed is the binding of drugs with sulfhydryl-containing amino acids present in hair.

There is an abundance of amino acids such as cystine in hair which form cross-linking S-S bonds to stabilize the protein fiber network. Drugs diffusing into hair cells could be bound in this way. From various studies, it has been demonstrated that after the same dose, black hair incorporates much more drugs than blond hair. This has resulted in discussions about a possible racial bias in hair analysis and is still under evaluation. After incorporation in the hair shaft, organic substances are capable of surviving for hundreds of years under favorable conditions (protected from light and humidity). For example, the cocaine metabolite, benzoylecgonine, was detected in the hair of ancient, mummified human remains of 163 individual samples. These samples were obtained from populations living in northern Chile in the past 4000 years.

METHODOLOGY

Collection procedures for hair analysis for drugs have not been standardized. In most published studies, the samples are obtained from random locations on the scalp. However, hair is best collected from the area at the back of the head, called the *vertex posterior.* Compared with other areas of the head, this area has less variability in the hair growth rate, the number of hairs in the growing phase is more constant, and the hair is less subject to age- and sex-related influences. The sample size varies considerably among laboratories and depends on the drug to be analyzed and the test methodology. Sample sizes reported in the literature range from a single hair to 200 mg. When sectional analysis is performed, the hair is cut into segments of about 1, 2, or 3 cm, which corresponds approximately to about 1, 2, or 3 months' growth. When scalp hair is not available, other types of hair (pubic hair, arm hair, or axillary hair) can be collected as an alternative source for drug detection. Contaminants of hair would be a problem if they were drugs-of-abuse or their metabolites or if they interfered with the analysis and interpretation of the test results. It is unlikely that anyone would intentionally or accidentally apply anything to their hair that would contain a drug of abuse.

The most crucial issue facing hair analysis is the avoidance of technical and evidentiary false-positives. Technical false-positives are caused by errors in the collection, processing, and analysis of specimens, whereas evidentiary false-positives are caused by passive exposure to the drug. Various approaches for preventing evidentiary false-positives due to external contamination of the hair specimens have been described. The majority of laboratories use a wash step; however, there is no consensus or uniformity in the washing procedures. Among the agents used in washing are detergents such as Prell shampoo, surgical scrubbing solutions, surfactants such as 0.1% sodium dodecylsulfate, phosphate buffer, or organic solvents such as acetone, diethyl ether, methanol, ethanol, dichloromethane, hexane, or pentane of various volumes for various contact times.

Generally, a single washing step is utilized, sometimes a second identical wash is performed. If external contamination is found by analyzing the wash solution, the washout kinetics of repeated washing may demonstrate that contamination has been removed. Detection of drug metabolite(s) in hair, whose presence could not be explained by hydrolysis or environmental exposure, would unequivocally establish that internal drug exposure had occurred. Cocaethylene and nor-cocaine

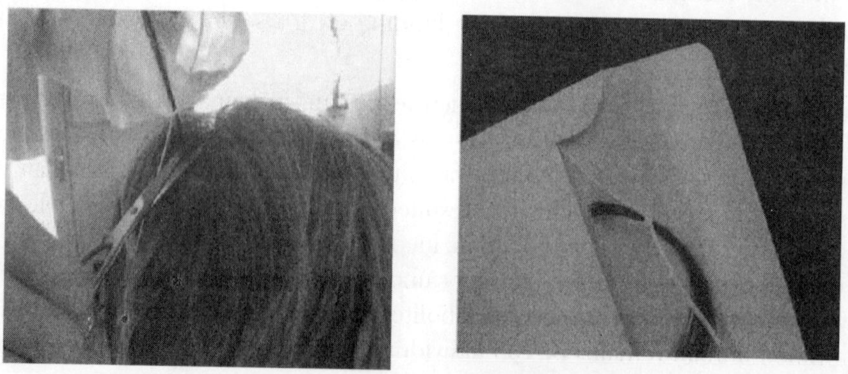

Fig. 12.1. Hair collection procedure.

would appear to meet these criteria, as these metabolites are only formed when cocaine is metabolized. Because these metabolites are not found in illicit cocaine samples, they would not be present in hair as a result of environmental contamination, and thus their presence in hair could be considered a marker of cocaine exposure. This concept may be extended to other drugs, such as THC-COOH for cannabis. However, there is still controversy about the potential risk of external contamination, particularly for crack, cannabis, and heroin when smoked as several authors have demonstrated that it is not possible to fully deduct due to contamination.

In conclusion, although it is highly recommended to include a decontamination step, there is no consensus on which procedure performs best. Therefore, each laboratory must validate its own technique. After decontamination, the hair sample can be pulverized in a ball-mill or cut into segments before the hydrolysis step or dissolved in alkaline medium (NaOH) to enhance drug solubilization. Finally, the xenobiotics are extracted or purified from the incubation medium before the analysis. The first publication describing the analysis of morphine in hair for determining opiate abuse histories reported on the analysis with RIA. This paper was followed by a large number of studies, which mostly included RIA and/or GC/MS. Chromatographic procedures seem to be a more powerful tool for the identification and quantification of drugs in hair, owing to their separation ability and high sensitivity.

Chromatographic techniques [GC or high-performance liquid chromatography (HPLC)] coupled to MS or tandem MS represent the gold standard in hair analysis for xenobiotics. The Society of Hair Testing has published recommendations for hair testing in forensic cases that include cut-off concentrations. Numerous papers have been published regarding the analytical aspects of drug detection. These papers include methods to test for opioids, cocaine, cannabis, amphetamines, phencyclidine, *gamma-hydroxybutyric acid* (GHB), or benzodiazepines.

APPLICATIONS

Correlative Testing

There are essentially three problems with drug testing in urine: false-positives, instability of some compounds, and evasive maneuvers on the part of the subject such as adulteration and substitution. These problems can be reduced or eliminated using hair analysis. It is possible to obtain a fresh, identical hair sample if there is a claim of specimen mix-up or breach in the chain of custody. When head hair is missing (shaved head, bold subject), it is possible to collect alternative sources of hair, such as pubic or arm hair. This makes hair analysis essentially fail-safe in contrast to urinalysis, as an identical urine specimen cannot be obtained at a later date. To the best of the author's knowledge, no claims of external contamination have been addressed in France in any forensic situation. Another potential use of hair analysis is to verify accidental or unintentional ingestion of liquid or food laced with drugs.

In the case of single use, the hair will not test positive. Ingestion of poppy seeds appears to be sufficient for the creation of a positive opiate result in urine, whereas ingestion of up to 30 g of poppy seeds did not result in a positive hair result. Its greatest use, however, may be in identifying false-negatives from urine testing, as neither abstaining from a drug for a few days nor trying to "*beat the test*" by diluting urine will alter the concentration in hair. Urine testing does not provide information relating to the frequency of drug use in subjects who might deliberately abstain for several days before biomedical screenings. While analysis of urine specimens cannot distinguish between chronic use or single exposure, hair analysis can make this distinction.

Table 12.1 Comparison between urine and hair

Parameters	*Urine*	*Hair*
Major compound	Metabolites	Parent drug
Detection period	2–5 days	Weeks, months
Type of measure	Incremental	Cumulative
Screening	Yes	Difficult
Invasiveness	High	Low
Storage	-20°C	Ambient temperature
Risk of false-negative	High	Low
Risk of false-positive	Low	Undetermined
Risk of adulteration	High	Low
Control material	Yes	Needed

Effects of Long Exposure of Drug

By providing information on exposure to drugs over time, hair analysis may be useful in verifying self-reported histories of drug use in any situation in which a history of past rather than recent drug use is desired, such as in pre-employment and employee drug testing. Hair drug testing will detect use in an addict. In the case of an addict who takes drugs only every few days, a urine and/or blood test may be negative even when the tests are repeated. Hair analysis can also provide a retrospective history of an individual's drug use. For this work, multi-sectional analysis is required and involves taking a length of hair and cutting it into sections to measure drug use during shorter periods of time.

The hair must be cut as close as possible to the scalp and particular care is also required to ensure that the individual hairs in the cut-off tuft retain the position they originally had beside one another. For example, one can perform multi-section analysis for people who test positive on an initial screen. This information can be used to validate an individual's claim of prior drug use or abstinence during the most recent several months. The most extensive studies on sectional analysis for drugs-of-abuse involved patients in rehabilitation centers. Segmental hair analysis was used to verify both previous drug history and recent enforced abstinence.

In the instance of drug replacement therapy, the lowest drug of abuse concentrations were measured in the segments nearest the root, thus confirming decreased drug use over time or recent abstinence. As illicit heroin samples contain codeine or codeine derivatives, codeine may also be detected in cases of heroin abuse. Morphine is a metabolite of codeine and can be detected when codeine is abused. The differentiation of heroin users from individuals exposed to other sources of morphine alkaloids can be achieved by identifying heroin, 6-acetylmorphine, or acetylcodeine.

Fetal Drug Exposure

The increasing number of people consuming drugs result in an increase in pregnant women under the effect of these drugs. Because of the immediate and long-term health problems, newborns exposed to drugs during pregnancy should be identified soon after birth so that appropriate intervention and follow-up can be performed. Current methods to verify drug abuse include maternal self-reported drug history (self-reports are generally unreliable), maternal urinalysis (risk of false-negative results due to the short elimination half-life of the drugs and positive results only reflects exposure during the preceding 1-3 days), and analysis of amniotic fluid, urine, or meconium of the baby at time of delivery (qualitative test at the moment of delivery, risk of false-negative results due to abstinence during the preceding 1 or 3 days or limitations of testing technology). Testing the hair of a newborn allows an increase in the window of detection (from weeks to months) and may provide information concerning the degree and pattern of the mother's drug use.

Maternal drug abuse is a health hazard for the fetus, and the effects of cocaine, phencyclidine, nicotine, and other compounds are well documented. In 1987, Parton and colleagues first reported quantitation of fetal cocaine exposure by RIA of hair obtained from 15 babies. Graham and colleagues detected benzoylecgonine (range 0.2 to 27.5 ng/mg) in neonatal hair from seven infants whose mothers were known cocaine users. Other studies have demonstrated placental transfer of maternal drug in neonatal hair.

Excessive Alcohol Consumption

Considering the large number of alcohol-associated problems, the diagnosis of excessive alcohol consumption is an important task from a medical point of view. The methods used for this purpose are based on indirect alcohol markers such as increased liver enzyme activity, increased erythrocyte mean cell volume, or presence of carbohydrate deficient transferrins. These markers may also originate from pathological conditions. Markers of ethanol consumption are ethyl glucuronide, and phosphatidylethanol or *fatty acid ethyl esters* (FAEE). The first investigation of a marker of alcohol consumption in hair was reported by Sachs and colleagues and focused on ethyl glucuronide; however, recent examination of the presence of this ethanol metabolite in hair was rather discouraging.

Detection of ethyl glucuronide in hair is always associated with alcohol consumption, whereas a negative result does not exclude alcohol abuse. FAEE were used by Yeggles and colleagues to monitor alcohol consumption. FAEE are formed in the presence of ethanol and free fatty acids, triglycerides, lipoproteins, or phospholipids, by a FAEE synthase found in the liver but also in hair roots. FAEE determination is of interest as they appear responsible for alcohol-induced organ damage. In blood, FAEE can be used as markers of recent alcohol intake for at least 24 h after cessation of drinking. Hair concentrations of four FAEE (ethyl myristate, ethyl palmitate, ethyl oleate, and ethyl stearate) found in hair of children, adult teetotalers, and social drinkers compared with FAEE concentrations found in hair of alcoholics led the authors to conclude that FAEE are suitable markers for the detection of heavy alcohol consumption. Segmental hair analysis in a case of alcohol withdrawal treatment showed a decrease in FAEE content from the distal to the proximal root segment. Further studies are in progress to examine the applicability of FAEE determination in clinical practice.

Detection of Doping Habits

Athletes use both endogenous (testosterone and DHEA) and exogenous anabolic steroids (nandrolone, stanozolol, and mesterolone) because it has been claimed that they increase lean body mass, strength, and aggressiveness and lead to a shorter recovery time between workouts. Hair testing may be used to identify false-negative results due to recent abstinence for a drug a few days before a competition. Hair can also indicate the history and frequency of drug intake as repetitive abuse can be demonstrated by segmental analysis along the hair shaft in contrast to urinalysis. A search of the international literature demonstrated a lack of articles concerning the identification of anabolic steroids in human hair.

As a complement to testosterone determination, the identification of unique testosterone esters in hair supports a charge of doping because the esters are exogenous substances. In 2000, Thieme and colleagues published a complete analytical strategy for detecting anabolics in hair. After extensive clean-up procedures, drugs were identified either by GLC/MS/MS or GC with *high-resolution mass spectrometry* (HRMS). Metandienone, stanozolol, mesterolone, metenolone enantate, nandrolone decanoate, and several testosterone esters were identified in the hair of several bodybuilders. Another advantage of hair analysis is the possible discrimination between nandrolone and other 19 norsteroids (norandrostenedione and norandrostenediol) abuse that lead to the same urinary metabolites (norandrosterone and noretiocholanolone). Discrimination is not possible in urine, but hair can identify the parent compound. The analysis of a hair strand obtained from an athlete who tested positive for nor-androsterone in urine revealed the presence of 19-norandrostenedione, distinguishing the result from nandrolone doping. Corticosteroids or β-adrenergic drugs can also be identified in hair.

Influence of Psychoactive Drugs

The major practical advantage of hair testing compared with urine and blood testing for drugs is the larger detection window, weeks to months, depending on the length of the hair shaft, compared with a few days for urine. Italy and Germany are using this approach in cases of driving license regranting. Individuals, whose driving license has been refused or suspended for addiction to psychoactive drugs or for driving "under the influence," may claim they have stopped using drugs. They can obtain a license after a medical committee has confirmed abstinence from illicit drugs and excluded any additional risk of relapse. To provide objective evidence of abstinence from drugs with an acceptable chronological window to support the clinical decision of this medical committee, hair analysis has been included in a panel of clinical and laboratory tests aimed at retrospectively investigating the toxicological behavior of subjects. Comparison between hair analysis and urinalysis demonstrates higher diagnostic sensitivity for hair tests.

Modification of Behavior

The use of a drug to modify a person's behavior for criminal gain is not a recent phenomenon. However, the recent increase in reports of drug-facilitated crimes (sexual assault and robbery) has caused alarm in the general public. Drugs involved can be pharmaceuticals, such as benzodiazepines, hypnotics, sedatives or anesthetics; drugs-of-abuse, such as cannabis, ecstasy, or LSD or more often ethanol. Because of low dosing, except for GHB, surreptitious administration into beverages such as coffee, soft drinks, or alcoholic cocktails is relatively simple. Most of these substances possess amnesic properties and therefore, the victims are less able to accurately recall the circumstances under which the sexual offence occurred.

As the drugs are generally short-acting, they impair an individual rapidly. In these situations, blood or even urine may not be helpful. It has been suggested that hair be tested since delays in reporting result in the removal of drug from the routine testing specimens due to metabolism. Although there are many published reports on the identification of drugs (mainly drugs-of-abuse) in hair following chronic use, those addressing controlled single dose are scarce. The ability to differentiate between a single exposure and long-term use can be documented by multi-sectional analysis. If it is assumed that drug does not migrate along the hair shaft, a single point of exposure must be present in the segment corresponding to the period of the alleged event, using a growth rate for scalp hair of 1 cm/month. As this growth rate may vary from 0.7 to 1.4 cm/month, the length of the hair section must be calculated accordingly.

A delay of 4-5 weeks between the offense and hair collection for sectional analysis (2-cm segments) can be considered as satisfactory to ensure that the hair shaft includes the time of the alleged exposure. The hair must be cut as close as possible to the scalp. Particular care is also required to ensure that the individual's hair in the strand retains the position it originally had beside another. After decontamination, hair is then segmented as follows: 0-2 (segment corresponding to the period of alleged crime), 2-4, and 4-6 cm (which should be drug-free). Drug exposure was demonstrated by hair segmentation in several cases. It was observed that the target concentrations in hair after a single exposure are generally in the range of few pg/mg. To obtain the low limits of detection required for such testing mandates the use of highly sensitive instrumentation such as tandem MS. It appears that the value of hair analysis for the identification of drug users is steadily gaining recognition. This can be seen from its growing use in preemployment screening, in forensic sciences, and in clinical applications.

Hair analysis may be a useful adjunct to conventional drug testing in toxicology. Specimens can be more easily obtained with less embarrassment, and hair can provide a more long-term history of drug use. Although there remains controversy on how to interpret the results, particularly concerning external contamination, cosmetic treatments, ethnic bias, or drug incorporation, analytical work in hair analysis has reached a plateau, with almost all the analytical problems solved. Although GLC/MS is the method of choice, in practice, GC- or LC-MS/MS technology is used today in many laboratories, even for routine cases, particularly to target low-dose compounds.

13

Ethanopharmacology

The aim of this book is to provide a conceptual framework for medicinal chemistry. It is now necessary to develop a method of designing drug molecules to fit into receptor molecules. The *multiphore method* of drug design is such a method. The multiphore method conceptualizes a drug as being constructed in a modular fashion from bioactive subunits, or *biophores*. Since a drug is invariably composed of many biophores, it is a *multiphore*. The most important biophore within the drug structure is the *pharmacophore*, the subset of atoms within the drug that permits energetically favorable binding to the receptor site with the elucidation of a subsequent beneficial biological response. Other portions of the molecule determine the metabolic and toxicological properties of the drug; these are the *metabophores* and *toxicophores*, respectively.

In the design of drugs using the multiphore method it is important to remember that there is nothing special about any particular drug molecule. A successful drug molecule is merely a collection of "*hetero-atom rich*" functional groups appropriately positioned on the three-dimensional space of a hydrocarbon framework in a fixed geometrical relationship that enables a desirable interaction with a receptor macromolecule. When the medicinal chemist knows the *bioactive zone* of the receptor macromolecule, he or she identifies multiple functional groups, usually within 15 Å of each other, within that bioactive zone. The selection of these receptor-based functional groups is a crucial step. For example, if the receptor zone were within the brain, it would be inadvisable to select many charged (anionic or cationic) functional groups, since a drug capable of binding to these receptor-based functional groups via electrostatic interactions would be too polar to diffuse across the blood-brain barrier and enter the brain.

Next, complementary functional groups capable of energetically favorable intermolecular interactions with the receptor-based functional groups are selected. These complementary functional groups will ultimately form part of the drug that is being designed and will constitute the pharmacophore. These drug-based functional groups are then "*clicked together*" in three-dimensional space by being covalently attached to a relatively rigid hydrocarbon frame. The number of functional groups determines the number of contact points between the drug molecule and the receptor macromolecule. A *three-point pharmacophore* will have three different intermolecular interactions between the drug and the receptor. The number of points of contact is also an important consideration. A large number of points of contact is favorable from a pharmacodynamic perspective since it enables a more specific and unique drug–receptor interaction, concomitantly decreasing the likelihood of toxicity. However, a large number of points of contact is unfavorable from a pharmacokinetic perspective, since the resulting increased polarity of the drug molecule tends to decrease the pharmacological half-life and also to decrease the ability of the drug to diffuse across membranes during its distribution throughout the body.

In general, most neuroactive drugs have 2–4 points of contact, while most non-neuroactive drugs have 3–6 points of contact. Once the pharmacophore has been designed, the remainder of the molecular fragments (individually composed of metabophores or toxicophores or inert bioinactive spacers, but collectively referred to as *molecular baggage*) are assembled. One of the primary goals of the molecular baggage component is to hold the pharmacophore in a desired conformation such that it can interact with its receptor. However, these additional molecular fragments also serve other functions. For instance, a metabophore can be inserted into this portion of the molecule. If one is designing an intravenous drug with a short half-life, one may want to include an ester moiety. This would constitute a metabophore since the ester would be hydrolyzed, resulting in rapid inactivation of the whole molecule. Once the prototype drug molecule has been prepared and biologically evaluated, various toxicities may become apparent during preliminary testing in animals.

Fragments of the molecule that are responsible for these toxicities (i.e., toxicophores) can then be deduced. If the toxicophore is separate and distinct from the pharmacophore, a new toxicity-free molecule can be engineered. If there is too much overlap between the toxicophore and the pharmacophore, it may not be possible to "design out" the toxicity. It is important to emphasize that a drug molecule may have many different toxicophores, reflecting different toxicities. A toxicophore is merely a pharmacophore that permits an undesirable interaction with an "untargeted" receptor. The multiphore method is versatile and is not restricted to *de novo* drug design, as the above discussion might imply. For example, if the drug molecule is discovered by accident or in a random screening process, the multiphore conceptualization is still applicable. Through structure-activity studies it is still possible to discern fragments that constitute the pharmacophore and potential toxicophores, and thus it is still possible to re-engineer the molecule for improved performance.

The strength of the multiphore method is its treatment of drug molecules as collections of bioactive fragments. If one fragment is giving problems, it is possible to simply insert another biologically similar fragment (*bioisostere*) that will hopefully overcome the identified problem. Clearly, the multiphore method of drug design is an iterative process. It takes repeated rounds of re-evaluation and redesign before a final candidate drug molecule is developed. This iterative process has the following five steps:

1. Design or identification of a lead (prototype) compound
2. Synthesis and initial biological evaluation of the lead compound
3. Optimization of the lead compound for the pharmacodynamic phase

4. Optimization of the lead compound for the pharmacokinetic and pharmaceutical phases
5. Pre-clinical and clinical evaluation of the optimized lead compound analog

PROTOTYPE DRUGS

Lead compounds are still a sine qua non of drug design. A lead compound is invariably an organic molecule that acts as a *prototype drug* around which future optimization is centered and focused. Identifying a lead compound is the key to starting the drug discovery engine. There are several well-tested methods for uncovering or identifying lead compounds as prototype agents around which to design and optimize a drug molecule:

1. Serendipity
2. Endogenous sources (drug candidates from molecules within us; e.g., insulin for diabetes)
3. Exogenous sources: ethnobotany (drug candidates from natural product sources)
4. Rational drug design
5. High throughput screening programs
6. Genomics/Proteomics

Elegant though some of these may sound, serendipity has historically been the most successful discoverer of drugs.

Work Done

"How have drugs historically been discovered? Can these traditional techniques be further exploited to help discover new and better drugs?" The answer to the latter question is probably "no"! Since the dawn of humankind, efforts have been made to discover remedies for the ailments of life. Although there are numerous examples of the trials and tribulations associated with these efforts, the story of epilepsy affords many instructive anecdotes. The failure of premodern physicians to develop adequate therapies reflected their inability to gain a viable mechanistic understanding of epilepsy. In primitive times, surgical "*therapies*" for epilepsy included *trephining* holes through the patient's skull in order to release "*evil humours* and *devil spirits.*" The ancients also employed a varied and bizarre assortment of ad hoc "medical" therapies, ranging from rubbing the body of an epileptic with the genitals of a seal, to inducing episodes of sneezing at sunset.

In early Roman times human blood was widely regarded as curative, and people with epilepsy frequently sucked the blood of fallen gladiators in a desperate attempt to find a cure. Charlatans would also offer (for a generous fee, of course) to massage the heads of people with epilepsy, thereby realigning the bone plates of the skull, taking pressure off the brain, and alleviating the curse of epilepsy. By the Middle Ages, alchemy and astronomy formed the scientific foundations of epilepsy therapy. The use of "*magical prescriptions*" flourished. These remedies ranged from grotesque therapies, such as the ingestion of dog bile or human urine, to the use of somewhat more innocuous precious stone amulets. During the Renaissance, these magical treatments were rejected by the medical profession in favor of "*rational and scientific*" Galenic therapies. These treatments relied extensively upon forced vomiting and bowel purging with concomitant oral administration of peony extracts. Also, during this time, the notion of epilepsy being secondary to hypersexuality emerged, and castration, circumcision, or clitoridectomy were widely advocated. Occasionally, various pharmacologically active organic molecules (e.g., strychnine, curare, atropine, valerian, picrotoxin, and quinine) were also used, but all ultimately failed–sometimes killing the unfortunate patient.

Table 13.1 Drug discoveries

1842	Long introduces ether as an anesthetic
1857	Locock accidentally discovers bromides as anti-convulsants
1867	Lister pioneers use of phenol as a surgical antiseptic
1869	Liebreich discovers hypnotic effects of chloral hydrate
1876	Stricker uncovers analgesic properties of salicylic acid
1882	Guthzeit and Conrad synthesize a series of barbiturates
1891	Erlich pioneers concepts of "receptor" and "chemotherapy"
1899	Meyer and Overton discover effect of lipid solubility on anesthetic action
1903	Fischer and von Mering identify hypnotic properties of barbiturates
1906	Hunt and Taveau synthesize and study acetylcholine
1912	Hauptmann accidentally discovers barbiturates as anticonvulsants
1921	Loewi demonstrates that acetylcholine is a neurotransmitter
1922	Banting and Best purify insulin as treatment for diabetes
1927	Szent-Gyorgyi isolates ascorbic acid (Vitamin C)
1929	Fleming serendipitously discovers antibacterial properties of penicillin
1932	Mietzach, Klarer, Domagk introduce first anti-streptococcal drug
1934	Ruzicka first synthesizes progesterone
1938	Merritt and Putnam use screening to identify hydantoins as anticonvulsants
1940	Chain and Florey introduce manufactured penicillin
1942	Ehrhard and Schauman produce synthetic analgesics (meperidine, methadone)
1945	Woodward and Doering synthesize quinine
1947	Lands introduces isoproterenol as a bronchodilator
1952	Charpentier identifies tricyclic phenothiazines as antipsychotics
1953	Watson and Crick deduce structure of DNA
1959	Beecham Laboratories develops semisynthetic penicillins
1959	Searle introduces the birth control pill
1960	Hoffmann-La Roche tests benzodiazepines as anxiolytics (Librium, Valium)
1962	Hansch develops principle of quantitative structure–activity relationships
1962	Pullman introduces quantum mechanics to drug design
1967	Cotzias pioneers the use of L-DOPA for treatment of Parkinson's disease
1975	Biochemically driven rational drug design begins to flourish as method
1985	Improved computers enable computer-aided drug design to advance
1995	Advances in combinatorial chemistry advance high throughput screening
2000	Widespread use of cholinesterase inhibitors for symptomatic treatment of Alzheimer's disease

Inorganic salts were also considered as putative therapies during the late Renaissance. Copper-based therapy flourished during the eighteenth and nineteenth centuries. Aretaios had introduced the use of antiepileptic copper salts in the first century AD, and had described significant therapeutic successes. These reported successes with copper therapy were embraced during the 1700s, leading to other therapeutic attempts with lead, bismuth, tin, silver, iron, and mercury, thus giving rise to *metallotherapy*. The subsequent widespread failure of metallotherapy, due to lack of efficacy and excessive toxicity, led to its abandonment during the late pre-modern era. Thus, in the millennia extending from antiquity to the mid-nineteenth century, epilepsy remained a medical condition surrounded by mystique–permitting charlatanism, superstition, and quackery to prosper. In general, the therapies of this time were without merit, as demonstrated by the detailed but disturbing description of King Charles II's death, which provides a comprehensive summary of the complexity and futility of seizure therapy during the pre-modern era. In 1685, the king fell backward and had a violent convulsion.

Treatment was begun immediately by a dozen physicians. He was bled to the extent of 1 pint from his right arm. Next, his shoulder was incised and cupped, depriving him of another 8 oz. of blood. After an emetic and 2 purgatives, he was given an enema containing antimony, bitters, rock salt, mallow leaves, violets, beet root, chamomile flowers, fennel seed, linseed, cinnamon, cardamom seed, saffron and aloes. The enema was repeated in 2 hours and another purgative was given. The king's head was shaved and a burn blister was raised on his scalp. A sneezing powder of hellebore root and one of cowslip flowers were administered to strengthen the king's brain. Soothing drinks of barley water, licorice and sweet almond were given, as well as extracts of mint, thistle leaves, rue, and angelica. For external treatment, a plaster of Burgundy pitch and pigeon dung was liberally applied to the king's feet. After continued bleeding and purging, to which were added melon seed, manna, slippery elm, black cherry water, and dissolved pearls, the king's condition did not improve and, as an emergency measure, 40 drops of human skull extract were given to allay convulsions.

Finally, bezoar stone was forced down the king's throat and into his stomach. As the king's condition grew increasingly worse, the grand finale of Raleigh's antidote, pearl julep, and ammonia water were pushed into the dying king's mouth. During an era in which Isaac Newton and Charles Darwin were making colossal advances in physics and biology, medical therapy and drug development was still mired in primitive, unprogressive and cruel rituals. Fortunately, by the mid 1 800s times began to change, and over the subsequent 150 years substantial progress was made. Although advances were obviously being made, much of this drug discovery relied on serendipity rather than rational drug design.

Advances in Drug Discovery

Many of the significant advances in drug discovery over the past one and a half centuries owe their success to serendipity; the discovery of penicillin is a legendary example of its importance. However, many other drugs were also discovered by accident.

Classes of drugs

The benzodiazepines are one of the most widely used (and abused) classes of drugs; they are widely used in the treatment of epilepsy, insomnia, anxiety, movement disorders, and a variety of other neurological disorders. The discovery of the benzodiazepines is a good example of the importance of serendipity. In the early 1950s, following the accidental discovery of phenothiazines as antipsychotic agents, interest in tricyclic molecules as potential therapeutic agents for neurological and psychiatric disorders became widespread.

Accordingly, in 1954, Sternbach of Hoffmann–La Roche laboratories decided to reinvestigate the synthetic chemistry of a series of tricyclic benzheptoxdiazine compounds upon which he had worked more than twenty years earlier. By reacting an alkyl halide with a variety of secondary amines, he prepared forty analogs, all of which were inactive as muscle relaxants and sedatives. When additional chemical studies revealed that these compounds were really quinazoline-3-oxides rather than benzheptoxdiazines, the project was summarily abandoned in 1955. Nearly two years later, a colleague at Hoffmann–La Roche discovered an untested crystalline sample while cleaning and tidying cluttered laboratory benches. Although many other compounds had been simply discarded, this compound, later called chlordiazepoxide, was submitted for biological evaluation. Chlordiazepoxide (7-chloro-2-(methylamino)-5-phenyl-3H-1,4-benzodiazepine-4-oxide) demonstrated profound anti-anxiety properties. Additional chemical studies revealed that this single compound was a 1,4-benzodiazepine, unlike its forty quinazoline-3-oxide predecessors–the chance use of methylamine, a primary rather than a secondary amine, had resulted in a different synthetic pathway.

Numerous analogs, including diazepam, were soon prepared and the anticonvulsant properties of this class of compounds were quickly discerned. The benzodiazepines soon emerged as one of the most important, and lucrative, classes of drug molecules. The discovery of benzodiazepines is a story of serendipity and certainly one that is difficult to predictably reproduce as part of a drug discovery program. Regrettably (or fortuitously), this story of the benzodiazepines is not an isolated example. Valproic acid, an agent used to treat epilepsy, migraine, chronic pain, and bipolar affective disorder, was also discovered by accident.

Khellin derivatives

In postwar France, the Berthier Pharmaceutical Company in Grenoble began to pursue a sideline project of producing soothing liquid bismuth preparations for acute tonsillitis. Being dissatisfied with the commonly used oils, they elected to use the supposedly physiologically inert valproic acid as a solvent for their bismuth compounds. In doing so, they gained immense practical experience with the handling and manipulation of this solvent. In 1962, Pierre Eymard, a graduate student at the University of Lyon, synthesized a series of khellin derivatives. Khellin is a biologically active substance that occurs in the fruit of the wild Arabian *Khell* plant and which has been used for centuries by herbalists for the treatment of kidney stones. Eymard arranged to have his new compounds biologically evaluated at the École de Medecine et de Pharmacie in nearby Grenoble. When attempts to produce a solution of these khellin compounds failed, advice was sought from H. Meunier of the nearby Laboratoire Berthier.

In view of Berthier's recent peripheral interest in valproic acid as a solvent for bismuth compounds, Meunier recommended valproic acid as a nontoxic inert solvent. Eymard's khellin derivatives were dissolved in valproic acid and, following the practice of submitting all such compounds for evaluation in an antiepileptic screening model, they were studied for anticonvulsant activity. These preliminary studies revealed profound anticonvulsant activity. Shortly after this,

Meunier serendipitously decided to use valproic acid as a solvent for an unrelated coumarin compound and, although chemically dissimilar to Eymard's khellins, this coumarin exhibited identical anticonvulsant properties. The fact that both compounds had been dissolved in the same solvent was realized immediately. The antiepileptic action of valproic acid was thus discovered completely by accident, with the first successful clinical trial occurring in 1963. Although serendipity has been quite successful in drug design, it is a method that is difficult to reproduce. Accordingly, over the past fifty years, a variety of other drug discovery methods have been pioneered.

Methods for Discovering Lead Compounds

In the attempt to identify logical or rational methods for designing and discovering lead compounds, the notion of exploiting endogenous molecules quickly comes to mind. Human disease arises from perturbations of normal biochemical processes. A logical therapeutic approach involves the administration of one or more of these naturally occurring endogenous biochemical molecules, or analogs thereof. In addition, certain human diseases seem to arise from a deficiency of a certain endogenous molecule. It is reasonable to assume that such diseases could be cured or at least helped by the administration of the missing molecule. Medicinal chemistry has many examples of the development of successful therapeutics based on an exploration of endogenous compounds. The treatment of *diabetes mellitus*, for example, is based upon the administration of insulin, the hormone that is functionally deficient in this disease.

The current treatment of Parkinson's disease is based upon the observation that the symptoms of Parkinson's disease arise from a deficiency of dopamine, an endogenous molecule within the human brain. Since dopamine cannot be given as a drug since it fails to cross the blood-brain barrier and enter the brain, its biosynthetic precursor, L-DOPA, has been successfully developed as an anti-Parkinson's drug. Analogously, the symptoms of Alzheimer's disease arise from a relative deficiency of acetylcholine within the brain. Current therapies for Alzheimer's-type dementia are based upon the administration of cholinesterase enzyme inhibitors that prolong the effective half-life of remaining acetylcholine molecules within the brain. The human body contains many different molecules and thus offers many opportunities for the discovery of lead compounds based on endogenous molecules. Nowhere is this opportunity more apparent than in the area of peptide neurotransmitters and peptide hormones. Neurotransmitters and hormones are endogenous messengers, controlling diverse biochemical processes within the body. Not surprisingly, they have the capacity to be ideal starting points in the drug discovery process. However, there are a number of major problems that must be confronted when exploiting peptides or proteins as lead compounds for drug discovery. Peptidomimetic chemistry is an attempt to address these problems.

Biochemical characteristics

Although they are potent endogenous bioactive molecules, peptides rarely make good drugs. There are several reasons for the failings of peptides as drugs:

1. Peptides are too big (molecular weight frequently over 1000 dalton).
2. Peptides are often too flexible (thus binding with too many receptors, leading to toxicity).
3. Peptides contain amide bonds that can be metabolized by hydrolysis.

4. Peptides cannot be given orally as drugs (they tend to be digested).
5. Peptides do not readily cross the blood–brain barrier to enter the brain.

Despite these obvious deficiencies, peptides have a number of properties that make them attractive as starting points in drug design:

1. Peptides contain numerous stereogenic (chiral) centers (an excellent starting point when designing stereoselective drugs).
2. Peptides contain many functional groups (e.g., carboxylate, ammonium, hydroxyl, thiol) that can readily constitute functional groups within a pharmacophore; since receptors are usually proteins, peptides are good starting points for designing a molecule to interact with a receptor, owing to the energetically favorable nature of peptide–peptide interactions.
3. Peptides are easily synthesized and many analogs can be readily produced.
4. Peptides can have their conformation and geometries easily optimized by energy minimization calculations using current computational methods (e.g., molecular mechanics); this makes subsequent modeling studies easy.
5. Peptides function as neurotransmitters and hormones and thus are good starting materials when designing bioactive molecules.

Since peptides are ideal starting molecules that cannot be turned into successful peptidic drugs, the specialty area of *peptidomimetic chemistry* has emerged. The goal of peptidomimetic chemistry is to design small, conformationally constrained, non-peptidic organic molecules that possess the biological properties of a peptide. Hopefully, this will retain the strength of the peptide as a putative drug while eliminating the problems. There are two approaches whereby peptidomimetic chemistry can achieve this design goal. This approach uses

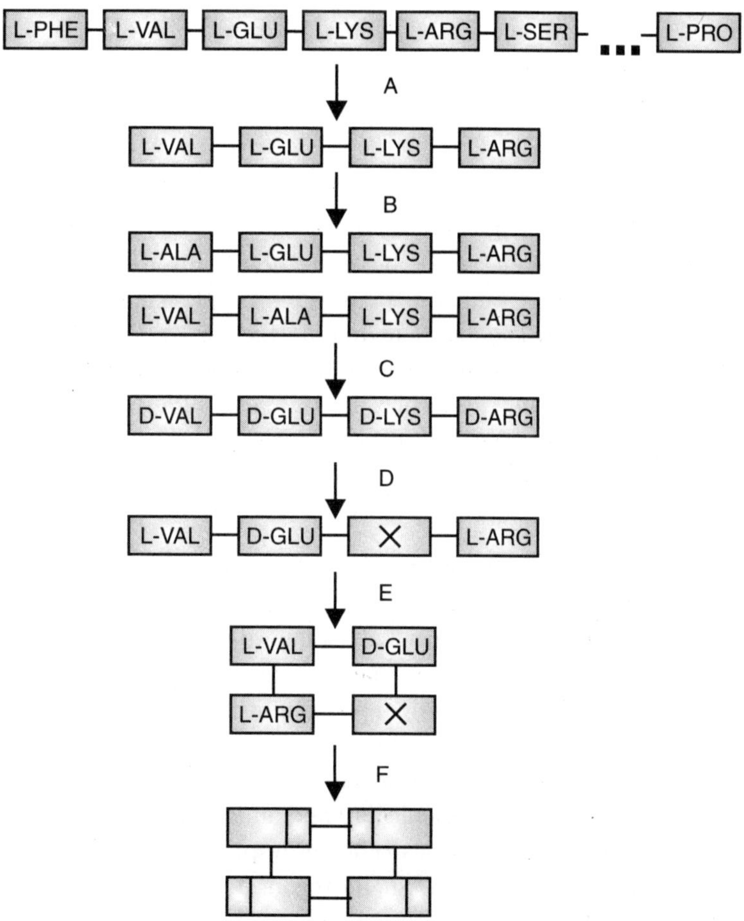

Fig. 13.1. Peptidomimetic chemistry attempts to produce a non-peptidic drug to mimic a bioactive peptide.

various techniques (e.g., alanine scanning) to identify the smallest peptide segment with biological activity within the overall peptide. This "*minimal bioactive segment*" may be cyclized or have its stereochemistry altered in order to attain restriction of conformational freedom and to render the molecule less like a naturally occurring peptide.

Next, this segment is then rebuilt isosteric fragment by isosteric fragment, gradually replacing each portion of the molecule in a stepwise fashion. For example, the amide bond may be replaced by a bioisosterically equivalent amide bioisostere. In this fashion, an equivalent but nonpeptidic organic molecule drug eventually emerges. An alternative approach is a little less plodding and perhaps a little more elegant. The three-dimensional structure of the peptide is determined using either theoretical (molecular mechanics, molecular orbital calculations) or experimental (X-ray crystallographic, NMR spectroscopic) methods. Next, an educated guess (hopefully based on some experimental data) is made to suggest which portion of the peptide is the pharmacophore.

The geometries of the functional groups within the pharmacophore are then measured from the theoretical and experimental studies of the peptide's geometry and conformation. For example, these data may show that the peptide pharmacophore contains a carboxylate group located 4.6 Å from a hydroxyl group, which in turn is 5.1 Å from a phenyl group. Using these precise data, databases of known organic molecules are then computationally searched to identify an organic molecule with similar functional groups held in the same position in three-dimensional space. Hopefully, this will yield a non-peptidic but bioactive organic molecule drug.

Other features of endogenous drugs

Although peptides have been studied the most extensively, there are other endogenous molecules within the human body worthy of exploitation as drug discovery platforms. These include nucleic acids, lipids, and carbohydrates, which are discussed in detail in chapter 8. These molecules share the same potential strengths and weaknesses as do peptides. Likewise, there is a need to develop small organic molecules as mimetics of these other endogenous molecules. Although not as clearly defined as peptidomimetic chemistry, ultimately, "*nucleotidomimetic*" or "*carbohydromimetic*" chemistries may eventually emerge as new design strategies for lead compound identification.

Lead Compound Identification from Exogenous Sources

The utility of naturally occurring molecules, which are endogenous to the human body, was discussed as a source of bioactive lead compounds. An alternative is to exploit molecules that are endogenous to other life forms (animal or plant) but do not naturally occur within humans. Such molecules would be classed as exogenous from the perspective of drug design for humans. Historically, plant-based natural products have been a source of useful drugs. The analgesic opiates come from the poppy plant. Digitalis for congestive heart failure was first isolated from the foxglove plant. Various antibiotics (penicillin) and anticancer agents (taxol) are derived from natural product sources. There are numerous other examples. There is good reason to be optimistic about the potential future usefulness of such exogenous compounds as a continuing source of potential lead compounds. With many thousands of years of trial-and-error by evolution on her side, Mother Nature is a vastly superior experimentalist to any mere human organic chemist.

Insect evolution has permitted the biosynthesis of anticoagulant molecules to ensure that a bite will provide a good supply of blood to the biting insect; such compounds could be useful in the treatment of stroke secondary to blood clots within the brain. Amphibian evolution has enabled the biosynthesis of antibacterial peptides on the skins of frogs so that they can avoid infections as they swim through stagnant swamp waters; peptides such as these could be a good starting point for the peptidomimetic design of novel antibacterial agents. Reptile evolution has culminated in the biosynthesis of neuroactive venoms for purposes of hunting and defense; these molecules have been fine-tuned by evolution as agents specific for neurotransmitter receptors.

Plant evolution has culminated in a wide variety of biomolecules that affect any animal that may choose to eat them: it is biologically advantageous for some plants to be eaten so that their seeds can be dispersed in the stool of the animal that ate them; conversely, it is biologically advantageous for other plants to produce noxious chemicals to decrease the likelihood of their being eaten. Because of these diverse biological activities, any of these non-human biosynthetic molecules could, in principle, be a lead compound for human drug discovery. Another promising feature of animal- or plant-based natural products is that they are a superb source of molecular diversity. As a synthetic chemist, Nature is much more creative and is not constrained to the same finite number of synthetic reactions typically employed by human synthetic organic chemists.

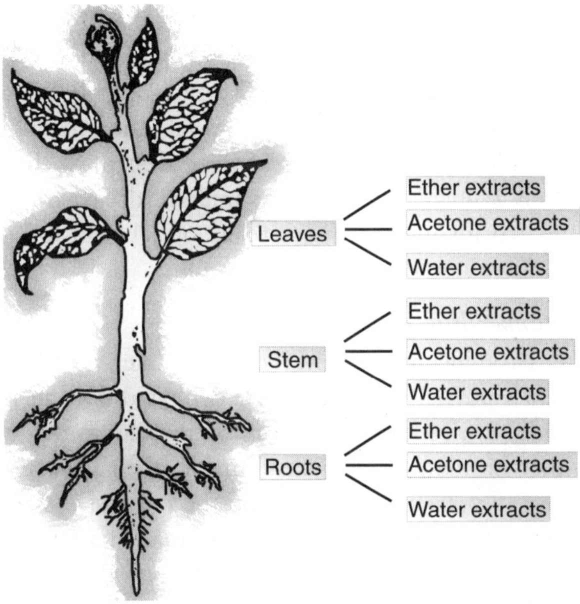

Fig. 13.2. Drugs from natural sources: different molecules can be isolated from the leaves, stem, and roots.

When designing new innovative therapies, molecular diversity is important. Furthermore, when developing compound libraries for high throughput screening, it is important to have libraries that capture molecular diversity and are not merely large collections of structurally similar analogs. Although ethnopharmacology, the scientific investigation of natural products, folk medicine, and traditional remedies, has led to some bona fide drugs (e.g., reserpine, quinine, ephedrine), it has not proven to be a reliable or efficient source of leads. However, natural products have always been and still are an inexhaustible source of drug leads as well as drugs. Renewed interest in natural products and the novel structures they provide is especially noticeable in marine pharmacology, a practically virgin territory. Several research institutes and well-established groups are producing some very promising results in this field.

The isolation of prostaglandins from a coral was one of the more startling recent discoveries in marine pharmacology. An extension of natural products chemistry is the biochemical information derived from the study of metabolic pathways, enzyme mechanisms, and cell physiological phenomena; this research has revealed exploitable differences between host and parasite (including malignant cells), and between normal and pathological function in terms of these parameters. The large and fertile area of *antimetabolite* (metabolic inhibitors) and *parametabolite* (metabolic substitutes) chemistry is based on such stratagems, and has found use in the field of enzyme inhibition and in conjunction with nucleic acid metabolism. The design of drugs based on biochemical leads remains a highly sophisticated endeavor, light-years removed from the random screening of sulfonamide dyes in which it has its origin.

Methods of Lead Compound Discovery

The previous two methods of lead compound discovery exploited naturally occurring endogenous or exogenous compounds. However, of the approximately 10^{200} "small" organic molecules that could theoretically exist in our world (10^{52} of which are drug- like molecules), many would be purely synthetic substances that do not occur naturally. The lead compound discovery methods afford an opportunity to explore these non-naturally occurring synthetic compounds as potential lead compounds. The concept of rational drug design implies that the disease under consideration is understood at some fundamental molecular level and that this understanding can be exploited for purposes of drug design. Such an understanding would facilitate the design of purely synthetic molecules as putative drugs. Although this ideal of rational drug design has been pursued for many years, it really only began to emerge as a viable drug discovery strategy in the last three decades of the 20th century.

Rational drug design

Bromine was discovered in seawater in 1826. Recognizing its chemical similarity to iodine, French physicians immediately exploited it as an iodine alternative for the treatment of numerous conditions, including syphilis and thyroid goitre. Although no beneficial effects were reported for either bromine or its potassium salt, their widespread use persisted and eventually the depressant effect of potassium bromide on the nervous system, so-called *ivresse bromurique*, was recognized. However, it was a report in the German literature concerning bromide's ability to induce impotence and hyposexuality, rather than *ivresse bromurique*, which lead to its discovery as an anticonvulsant. In 1857, Sir Charles Locock, the physician accoucheur to Queen Victoria, ascribing to the then prevalent view that epilepsy arose from excessive sexuality, introduced bromide as an anaphrodisiac to suppress the supposed hypersexuality of epileptics.

The bromide salts (e.g., potassium bromide, sodium bromide) were administered in substantial doses, ranging from 0.3 g/day in children to a staggering 14 g/day in adults. Although side effects had been considerable (and included psychoses and serious skin rashes), bromides were successful in 13 of the 14 patients treated. On 11 May 1857, at a meeting of the Royal Medical and Chirurgical Society, Locock proudly reported his success in treating "*hypersexual*" epilepsies with bromides. He argued that logical and rational drug development had finally been achieved for the time: epilepsy arises from excessive sexuality; potassium bromide suppresses sexuality; therefore, potassium bromide successfully treats epilepsy. In principle it seemed like a major success of rational drug design. In reality, it was little more than yet another serendipitous discovery, since hypersexuality has absolutely nothing to do with epilepsy. Regardless of the flawed reasoning, bromides were a major step forward in the treatment of epilepsy and their use persisted until the introduction of phenobarbital in 1912.

Improved rational drug design

Thankfully, the science of rational drug design has improved substantially since Victorian times. In modern times, rational drug design is defined as a method whereby a disease is understood at a molecular level such that the biological macromolecules involved in the disease process have been identified, purified, characterized, and have had their three-dimensional structure elucidated, thereby enabling the intelligent and insightful engineering of organic molecules to dock to that macromolecule and alter its functional properties in a therapeutically beneficial manner. This is a design method that is lofty and difficult to attain. Rational drug design is an iterative process, dependent upon feedback loops and new information. When the drug designer makes the first prototype molecule, this molecule becomes a probe with which to test the drug design hypotheses.

The molecule can then be further designed and refined to better improve its ability to dock with the receptor site and elicit a biological response. This cycle of "*design-test-redesign-retest*" can go on for several iterations until the optimized molecule is achieved. Rational drug design is a very intellectually satisfying activity. The successful rational design of a drug is similar to solving a major puzzle using your wits and wisdom. The macromolecules involved in the disease have been determined; the structures of these macromolecules have been ascertained using X-ray crystallography and/or computer-aided molecular design; a small organic molecule capable of binding to the macromolecule has been cleverly designed; a synthesis for this small organic molecule has been devised; and biological testing has confirmed the bioactivity of the small organic molecule. Despite protestations from the "*naysayers*" who despondently claim that all drugs are discovered by serendipity, there is an increasing number of examples that exemplify the successes and practical utility of rational drug design.

Perhaps one of the earliest examples is the discovery of cimetidine, an H2-antagonist drug used for the treatment of peptic ulcer disease. Even though the complete structure of the receptor was not fully appreciated, the careful manipulation of the molecule's physicochemical properties led to the discovery of cimetidine. More recently, the design of drugs for the treatment of AIDS has provided superb examples of rational drug design. The HIV virus encodes for a protease, called HIV-1/PR, that is required for its replication.

Using X-ray crystallographic studies of recombinant and synthetic HIV-1/PR helped to identify the active site of the protein. Next, a complex between the HIV-1/PR protein and a prototype inhibitor was also structurally solved using X-ray crystallography. These structural studies greatly facilitated the process of rational drug design, ultimately leading to six rationally designed therapeutics: amprenavir, indinavir, nelfinavir, ritonavir, saquinavir, and lopinavir. As evidenced by the aforementioned examples, structural chemistry is front and center in enabling rational drug design.

Molecular modeling, also called quantum pharmacology, has been instrumental to many of the advances in rational drug design. Some cynics are quick to pontificate that there are no drugs that have been designed by computers. Strictly speaking, this is true; likewise, no drugs have been designed by a nuclear magnetic resonance spectrometer. Computers do not design drugs, people do. However, computers are immensely valuable in advancing and progressing the art and science of rational drug design.

Pharmacological calculations

Quantum pharmacology calculations have, are, and will continue to revolutionize rational drug design. Until the dawning of the 20th century, it was believed that all reality was eminently describable through Newton's Laws. These laws, the foundation of Newtonian mechanics, appeared able to rigorously represent the physical properties of reality, including natural objects as large as planets or as small as billiard balls. However, through the work of Bohr and others, it was eventually realized that classical Newtonian mechanics failed at the atomic level of reality–atoms did not behave like billiard balls. An alternative approach was needed for the quantitative evaluation of molecular phenomena. Molecular quantum mechanics was to be such an approach. In the first three decades of the 20th century, there occurred many significant advances in theoretical physics and physical philosophy.

Planck showed that energy is emitted in the form of discrete particles or quanta; Einstein expanded upon this theory with the proposal that an atom emits radiant energy only in quanta, and that this energy is related to the mass and to the velocity of the light; Schrodinger incorporated these evolving ideas of the new quantum theory into an equation that described the wave behavior of a particle (wave mechanics); Heisenberg formulated a complete, self-consistent theory of quantum physics, known as matrix mechanics; and Dirac showed that Schrodinger's wave mechanics and Heisenberg's matrix mechanics were special cases of a larger operator theory. The capacity for a robust, mathematical description of molecular-level phenomena seemed to be at hand.

Since the Schrodinger equation (which lies at the mathematical heart of quantum mechanics) permitted quantitative agreement with experiment at the atomic level, the physicists of the 1930s predicted an end to the experimental sciences, including biology, suggesting that they would merely become a branch of applied physics and mathematics. These hopes were excessively optimistic and soon proved groundless. Although in principle the Schrödinger equation afforded a complete description of Nature, in practice it could not be solved for the large molecules of medical and pharmacological interest. Early hopes that quantum mechanics would solve the problems of drug design were dashed in despair. Over the past thirty years, however, three advances have changed the practical usefulness of molecular quantum mechanics:

1. The advent of semi-empirical molecular orbital calculations and density functional theory, which employ mathematical assumptions to simplify the application of quantum mechanics to drug molecules of intermediate to large size.
2. The development of molecular mechanics, which incorporates quantum mechanical data into a simplified mathematical framework derived from the classical equations of motion to permit reasonable calculations on biomolecules of large size.
3. The construction of "*supercomputers*" capable of performing the massive calculations necessary for considering very large biomolecules. Accordingly, quantum pharmacology has become an attainable goal and calculational computer modeling permits large molecules to be studied meaningfully.

Computer-assisted molecular design (CAMD) employs these powerful computational techniques to engineer molecules for desired receptor site geometries. CAMD had gained widespread acceptance and is beginning to prove its usefulness in many realms of pharmacological endeavor. It has the potential to profoundly influence the future of rational drug design. CAMD enables rigorous modeling of drug molecules, of receptor macromolecules, and of complex drug–receptor interactions. All of these calculations are immensely important to rational drug design. The most important aspect of computer-assisted modeling is the capability to perform three-dimensional docking experiments, a tool used by all major drug companies and biotechnology firms. Starting with the X-ray structure of the macromolecule, a space-filling molecular model is created, including hydrate envelopes around it.

By separately generating the three-dimensional model of a hypothetical drug, a modeler can manipulate the two by modern fast computers and can directly examine the fit of the ligand in the active site; the investigator can change the substituents, conformation, and rotamers of the drug on the screen, and can repeat the docking. This enormous progress in computer hardware and software, elucidation of macro-molecular structure and ligand–receptor interactions, crystallography, and molecular modeling is hopefully bringing us to the threshold of a breakthrough in drug design. We are now able to design lead compounds de novo on the basis of the structure of the receptor macromolecule. However, computerized drug design is still only an instrument that reduces empiricism in an experimental science; the inherent approximations of innumerable conformers and molecular parameters of drug and receptor, and the methodological inaccuracies and difficulties of comparison, will never allow the elimination of insight and trial.

High Throughput Screening

Random screening, while seemingly wasteful, has an important place in developing lead compounds in areas in which theory lags. Screening for antitumor activity has been carried on for more than 30 years by the U.S. National Cancer Institute, with tens of thousands of compounds being tested on tumors *in vivo* and *in vitro.* More recently, a computerized prescreening method has been applied to this process, saving time and expense, and hence the screening is not as random as it used to be. A successful random search for antibacterial action was conducted by several pharmaceutical companies in the 1950s. They tested soil samples from all over the world, which resulted in the discovery of many novel structures and some spectacularly useful groups of antibiotics, notably the tetracyclines.

In fact, microbial sources have supplied an enormous number of new drug prototypes, sometimes of staggering complexity. Recently, the large-scale automated testing of microbial mutants has been realized and combined with recombinant DNA techniques to speed up the efficient discovery and production of new antibiotics. Some would argue that drug discovery through screening provides the “irrational” counterpart to rational drug design. This remark is unjustifiably harsh and is somewhat facetious. As mentioned, screening of compounds has a long and rather illustrious history and has produced many useful anticancer and antibiotic drugs. The discovery of the anticonvulsant drug phenytoin provides an early example of drug discovery through screening.

Screening and drug discovery

As a chemical species, the hydantoins have been known since the 1860s. By the latter half of the nineteenth century numerous hydantoin analogs had been synthesized, but only one, 5-ethyl-5-

phenylhydantoin (nirvanol), demonstrated any clinical utility. Wernecke introduced nirvanol in 1916 as a "*less toxic hypnotic*"; however, enthusiasm rapidly waned when its chronic toxicity became recognized. Not surprisingly, a second hydantoin, 5,5-diphenylhydantoin (phenytoin), which had long remained on the laboratory shelf, appeared doomed to obscurity; phenytoin had been first synthesized by Biltz in 1908, through a condensation of urea with benzil which exploited a pinacolone rearrangement. In the late 193 0s, T. Putnam initiated a screening programme to search for new anticonvulsants, using protection against electroshock-induced convulsions as a selection criterion.

A makeshift apparatus to execute these experiments was assembled using a commutator salvaged from a World War I German aircraft. Having studied the structure of phenobarbital, Putnam randomly requested a diverse selection of heterocyclic phenyl-containing compounds from a variety of chemical manufacturers. He also communicated with a number of pharmaceutical companies. The Parke-Davis Company provided nineteen heterocyclic phenyl-substituted compounds that had been deemed "*worthless hypnotics*." Phenytoin was one of these nineteen compounds. Putnam screened hundreds of compounds but only phenytoin combined high activity with low toxicity. In 1936, Putnam's colleague, Houston Merritt, initiated a clinical evaluation of phenytoin, which soon led to its widespread marketing as an anticonvulsant drug.

The pioneering screening techniques that heralded the discovery of phenytoin profoundly influenced subsequent antiepileptic drug discovery. Hundreds of hydantoin analogs were synthesized and screened for biological activity; hundreds of other penta-atomic heterocyclic compounds (e.g., succinimides, oxazolidinediones) were likewise synthesized and screened for biological activity. Many of these new compounds found their way into the market place, with varying degrees of therapeutic success.

Advanced method of drug discovery

Thankfully, the science of drug discovery by screening has advanced since the time of Merritt and Putnam. Modern drug discovery by screening is more of a systematic technological tour de force than a hit-or-miss gamble. The reasons for these advances are obvious. Although rational drug design is elegant, it is also slow and thus time-inefficient. It takes a long time to identify the proteins that are involved in a disease, then crystallize them and design drugs to bind to them. Worse, some proteins, especially membrane- bound proteins, seem to defy crystallization, while some diseases do not even have identifiable proteins involved in their pathogenesis and etiology. Screening methods attempt to address all of these deficiencies in drug discovery.

If a reliable bioassay is available, it is possible to screen thousands or even millions of compounds against this bioassay. The crystal structures of key protein receptors do not have to be known; indeed, the proteins do not even have to be identified. If the bioassay is fast and efficient, and if the library of compounds being screened is diverse and comprehensive, then in principle it should be possible to identify a lead compound years before the practitioner of rational drug design. However, the key to success lies in the "goodness of the library of compounds" (i.e., a combinatorial chemistry library) and in the "goodness of the screening bioassay" (i.e., high throughput screening methods).

Combinatorial chemistry

A key to success in drug discovery by screening is the availability of a large and structurally diverse library of compounds. If the library contains a million compounds that are all analogs of each other, then it may be large but it is probably not sufficiently diverse. The library should have the full range of functional groups (cations, anions, hydrogen bond donors, hydrogen bond acceptors, lipophilic, aromatic, etc.) displayed in all possible permutations and combinations in three-dimensional space. Creating such a library is not a trivial task. *Combinatorial chemistry* is both the philosophical and the practical method with which to create structurally diverse compound libraries. Combinatorial chemistry is defined as that branch of synthetic organic chemistry that enables the concomitant synthesis of large numbers of chemical variants in such a manner as to permit their evaluation, isolation, and identification. Combinatorial chemistry affords techniques for the systematic creation of large but structurally diverse libraries. From a technical perspective, there are several avenues of approach to library creation:

1. Libraries of oligomers of naturally occurring monomers
   (a) Oligopeptide libraries
   (b) Oligonucleotide libraries
2. Libraries of oligomers of non-naturally occurring monomers
   (a) Oligocarbamate libraries
   (b) Oligourea libraries
   (c) Oligosulfone libraries
   (d) Oligosulfoxide libraries
3. Libraries of monomers with multiple sites for substituents
   (a) Synthetic ease privileged structure
       Dioxapiperazines
   (b) Pharmacological activity privileged structures
       Benzodiazepines
       Dihydropyridines
       Hydantoins
   (c) Novel template structures
       Dihydrobenzopyrans

Historically, the first major libraries were oligomers of naturally occurring monomers. A good example would be a library of all possible tripeptides. Using the twenty naturally occurring amino acids, it is possible to produce 8000 different tripeptides. If atypical amino acids and amino acids in the unnatural D configuration are included, it is possible to achieve 125,000 different compounds with relative ease. Peptide libraries are easy to synthesize and, since amino acid side chains possess a wide variety of different functional groups, it is possible to achieve a good measure of structural diversity. However, in general, peptides are not drugs and a peptide lead would have to be modified into a drug-like molecule. In addition to oligopeptides, other naturally occurring oligomeric libraries are possible, including oligonucleotide libraries. A step beyond the naturally occurring oligomeric libraries is to create libraries from non-naturally occurring monomeric building blocks.

The medicinal chemistry literature contains a fair number of examples of such libraries, including oligocarbamates and oligoureas. Although these libraries overcome the limitations of naturally occurring oligomeric libraries, most drugs are not polymers. To address this problem, new libraries emerged in which the central moiety was a small organic molecule. The diversity library was then constructed by attaching many different substituents to this central moiety. Some of these moieties were selected because they were very simple to synthesize. For example, dioxapiperazines are cyclic dipeptides and thus are relatively trivial to prepare.

Other monomers were selected because they had a good track record for being drug-like molecules. Benzodiazepines are a good example of such libraries. In preparing these various libraries, extensive use is made of *solid phase synthetic methods.* These methods are all derived from the *solid phase peptide synthesis* (SPPS) method developed by Merrifield in 1963. When performing a large number of syntheses, it is preferable to perform the synthetic steps on a solid bead rather than completing the entire synthesis in the solution phase. The solid-phase technique makes byproduct removal and final compound purification easier. The organic chemistry literature contains a wealth of different types of solid-phase supports and novel linkers for attaching the synthetic substrate to the bead.

Chemically diverse library

If a large, chemically diverse library is available, the next problem is to evaluate these compounds in a time-efficient manner. If a 200,000 compound library is available, the biological evaluation assay must be rapid and reliable. If the assay were capable of testing five compounds per day, it would take 110 years to evaluate the entire library. Clearly, this is not the time for elaborate *in vivo* testing. Fast, efficient *in vitro* assays are required. The ability to inhibit an enzyme is a good example of a potentially useful assay for high throughput screening. A variety of high throughput assays have been developed and perfected over the past 10–20 years. These include the following basic types of assay:

1. Microplate activity assays (assay is in solution in a well; the result of the assay, such as enzyme inhibition, is linked to some observable, such as color change, to enable identification of bioactivity)
2. Gel diffusion assays (biological target is mixed in soft agar and spread as a thin film; the compound library is spread on the surface of the film; after allowing for compound diffusion, an appropriate developing agent is sprayed on the agar surface and areas in which bioactivity has occurred will show up as distinct zones)
3. Affinity selection assays (compound library is applied to a protein target receptor; all compounds that do not bind are removed; compounds that do bind are then identified)

Of these, microplate assays are probably the most widely used. Screening combinatorial libraries in 96- or even 384-well microplates is time and cost efficient. Using modern robotic techniques, it is possible to perform more than 100,000 bioassays per week in a microplate system (permitting the above-described 200,000 compound library to be screened in two weeks, rather than over a century). In addition to selecting an appropriate assay, it is also necessary to have a pooling strategy. It is more efficient to test many compounds per well on the microplate, rather than one. If one could test 100 compounds per well, then the standard 96-well plate would enable almost 10,000 compounds to be evaluated in one experiment. (Currently, multiwell plates containing more than 96 wells are routinely being used.)

To facilitate effective pooling, the library of compounds is usually divided into a number of nonoverlapping subsets. The synthetic strategy employed during the combinatorial syntheses can be used to assist in determining these pooling strategies. In *random incorporation syntheses*, a single bead could contain millions of different molecular species. In *mix and split syntheses* (also called *pool and divide syntheses* or *one bead-one compound syntheses*) only one compound is attached to any given solid-phase synthetic bead. The evolution of methods for combinatorial syntheses and high throughput screening will be necessary to address the explosion of druggable targets soon to be identified by the genomics and proteomics revolutions. Genomics and proteomics represent the future of lead compound identification.

Genetic Factors

Conservative estimates suggest that more than 4000 human diseases are influenced by distinct and potentially targetable (or druggable) genetic factors. Current drug design strategies are struggling with fewer than 500 druggable receptor proteins. Endeavoring to identify lead compounds for an additional 3500 targets will overwhelm present-day drug design technologies. Drug design has come a long way since the unfortunate dêmise of King Charles II, but clearly it has a very long way to go. Genomics and proteomics represent a possible pathway to enhanced future drug discovery.

Genetic engineering and lead compound

Genomics is the study of genes and their functions. On June 26, 2000-the dawning of the present century-a historic milestone in genomic science was attained when researchers involved with the Human Genome Project jointly announced that they had sequenced 97-99% of the human genome-the all-encompassing collection of human genes. The human genome consists of 23 pairs of chromosomes with three billion base pair codes for approximately 24,000-30,000 functional genes (original estimates of 100,000-120,000 genes seem to have been incorrectly high). The genomic gold rush that followed the human genome project produced a flood of 50,000-75,000 sequences as potential targets for future drug design. Despite the size of this flood, its flow has not filled the drug discovery pipeline with winning candidates. Determining gene structure and function through genomics definitely does illuminate the path for deciphering human biochemistry and for linking specific genes to specific diseases.

Although genomics did deliver phenomenal masses of raw information, the genomics technologies have so far failed to deliver the more than 10,000 anticipated druggable targets predicted by the early hyperbole of the genomics era. Such success will require post-genomics technologies. Taking genomics one step further for the purpose of drug discovery will require linking specific proteins to those specific genes. Clearly, there exists a vast gap between genomics and drug discovery. Bridging this gap will ultimately be a daunting task that lies within the domain of proteomics.

Proteomics

Proteomics is a protein-based science that seeks to provide new, fundamental information about proteins on a genome-wide scale; to date, proteomics is still more of a concept that a neatly defined smooth-running biotechnology. More concretely, proteomics is the molecular biology discipline that seeks to elucidate the structure and function profiles of all proteins encoded within a specific genome; this collection of proteins is termed the *proteome*. The proteomes of multicellular organisms present an immense challenge in that more than 75% of the predicted proteins have no apparent cellular function. Furthermore, although the human proteome has more than 100,000 proteins, only a fraction of these proteins are expressed in any individual cell type.

If specific diseases are to be linked to specific proteins, it is imperative that ways be developed to deduce which individual protein is expressed in which individual cell. Since protein and mRNA concentrations tend to be correlated, *DNA microarray technology* is a powerful technique with which to monitor the relative abundance of a specific mRNA in an individual cell and to correlate this with a specific protein. (Regrettably, since mRNA and protein levels do not perfectly correlate, a direct measurement of protein abundance would be preferable.)

In addition to these microarray technologies, many other technologies will be required if proteomics is to deliver the drugs promised by genomics. For example, drug design requires much more than merely knowing the primary amino acid sequence of a protein; it requires a precise knowledge of the protein's three-dimensional structure, down to the level of the ångström. To date, science has no technology that enables one to use the information coded in a protein's primary amino acid sequence to deduce the overall tertiary structure of the protein. This is the *multiple minima problem* (also called the *protein folding problem*).

The need to solve this problem has given rise to the subdiscipline of *structural proteomics*, a technology that is based upon the principle that structure underlies function and that endeavors to provide three-dimensional structural information for all proteins. Another evolving subdiscipline is *interaction proteomics*. Protein–protein interactions are a key element of almost all cellular processes. These interactions underlie the events of cell-cycle regulation, cellular architecture, intracellular signal transduction, nucleic acid metabolism, lipid metabolism, and carbohydrate metabolism.

A rigorous characterization of the complexities of cellular protein–protein interactions will afford a robust understanding of the integrated networks that drive cellular function. Furthermore, many human diseases, including cancer and neurodegenerative diseases, seem to arise from aberrant protein–protein association mechanisms. Interaction proteomics seeks to elucidate the complete set of interactions that define protein-protein associations. Even when the technologies of structural proteomics and interaction proteomics have evolved to maturity, the pathway to the awaiting plethora of drugs is still not paved and perfect. Drugs are small organic molecules. Obtaining these drug molecules will require yet another step in the "from genomics–to proteomics–to disease" cascade. Just as proteomics is a crucial bridge uniting genomics to disease, so too will an equally crucial bridge be needed to unite proteomics with therapeutics. Hopefully, the bioinformatics/ cheminformatics spectrum will be that bridge.

Bioinformatics and cheminformatics in drug discovery

Bioinformatics and cheminformatics constitute an *in silico* science that endeavors to predict the phenomenology of cellular physiology and pharmacology at a molecular level using computational methods. Using databases of compounds and other theoretical molecular design techniques, bioinformatics and cheminformatics will attempt to identify novel molecules to alter the function of various proteins defined by the genome-based proteome. Bioinformatics/ cheminformatics will apply knowledge-discovery and pattern- recognition algorithms to the genome-wide and proteome-wide experimental data, thereby facilitating drug design.

If structural proteomics has identified the functional portion of an important protein, cheminformatics will search large databases of drug-like molecules to identify one that has the right shape and properties to dock with the protein. An interesting and recent advance in cheminformatics is *chemogenomics*. In conventional cheminformatics, a single drug is designed for a single protein target; in chemogenomics, multiple drugs will be designed to target multiple-gene families. Data gleaned for one protein can be applied to structurally similar proteins coded by the same gene family. Chemogenomics represents a new conceptual approach to target identification and drug development.

Spectrum of Pharmacogenomics

The spectrum of genomics/proteomics/bioinformatics/cheminformatics is defining the future of lead compound discovery and drug design; *pharmacogenomics* is defining the future beyond that! Conventional drug design attempts to discover drugs to treat particular diseases; pharmacogenomics attempts to design individualized drugs to treat particular people with particular diseases. On the basis of a variety of genetic testing, a physician would be able to predict how an individual patient would respond to a specific drug and if this patient will experience any specific side effects.

On the basis of person-to-person variability in pharmacokinetics and pharmacodynamics, pharmacogenomics will study how genetic variations affect the ways in which particular people respond to specific drug molecules. Traditionally, drug design has developed drugs for "everyone" with a given disease; pharmacogenomics will enable the tailor-made design of chemotherapies for specific populations or individuals with diseases. In attempting to achieve this lofty ideal, pharmacogenomics will rely upon genetic data such as *single nucleotide polymorphism maps*.

The assembly of a *single nucleotide polymorphism* (SNP) map for the genome will represent a set of characterized biomarkers spread throughout the human genome; this SNP map will highlight individual variations within particular genes, some of which may be associated with particular diseases, thus identifying the genetic variability inherent in human populations that is crucial to the task of individualized drug design.

The emergence of pharmacogenomics will also enhance the interaction of medicinal chemistry as a discipline with other disciplines, including the social sciences, ethics, and economics. Society has a difficult enough time paying for currently available drug therapies. Who will pay for individualized therapies? Will this further widen the chasm between "have" and "have-not" populations, between "developed" and "developing" nations?

SYNTHESIS OF TARGET MOLECULES

Although drug design and computer-aided molecular design are powerful techniques, ultimately the lead compound must be made and evaluated; it is imperative that we make the jump from *in silico* to *in vitro* and *in vivo*. This is the point at which medicinal chemistry overlaps heavily with synthetic organic chemistry. Organic synthesis (from the Greek, *synthetikos*, "to put together") is the preparation of complicated organic molecules from other, simpler, organic compounds. Because of the ability of carbon atoms to form chains, multiple bonds, and rings, an almost unimaginably large number of organic compounds can be conceived and created. In planning a synthetic route for the preparation of a desired molecule (termed the *target molecule*) the organic chemist devises a *synthetic tree*–an outline of multiple available routes to get to the target molecule from available starting materials. An organic synthesis may be either *linear* or *convergent*.

A linear synthesis constructs the target molecule from a single starting material and progresses in a sequential step-by- step fashion. A convergent synthesis creates multiple subunits through several parallel linear syntheses, and then assembles the subunits in a single final step. Since overall yield is a function of the number of steps performed, convergent syntheses have higher yields and are preferred over linear syntheses. In creating synthetic routes for the development of drug molecules, the synthetic chemist wants to create a molecular entity in which functional groups (carbonyls, amines, etc.) are correctly positioned in three-dimensional space; this will enable the creation of functional biophoric fragments such as the pharmacophore. The synthetic chemist has ten general classes of reactions available for such synthetic tasks:

1. Aliphatic nucleophilic substitution
2. Aromatic electrophilic substitution
3. Aliphatic electrophilic substitution
4. Aromatic nucleophilic substitution
5. Free-radical substitution
6. Additions to carbon–carbon multiple bonds
7. Additions to carbon–heteroatom multiple bonds
8. Elimination reactions
9. Oxidations and reductions

These nine reactions provide the capacity to construct a molecular framework and then to position functional groups precisely on this framework. Accordingly, these ten reactions permit two fundamental construction activities:

1. Creation of C-C/C=C/C-H bonds (for the purpose of building a structural framework)
2. Creation of functional groups (to give functionality to the framework)

Retrosynthetic Analysis

To use these 100 reactions (and additional ones) for the purposes of drug molecule synthesis requires an organized and systematic approach. The synthesis of a complicated drug molecule from simple starting materials must be approached in a rigorous and systematic fashion. The *synthon approach* provides such a scheme. This approach is based upon the notion that it is easier to work backwards from the target molecule (i.e., the drug to be synthesized) to the starting materials. This backwards-thinking process is referred to as *retrosynthetic analysis.* In retrosynthetic analysis, a procedure known as *disconnection* is used to dissect a molecule into progressively smaller and smaller fragments until readily available starting materials are obtained. A disconnection involves breaking a bond to a carbon atom.

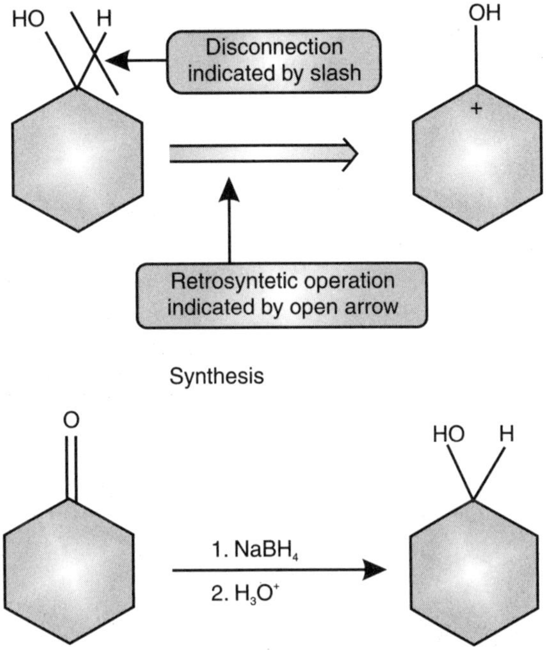

Fig. 13.3. The synthon approach: example of a retrosynthetic approach to the synthesis of cyclohexanol.

The fragments that result from this disconnection are referred to as *synthons.* Typically, a bond is broken and the electron pair is assigned to one of the fragments, resulting in a positively charged synthon and a negatively charged synthon. The next task is to find readily available starting materials that can actually be used as sources of these synthons. These available materials that are equivalent to a synthon are referred to as *synthetic equivalents*, and are generally commercially available compounds that represent the nucleophile and the electrophile that must react. Sometimes, prior to a disconnection, one functional group is converted to another "synthetically equal" functional group through the process of a *functional group interconversion* (FGI). This FGI may produce a molecule which is easier to disconnect and thus easier to synthesize. Cyclohexanol may be disconnected to a hydride ion and a hydroxycarbocation (the synthons). Sodium borohydride and cyclohexanone are the synthetic equivalents of these two synthons. Thus, reacting cyclohexanone with $NaBH_4$ will produce cyclohexanol. It is apparent that this system of disconnections can be applied to molecules of significant complexity to deduce a synthetic route.

Research Level Synthesis

Typically, medicinal chemists work in either academic or industrial research laboratories. Accordingly, they tend to develop "*research level*" syntheses. At the research level, it is acceptable for a synthetic scheme to employ unusual catalysts or large amounts of organic solvents. However, this is not acceptable at the "*scale-up level.*" For a drug molecule to be successful, it must also be affordable. If a drug costs $85 per milligram to synthesize and will be administered at 60 mg/day for many months, then it will not be a commercially viable drug. If a drug is to be successful it will have to be produced in large quantities in an industrial setting, necessitating the implementation of syntheses that are "*scalable,*" efficient, and environmentally friendly. A synthetic scheme with fewer steps and employing water as a preferred solvent has distinct advantages over a technically

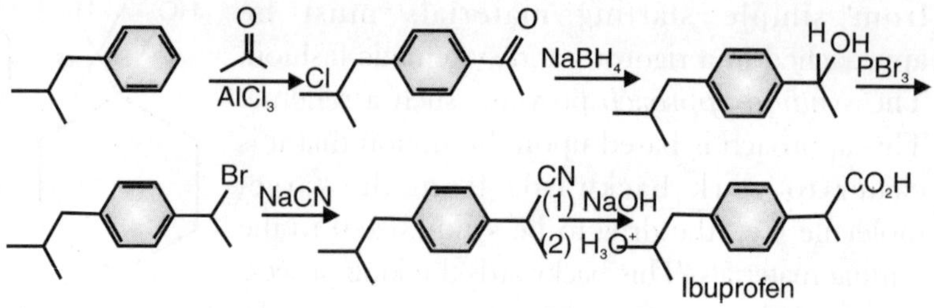

Fig. 13.4. The synthesis of ibuprofen is initiated by a Friedel-Crafts acylation of an alkyl-substituted benzene ring.

complicated, low-yield synthesis that uses large quantities of organic solvents that are difficult to dispose of. Thus, there are some differences between a classical synthetic organic chemist and a synthetic medicinal chemist. The classical synthetic organic chemist is proud of a complex multistep synthesis which may, regrettably, have a low yield. The synthetic medicinal chemist is pleased to design a molecule that can be synthesized in as few steps as possible, hopefully with a high yield and few by-products.

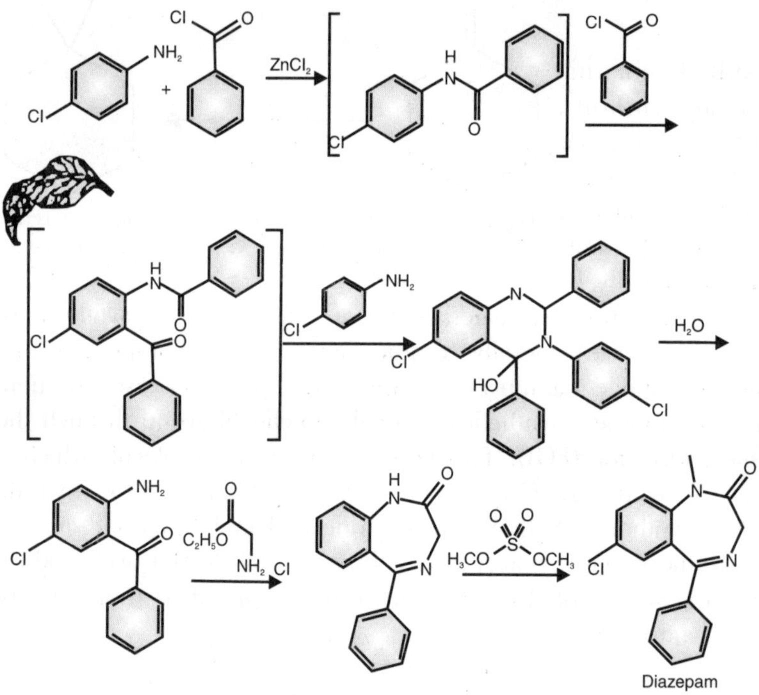

Fig. 13.5. The synthesis of diazepam is initiated by the double acylation of an aromatic amine with an armatic acid chloride.

Evaluation of Lead Compound

Once the lead compound has been synthesized, it is next necessary to biologically evaluate it. A number of very important characteristics go into making a biological assay useful. Ideally, the assay should be rapid, cost-effective, efficient, and easy to implement. This is particularly true if one is pursuing lead compound discovery by high throughput screening of millions of compounds. More importantly, the biological model should accurately reflect the human disease for which it is being used as a drug-screening tool. Just because an animal model produces a similar disease as is found in humans, there is no guarantee that it will truly reflect the corresponding human disease. For example, occluding the middle cerebral artery in a gerbil will produce a "*gerbil stroke*." However, are strokes in gerbils identical to strokes in humans?

A drug that successfully treats strokes in gerbils may not necessarily treat strokes in humans. Alternatively, there may be diseases that are unique to humans or primates, thereby making it difficult to develop meaningful biological assays in species such as rodents. Alzheimer's disease is a superb example of this dilemma. For years, drug design in Alzheimer's disease was delayed by the absence of a reasonable animal model. Modern advanced molecular biology techniques are addressing this issue by enabling the engineering of rats that over-express the β-amyloid protein that seems to cause Alzheimer's disease in humans. Biological assays for compound evaluation may be broadly categorized as follows:

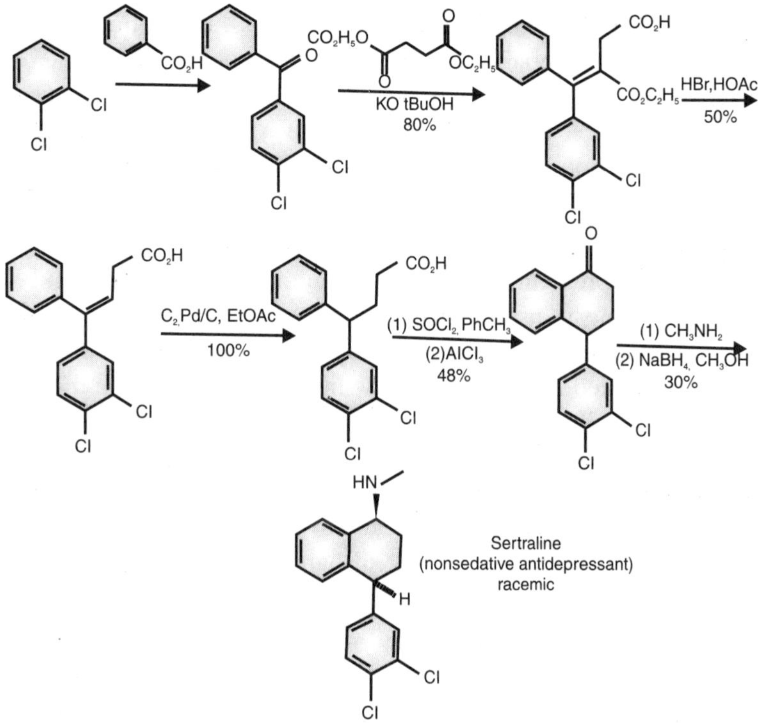

Fig. 13.6. The synthesis of sertraline.

1. *In silico*—theoretical evaluation achieved by computing simulation
2. *In vitro*— "*test-tube*" evaluation done without a whole animal
   (a) Binding studies (e.g., competition studies using radioactive ligands)
   (b) Functional assay studies (e.g., functional assays using enzymes)
3. *In vivo*—evaluation done with a whole animal

Each of these models has its strengths and weaknesses.

The *in silico* methods are the most rapid and cost effective. They are also the ones that are most divorced from reality. Consequently, *in silico* methods are acceptable for preliminary screens, but are completely unacceptable for the advanced assessment of a candidate compound. *In vitro* methods are a definite step up. Radiolabeled competitive binding studies can be used to ascertain whether a drug binds to a receptor. Such studies simply give information about the binding process; they do not tell if the drug is an agonist or antagonist. Functional *in vitro* assays with a measurable biological outcome are required to tell whether a compound is functioning as an agonist or an antagonist. By far the most useful assay is the *in vivo* assay. Regrettably but understandably, this assay is the most labor-intensive and costly. *In vivo* assays give the highest quality information about the efficacy of a lead compound. *In vivo* models are accurate animal models of a human disease. Ideally, a candidate drug molecule should not be advanced in the development process unless it demonstrates good to excellent efficacy in an appropriate *in vivo* model.

Optimization of Lead Compound

Once a biologically active lead compound has been identified (i.e., designed, synthesized, evaluated), the next decision is whether to proceed with the optimization of the lead compound. If the disease is important and if the lead compound has promising activity, then this decision may be relatively straightforward. Nevertheless, the optimization of a lead compound is a lengthy and expensive undertaking, fraught with a frighteningly high rate of failure. Even in the 21st century, drug design is more of a hope than an achievement. It means the application of previously recognized correlations of biological activity with physicochemical characteristics in the broadest sense, in the hope that the pharmacological success of a not yet synthesized compound can be predicted. Few drugs in use today were discovered entirely in this way.

One of the principal difficulties in this approach is that the available–and very sophisticated–methods for predicting drug action cannot foretell toxicity and side effects, nor do they help in anticipating the transport characteristics or metabolic fate of the drug *in vivo*. Although some practicing biologists and pharmacologists still regard efforts at drug design with some condescension and ill-concealed impatience, a slow but promising development gives renewed hope that progress in this area will not be less rapid than in the application of biology and physical chemistry to human and animal pathology. The explosive development of computer-aided drug design and bioinformatics promises to lead to the era of true rational drug design. Until the early 1960s, lead compound optimization was an intuitive endeavor based on long experience, keen observation, serendipity, sheer luck, and a lot of hard work.

The probabilities of finding a clinically useful drug were not good; it was estimated that anywhere from 3000 to 5000 compounds were synthesized in order to produce one optimized drug. With today's even stricter drug safety regulations, the proportions are even worse and the costs skyrocket, retarding the introduction of new drugs to an almost dangerous extent. The classical method usually applied in lead compound optimization was molecular modification–the design of analogs of a proven active "lead" compound. The guiding principle was the paradigm that minor changes in a molecular structure lead to minor, quantitative alterations in its biological effects. Although this may be true in closely related series, it depends on the definition of "minor" changes.

The addition of two seemingly insignificant hydrogen atoms to the Δ^8 double bond of ergot alkaloids eliminates their uterotonic activity, but replacement of the N-CH_3 substituent by the large phenethyl group in morphine increases the activity less than tenfold. Extension of the side chain of diethazine by only one carbon atom led to the serendipitous discovery of chlorpromazine and the field of modern psycho-pharmacology. There are two conclusions to be drawn from these random examples. First, a merely structural change in an organic molecule is meaningless as long as its physicochemical consequences remain unexplored and the molecular basis of its action remains unknown. Structure, in the organic chemical sense, is only a repository, a carrier of numerous parameters of vital importance of drug activity.

The second conclusion to be drawn from the above examples–and innumerable others–is that the discovery of qualitatively new pharmacological effects is often a discontinuous jump in an otherwise monotonous series of drug analogs and is hard to predict, even with fairly sophisticated methods. Although a beginning has been made, drug design is far from being either automatic or foolproof. The decision to optimize a proper lead compound–a necessity in drug design and development–is still based on experience, serendipity, and luck, given our basic ignorance of molecular phenomena at the cellular level. Now, however, in the 21st century we can at least have some confidence (thanks in no small part to computer-aided molecular design and bioinformatics) that the optimization of lead compounds and the corresponding discovery of new drugs will be able to keep pace with the progress of biomedical research.

PHARMACODYNAMIC PHASE

Once a lead compound has been identified by one of the techniques described above and the decision has been made to optimize this compound, the next task is to determine an approach to compound optimization. Typically, this approach is achieved via a two-step strategy:

1. Optimizing the compound for pharmacodynamic interactions
2. Optimizing the compound for pharmacokinetic and pharma-ceutical interactions

This two-step strategy enables the drug to be optimized first for its ability to bind to a receptor site and then for its ability to actually travel to that receptor site. Optimizing the lead compound for the pharmacodynamic phase is likewise approached via a two-step strategy:

1. Synthesizing analogs of the lead compound
2. Correlating analog structure with bioactivity via *quantitative structure-activity relationship* (QSAR) studies

This two-step strategy starts with the synthesis of numerous analogs of the lead compound. This is meant to explore the structural diversity of the lead compound. Next, QSAR studies are employed to correlate these structural diversities with bioactivity. The results of the QSAR studies are then used to design new and, hopefully, improved analogs.

Structural Modification

Classical structure modifications have been the mainstay of drug optimization since the earliest days. As emphasized earlier at several points, structural modifications expressed in organic chemical terms are really only symbols for modification of the physicochemical properties of various structures. Nevertheless, the medicinal chemist usually thinks in terms of structure, since that is the language of organic synthesis. It is therefore appropriate to deal with such an approach, provided one keeps in mind that it is somewhat obsolete because it is twice removed from the arena of drug-receptor interactions.

Homologation

The first approach is to vary substituents. The variation of substituents can follow many directions. It can be used to increase or decrease the polarity, alter the pK_a, and change the electronic properties of a molecule. Exploration of homologous series is one of the most often used strategies in this regard, because the polarity changes that are induced are very gradual. Homologation is a standard first approach to substituent variation; indeed, a "*standard joke*" in medicinal chemistry is to pursue the "methyl, ethyl, propyl, ... futile" series of analogs. Despite the somewhat facetious nature of this statement, there are many examples in which homologation is important to drug design. The case of the antibacterial action of aliphatic alcohols has been known in detail for many years; an increase in chain length leads to increased activity, with a sudden cutoff point at C_6–C_8, due to insufficient solubility of these homologs in an aqueous medium because of their high lipophilicity.

On the other hand, local anesthetics depend on lipid solubility in the membrane, and the duration of anesthesia produced by the nupercaine derivatives varies between 10 and 600 minutes for a series of alkyl substituents ranging from a simple -H to -n-pentyl. Another well-known example is the profound qualitative change in action between promethazine, an H_1-antihistaminic drug in which two $-CH_2-$ groups separate the ring and side-chain nitrogens, and promazine (an analog of promethazine), which has three methylene groups and predominantly exhibits tran-quilizing properties. Higher homologs can, on occasion, become antagonists of the lower members of a series.

Stereochemistry of molecules

The introduction of double bonds changes the stereochemistry of a molecule and decreases the flexibility of carbon chains. The E and Z isomers may show very different binding properties; for example, the Δ^1-double bond of prednisone increases its activity against rheumatoid arthritis over that of its parent compound cortisol by about 30-fold. Decreasing molecular flexibility is also important in drug design. If a drug molecule is too flexible, it will be able to fit into too many different receptors, leading to undesirable effects and toxicity.

Bewildering variations

Variations in ring structure are endless in drug synthesis, and are often used in the service of some other change or are introduced simply for patent-right purposes. Inspection of some of the *bewildering variations* of rings in the older H_1 antihistamines reveals them to be simply variations on the ethylenediamine structure, differing only quantitatively in their effect. Sometimes, however, relatively minor changes in ring structure lead to profound qualitative changes. The most famous example of this occurs in the transition from the neuroleptic dopamine-blocking phenothiazine drugs to the antidepressant dibenzazepines such as imipramine, in which the replacement of -S- by $-CH_2-CH_2-$ changes the molecular geometry. Ring opening or closure usually leads to subtle changes in activity, provided that nothing else changes. Three examples (among many possibilities) come to mind: incorporation of the N-methyl substituents of chlorpromazine (1.3) into a closed piperazine ring in prochlorperazine tremendously increases the antiemetic effect while the neuroleptic activity declines.

Of course, this may be due to the introduction of a new basic center. In thioridazine, the neuroleptic effect increases with the introduction of a closed ring, but extrapyramidal side effects (tremor, stooped posture, slow and shuffling gait) become noticeable. The inclusion of rings helps to conformationally constrain a molecule and make it less flexible. As discussed in the previous section, this is a desirable design strategy. Incorporating alkyl rings may change the solubility of the molecule, increasing lipophilicity. The incorporation of an aromatic ring may change the pharmacokinetics of the drug.

Pharmacophore

As noted earlier, the pharmacophore of a drug is usually confined to a few functional groups or parts of the whole molecule, which can be a large one. In the case of such complex natural products as alkaloids, which may be difficult or impractical to synthesize (e.g., tubocurarine), the first design attempt is usually directed at simplification of the molecule, pruning away those structural elements that are not part of the pharmacophore and do not serve to hold crucial binding groups in their appropriate positions. The most successful dissection of a molecule can be seen in the case of morphine. Starting with morphine, the oxygen bridge (i.e., the furan ring) is first removed, resulting in levorphanol, a morphinan.

By eliminating ring C, the benzomorphan series is obtained. Its most successful member is pentazocine, which retains only the two methyl groups from ring C and has a lower addiction liability. The simplest (and, incidentally, oldest) modification of the morphine molecule is seen in meperidine, a phenylpiperidine that has many congeneric analogs. Fentanyl is designed along similar lines and has some tremendously active analogs, such as sufentanil. Even in the methadone molecule, the remnants of the piperidine ring are discernible. On the basis of these and other analogs, the opiate pharmacophore consists of:

1. A nonbonding N electron pair
2. A phenyl ring, three carbons removed from the N
3. A quaternary carbon, next to the phenyl ring

Basically, the same criteria apply to the enkephalins. The addition of bulky substituents to a drug molecule often results in the emergence of antagonists, since it permits the utilization of auxiliary binding sites on the receptor. This trend is especially noticeable among the neurotransmitters. For example, the anticholinergics, β-adrenergic blocking agents, and some serotonin antagonists show this correlation. Large substituents often prevent enzymatic attack on a drug, thereby prolonging its useful life. This technique was used to impart resistance to β-lactamase to the semi-synthetic penicillins. The need for the proximity of the phenyl group to the lactam is quite interesting: phenylbenzyl penicillin is inactive as an enzyme inhibitor because the phenyl group no longer hinders access of the enzyme to the lactam bond.

Physicochemical characteristics

Alteration of the physicochemical characteristics within a drug series is, of course, a result of structural modification; it is just our point of view that changes. It is rather difficult to change only a single parameter with any specific modification, with the potential exception of lipophilicity, which increases with the addition of "inert" hydrocarbon groups. The degree of lipophilicity–so important in drug action and quantitative SAR (QSAR) investigations–is otherwise subject to change together with the Hammet σ-constant, a descriptor of the electron-donor or electron-acceptor capability of a substituent.

Isosteric replacement

The *isosteric replacement* of atoms or groups in a molecule is widely used in the design of antimetabolites or drugs that alter metabolic processes. Isosteric groups, according to Erlenmeyer's definition, are isoelectronic in their outermost electron shell. However, since their size and polarity may vary, the term isostere is somewhat misleading. Isosteres are classified according to their valence (i.e., the number of electrons in the outer shell):

Class I: halogens; OH; SH; NH_2; CH_3

Class II: O, S, Se, Te; NH; CH_2

Class III: N, P, As, CH

Class IV: C, Si, N^+, P^+, S^+, As^+

Class V: -CH=CH-, S, O, NH (in rings)

Thus, for instance, the exchange of OH for SH in hypoxanthine gives the antitumor agent 6-mercaptopurine. Fluorine, the smallest halogen, replaces hydrogen well, giving, for instance, fluorouracil, which is also an antitumor antimetabolite. Interchanges of -CH- and nitrogen are common in rings, as seen in the antiviral agent ribavirin. The oldest example of the use of "*nonclassical*" isosteres involves the replacement of the carboxamide in folic acid by sulfonamide, to give the sulfanilamides. Diaminopyrimidines, as antimalarial agents, are also based on folate isosterism, in addition to the exploitation of auxiliary binding sites on dihydrofolate reductase. This concept of nonclassical isosteres or *bioisosteres*–that is, moieties that do not have the same number of atoms or identical electron structure–is really the classical structure modification approach.

Analogous bioactivity

The approach involving the design of analogs of an active lead compound has remained unchanged for decades, and the expertise of the synthetic medicinal chemist is as much in demand as ever; however, the intuitive process of selecting structural modifications for synthesis becomes circumspect in this approach, and models based on multiple regression analysis and pattern recognition methods, using very powerful computer techniques, are increasingly being employed as aids. It is obviously much faster and cheaper to calculate the required properties of novel compounds from a large pool of data on their analogs than to synthesize and screen all such new compounds in the classical fashion. Only promising candidates are investigated experimentally. The results gained this way are incorporated into the database, expanding and strengthening the theoretical search. Eventually, sufficient material accumulates to aid in making a confident decision about whether the "best" analog has been prepared or whether the series should be abandoned. Quantitative structure-activity relationship studies represent a systematic approach to this correlation of structure with pharmacological activity.

Corelative Approach

The relationship between chemical structure and biological activity has always been at the center of drug research. In the past, drug structures were modified intuitively and empirically, depending on the imagination and experience of the synthesizing chemist, and were based on analogies. Surprisingly, the results were often gratifying, even if obtained only serendipitously or on the basis of the wrong hypothesis. However, this hit-or-miss approach, practiced even now, is enormously wasteful. Considering that only one of several thousand synthesized compounds will reach the pharmacy shelves, and that the development of a single drug can cost millions of dollars, it is imperative that rational short-cuts to drug design be found.

Quantitative structure-activity relationship (QSAR) studies represent this important rational short-cut. QSAR endeavors to elevate drug design from an art to a science. QSAR methods are in part retrospective as well as predictive, since a "*training set*" of compounds of known pharmacological activity must first be established. The purpose of such methods is to increase the probability of finding active compounds among those eventually synthesized, thus keeping synthetic and screening efforts within reasonable limits in relation to the success rate. There are three main classifications of QSAR methods:

1. 1D-QSAR (e.g., Hansch analysis)
2. 2D-QSAR (e.g., pattern recognition analysis)
3. 3D-QSAR (e.g., comparative molecular field analysis)

Each method has its own strengths and weaknesses.

Hansch analysis

Historically, this is the most popular mathematical approach to QSAR. The major contribution of Hansch analysis is in recognizing the importance of log*P*, where *P* is the octanol–water partition coefficient. Log*P* is perhaps the most important measure of a drug molecule's solubility. It reflects the ability of the drug to partition itself into the lipid surroundings of the receptor microenvironment. Introduced by Corwin Hansch in the early 1960s, Hansch analysis considers both the physicochemical aspects of drug distribution from the point of application to the point of effect and the drug–receptor interaction. In a given group of drugs that have analogous structures and act by the same mechanism, three parameters seem to play a major role:

1. The *substituent hydrophobicity constant,* based on partition coefficients analogs to Hammet constants:

$$\pi_X = \log P_X - \log P_H \qquad ...(1)$$

where P_X is the partition coefficient of the molecule carrying substituent X, and P_H is the partition coefficient of the unsubstituted molecule (i.e., substituted by hydrogen *only*). More positive *Jr* values indicate higher lipophilicity of the substituent. Since these values are additive, *P* values measured on standard molecules permit prediction of hydrophobicity of novel molecules.

2. The *Hammet substituent constant* σ
3. *Steric effects,* described by the Taft E_S values

The σ and π constants of substituents are often useful when correlated to biological activity in the statistical procedure known as multivariate regression analysis. As is well known from pharmacological testing of various drug series, such correlations can be either linear or parabolic. The linear relationship is described by the equation

$$\log 1/C = a\pi + bE_S + c\sigma + d \qquad ...(2)$$

where *C* is the drug concentration for a chosen standard biological effect, and *a, b, c,* and *d* are regression coefficients to be determined by iterative curve fitting. The parabolic relationship fits the equation

$$\log 1/C = -a\pi^2 + b\pi + cE_S + d\sigma + e \qquad ...(3)$$

The coefficients *a, b, c, d,* and *e* are fitted to the curve by the least-squares procedure, using regression methods for which computer programs are readily available. The extent of the fit is judged by the correlation coefficient *r* or the multiple regression coefficient r^2, which is proportional to the variance. A perfect fit gives $r^2 = 1.00$. Once the best fit has been achieved and *r* or r^2 has been maximized by using a reasonable number of known compounds (15-20 is an advisable number, depending on the number of variables tested, with even more compounds being even better), the curve can be used to predict the biological activity of compounds that have not been tested or, indeed, have not even been synthesized. This requires only the substitution of the optimized regression coefficient constants into the equation, and the use of π, σ, and E_S values, which are usually available for just about any substituent.

Naturally, independent variables other than π or σ–including ionization constants, activity coefficients, molar volumes, or molecular orbital parameters–can also be used. To achieve these various "*best fits*," statistical methods are employed. A regression analysis of the effects of various substituents on a molecule using the Hansch approach is very useful, saving much time and effort in the synthesis and testing of new drugs. Hundreds of examples of such analyses are available in the literature; many show positive predictive values for drug activity, whereas some other drug series cannot be interpreted by this method. Regression analysis is currently the most widely used correlative method in drug design. This is because it simplifies problems within a set of compounds by using a limited number of descriptors, notably the Hansch hydrophobic constant π, Hammet constants, or other electronic characteristics of substituents, and the Taft steric constant E_S.

Nevertheless, there are several difficulties and pitfalls in using the Hansch method. First, the inherent disadvantage of regression analysis is that one can obtain good fits ($r^2 > 0.9$) simply by manipulating the constants. Therefore, curve fitting must be done for a relatively large number of compounds to ensure that all predictors are considered. Second, the mode of action may change for drugs within a seemingly continuous series, invalidating the comparison of some compounds in the series with the predictor compounds. The Hansch method cannot anticipate such a change. Other problems with the Hansch method are that biological systems are often too crude as models for its application, or the electronic effects operative in a drug molecule are not sufficiently understood or precise.

Finally, the method requires considerable time and expense, even in the hands of an expert. Difficulties notwithstanding, the Hansch approach took both chemists and pharmacologists out of the dark age of pure empiricism and allowed them to consider simultaneously the effects of a large number of variables of drug activity–a feat unattainable with classical methods. Nevertheless, Hansch analysis revolutionized drug molecule optimization and directly led to two other strategies for molecule optimization: the Free-Wilson method and the Topliss decision tree.

Free-Wilson method

This method also assumes that biological activity can be described by the additive properties of the substituents on a basic molecular structure. In the Fujita-Ban modification of this method

$$\log 1/C = \Sigma a_i X_i + \mu_0 \qquad ...(4)$$

where C is the drug concentration for a standardized effect, a_i is the group contribution of the ith substituent to the pharmacological activity of the substituted molecule, X is unity if substituent i is present and zero otherwise, and $\mu_0 = 1/C$ for the parent compound. Regression analysis is used to determine a_i and μ. In the Fujita-Ban modification of the Free-Wilson method, no assumptions are made about the relevance of the model parameters to the biological activity of the molecule.

The effect of each substituent is considered to be independent of any other, and each makes a constant contribution to the overall activity of the molecule. Therefore the method is applicable to compounds with more than one variable group.

The result is a data matrix that shows the contribution of each substituent in each position to the overall biological effect of the molecule. The Free–Wilson equation bears close similarities to the linear Hansch equation, and the results of the two can be comparable. The Free–Wilson method, however, cannot predict the activities of compounds that have substituents not included in the matrix. Consequently, this method has found only limited application in drug series where many close analogs are already available but physicochemical data are lacking.

Topliss decision tree method

This method is quicker and easier to use than the Hansch method. The *Topliss scheme* is an empirical method in which each compound is tested before an analog is planned, and is compared in terms of its physical properties with analogs already planned. Like the Free–Wilson method, the Topliss decision tree is no longer extensively used. The 2D- and 3D-QSAR methods are gradually supplanting the 1D methods.

Pattern recognition analysis

2D-QSAR is a somewhat more advanced method for correlating activity and structure. The first step in performing a 2D-QSAR is to select the *training set*. This is a subset of molecules that are diverse in terms of both structure and bioactivity. Ideally, the compounds that are available cover the full spectrum of bioactivity, ranging from active (fully and partially, covering a 10^3-fold range in receptor binding affinities) to inactive. It is difficult to determine what makes a molecule bioactive (or conversely what makes a molecule bioinactive) if all of the compounds tested have similar bioactivities.

The more molecules the better, but a reasonable start can be made with as few as ten compounds. It is important not to use all available molecules, since another subset is held back and retained as a *test set*. This test set will ultimately be used to validate any prediction algorithm that is developed through the study of the training set. Next, every molecule in the training set, regardless of its pharmacological activity, is characterized by a series of *descriptors*:

1. Geometric descriptors Bond lengths

   Bond angles

   Torsional angles

   Interatomic distances

2. Electronic descriptors Charge densities on individual atoms

   Energy of the highest occupied molecular orbital

   Energy of the lowest unoccupied molecular orbital

   Molecular dipole

3. Topological descriptors

   Graph theory indices

Randic indices

Kier–Hall indices

Ad hoc indices

Number of rings in the molecule

Number of aromatic rings in the molecule

4. Physicochemical descriptors

Octanol–water partition coefficients

Log*P*

(Log*P*)^2

Hydrogen bonding number

Number of hydrogen bonding donor sites

Number of hydrogen bonding acceptor sites

These various descriptors may be calculated using various molecular mechanics and quantum mechanics approaches. The geometric descriptors reflect molecular geometry and are conceptually straightforward. Electronic descriptors reflect properties arising from variations in electron distribution throughout the drug molecule framework. Topological descriptors endeavor to describe molecular branching and complexity through the notion of *molecular connectivity.* The concept of molecular connectivity, introduced by Kier and Hall in 1976, describes compounds in topological terms. Branching, unsaturation, and molecular shape are all represented in the purely empirical connectivity index $^1\chi$, which correlates surprisingly well with a number of physicochemical properties including the partition coefficients, molar refractivity, or boiling point. These graph theory indices are useful to differentiate between an *n*-butyl substituent and a *tert*-butyl substituent.

The physicochemical indices reflect the ability of the drug to partition itself into the lipid surroundings of the receptor microenvironment. All of these descriptors are calculated for every compound within the training set. Next, a 2D data array is constructed. Along the vertical axis, all of the training set compounds are listed in descending order of bioactivity. Along the horizontal axis, all of the descriptors are arranged for every training set compound. This data array is then probed with statistical calculations to ascertain the minimum number of descriptors that differentiate active compounds from inactive compounds. In order to probe the data array, several methods are available. *Pattern recognition* and *cluster analysis,* two recent quantitative methods, make use of sophisticated statistics and computer software. Pattern recognition can be used to deal with a large number of compounds, each characterized by many parameters. First, however, these raw data must be processed by scaling and normalization–the conversion of diverse units and orders of magnitude from many sources–so that the chosen parameters become comparable.

Feature selection methods exist for weeding out irrelevant "*descriptors*" and obtaining those that are potentially most useful. By using "*eigenvector*" or "*principal component*" analysis algorithms, these multidimensional data are then projected two-dimensionally onto a plot whose axes are the two principal components or two (transformed and normalized) parameters that account for most of the variance; these are the two eigenvectors with the highest values. Previously unrecognized relational patterns between large numbers of compounds characterized by multidimensional descriptors will thus emerge in a new, comprehensible, two-dimensional plot. The projection of unknowns onto this eigenvector plot will determine their relationship to active and inactive compounds.

Cluster analysis is similar in concept to pattern recognition. It can define the similarity or dissimilarity of observations or can reveal the number of groups formed by a collection of data. The distance between clusters of data points is defined either by the distance between the two closest members of two different clusters or by the distances between the centers of clusters. Once the data array has been probed and the minimum number of descriptors that differentiate activity from inactivity has been ascertained, a *prediction algorithm* is deduced. This algorithm attempts to quantify the bioactivity in terms of the relevant descriptors. The predictive usefulness of this algorithm is then validated by being applied to the test set compounds. If the prediction algorithm is sufficiently robust, it can be used to direct the syntheses of optimized compounds.

Comparative molecular field analysis

Like other forms of QSAR, 3D-QSAR starts with a series of compounds with known structures and known biological activities. The first step is to align the molecular structures. This is done with alignment algorithms that rotate and translate the molecule in Cartesian coordinate space so that it aligns with another molecule. The work starts with the most rigid analogs and then progresses to conformationally flexible molecules that are aligned with the more rigid ones. The end result is that all the molecules are eventually aligned, each on top of another. Once the molecules of the training set have been aligned, a *molecular field* is computed around each molecule, based upon a grid of points in space.

Various molecular fields are composed of *field descriptors* that reflect properties such as steric factors or electrostatic potential. The *field points* are then fitted to predict the bioactivity. A *partial least-squares algorithm* (PLS) is used for this form of fitting. Based upon this PLS calculation, two pieces of information are deduced for every region of space within the molecular field about the molecule: the first piece of information states whether that region of space correlates with biological activity; the second piece of information determines whether the functional group on the molecule within that region of space should be bulky, aromatic, electron donating, electron withdrawing, and so on. The predictions from these molecular field calculations are then validated by being applied to a test set of compounds.

Pharmacophore identification

All drugs have pharmacological activity as a result of stereoelectronic interaction with a receptor. The receptor macromolecule "recognizes" the arrangement of certain functional groups in three-dimensional space and their electron density. It is the recognition of these groups rather than the structure of the entire drug molecule that results in an interaction, normally consisting of noncovalent binding. The collection of relevant groups responsible for the effect is the pharmacophore, and their geometric arrangement is called the pharmacophoric pattern, whereas the position of their complementary structures on the receptor is the *receptor map.* Over the years, many attempts have been made to define the pharmacophores and their pattern on many drugs. The first attempts were rather naive and simplistic, but the recent use of QSAR has contributed greatly to the evolution of sophisticated methods of practical significance.

The identification of the pharmacophore is a logical corollary of a QSAR calculation. If the minimum number of descriptors that differentiate activity from inactivity is known, it is possible to deduce the *bioactive face* of the molecule–that part of the molecule around which all of the relevant descriptors are focused. This bioactive face logically defines the pharmacophoric pattern of the bioactive molecules. When using QSAR calculations to optimize a drug for the pharmacodynamic phase, it is important to use relevant biological activities. If *in vivo* activities are used, the bioactivities will be influenced by pharmacokinetic and pharmaceutical factors. In order for QSAR calculations to reflect the pharmacodynamic phase, the bioactivities should be based on *in vitro* data–optimally, receptor binding studies. QSAR studies are not restricted to the optimization of biological activity at the pharmacodynamic phase. Since toxicity also arises from drug-receptor interactions, the QSAR method can be used to identify the *biotoxic* face of the molecule (i.e., the toxicophore), which could then be engineered out of the molecular structure. Once QSAR calculations have been used to optimize the pharmacodynamic interactions of the drug molecule, the next step is to optimize the pharmacokinetic and pharmaceutical phases of drug action.

PHARMACODYNAMIC PHASE

Once the lead drug molecule has been optimized for the pharmacodynamic phase, it must next be optimized for the pharmacokinetic and pharmaceutical phases. If a drug molecule cannot withstand the trip from the gut to the receptor microenvironment, it makes no difference whether the drug actually binds to the receptor. Many factors must be taken into consideration when optimizing for the pharmacokinetic and pharmaceutical phases. Will the drug be annihilated within the gastrointestinal tract? Will the drug be absorbed? Can the drug molecule be distributed throughout the body? Will the drug be destroyed in the liver? Will the drug's half-life be too short, or too long? Will the drug be excreted too rapidly? If the drug is destined for a brain-based receptor, can the drug cross the blood-brain barrier?

Pharmacokinetic Phase

When optimizing for the pharmacokinetic/pharmaceutical phases, considerations of ADME (absorption, distribution, metabolism, excretion) are among the most important. (Sometimes ADME is extended to ADMET because of the inclusion of toxicity.) The drug designer must optimize the drug so that it can remain structurally intact during its absorption and distribution. This can be a daunting task, since the body inflicts many metabolic chemical reactions upon the drug molecule during the processes of absorption and distribution. Understanding these metabolic reactions is crucial to the continuing optimization of the drug molecule.

Physiochemical reactions

The body uses its usual array of chemical reactions to attack the structural integrity of a drug molecule and to promote its excretion from the body. These reactions can be categorized into Phase I and Phase II reactions. Phase I transformations (oxidation, reduction, hydrolysis) initiate the chemical modification, frequently adding functional groups and increasing polarity; Phase II transformations (conjugations) promote the aqueous solubility of the drug metabolite so that it may be excreted from the body. A more detailed delineation of these transformations is as follows:

- Phase I transformations:
  - Oxidations
    - Oxidation of aliphatic carbon atoms
    - Oxidation of carbons adjacent to sp^2 hybridized atoms
    - Oxidation of C=C (alkene) systems
    - Oxidation of C-O systems
    - Oxidations of C-N systems
    - Oxidation of C-S systems
    - Oxidations of aldehyde and alcohol
    - Aromatic hydroxylations
  - Reductions
    - Carbonyl reductions
    - Nitro reductions
    - Azo reductions
    - Reductive dehalogenation
  - Hydrolyses
    - Ester hydrolysis
    - Amide hydrolysis
- Phase II transformations
  - Glucuronic acid conjugation
    - O-glucuronidation
    - N-glucuronidation
    - S-glucuronidation
    - C-glucuronidation
  - Sulfuric acid conjugation
  - Amino acid conjugation
  - Acetyl conjugation
  - Methyl conjugation
  - Glutathione conjugation
  - Hydration or water conjugation

Specific examples of physiochemical reactions

The previous section listed the various reactions that the body can inflict upon a drug molecule. This section presents brief examples of a number of these reactions.

Aliphatic carbon atoms

Oxidation at the terminal carbon atom of an alkyl substituent is ω-oxidation; oxidation of the carbon atom located second from the end is ω-1 oxidation. Unless specifically catalyzed by an enzyme, ω-1 oxidation tends to occur more frequently. The anticonvulsant drug ethosuximide is metabolized at both the ω and ω-1 position.

Alkene epoxidation

Alkenes may react to produce epoxides (alternatively, sometimes, the alkenes do not react and are metabolically stable). The epoxide is unstable and is subject to ring opening via a nucleophilic attack. The anticonvulsant drug carbamazepine is metabolized via epoxidation to yield carbamazepine-10, 11- epoxide; in turn, this is rapidly opened to yield carbamazepine-10,11-diol.

Carbons adjacent to sp^2 atoms

Carbon atoms that are situated adjacent to imine, carbonyl, or aromatic groups are frequently oxidized. This reaction appears to be catalyzed by the cytochrome P-450 enzyme system. Typically, a hydroxyl group is attached to the carbon as part of the oxidation process.

Aromatic hydroxylation

Since many drugs contain aromatic rings, this is a very common metabolic transformation. The process tends to be species specific, with human showing a strong tendency to hydroxylation in the *para* position. This reaction proceeds via an arene epoxide intermediate. The anticonvulsant drug phenytoin is metabolized by being *para*-hydroxylated in its aromatic rings.

Carbon–nitrogen systems

Primary amines may be hydroxylated at the nitrogen atom (*N-oxidation*) to yield the corresponding hydroxylamine. Alternatively, primary alkyl or arylalkyl amines may undergo hydroxylation at the α-carbon to give a carbinolamine that decomposes to an aldehyde and ammonia (through the process of *oxidative deamination*). Secondary aliphatic amines may lose an alkyl group first (*N-dealkylation*) prior to oxidative deamination.

Carbon–oxygen systems

Molecules containing ether linkages may undergo *oxidative O-dealkylation.* In this process, the carbon atom located α to the oxygen atom is hydroxylated, followed by cleavage of the C-O bond.

Alcohols and aldehydes

The oxidation of alcohols to aldehydes and of aldehydes to carboxylic acids is routine, and is catalyzed by alcohol dehydrogenase and aldehyde dehydrogenase, respectively.

Carbon–sulfur systems

The most common metabolic process that affects a C-S system is *S-oxidation.* The S atom is oxidized to a sulfoxide. In the case of thioketones, the C=S double bond is converted to a C=O bond. For thioethers, oxidative *S-dealkylation* is a possibility.

Carbonyl groups

The carbonyl group (-(C=O)-) is reduced through a reaction that is catalyzed by an aldo-keto reductase requiring NADH as a cofactor. A large number of aromatic and aliphatic ketones are reduced to the corresponding alcohols; these reductions are frequently stereospecific. α,β-Unsaturated ketones are typically metabolized to saturated alcohols.

Nitro groups

Nitro reduction is catalyzed by the cytochrome P450 system in the presence of the NADPH cofactor; it is a multistep process in which the reduction of the nitro group to a nitroso group is a rate-limiting step. The metabolic conversion of the nitro group in clonazepam to an amine is a representative example.

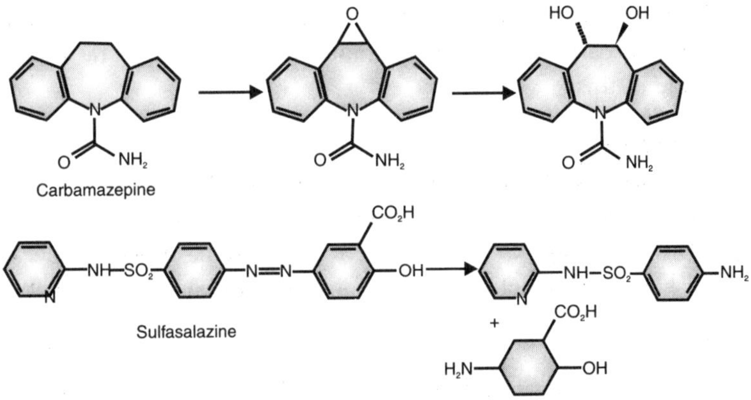

Fig. 13.7. Example of drug metabolism by oxidative and reductive pathways.

Esters and amides

The plasma, liver, kidney, and intestines contain a wide variety of nonspecific amidases and esterases. These catalyze the metabolism of esters and amides, ultimately leading to the formation of amines, alcohols, and carboxylic acids. Kinetically, amide hydrolysis is much slower than ester hydrolysis. These hydrolyses may exhibit stereoselectivity.

Conjugations

Conjugation reactions are Phase II metabolic reactions that are enzymatically catalyzed and involve the attachment of small polar molecules (glucuronic acid, sulphate, amino acids) to the drug. This, in turn, makes the drug more water soluble and thus more readily excreted in the urine or bile.

The enzymes that catalyze conjugations are transferases such as glucuronosyltransferase, sulfotransferase, glycine N-acyltransferase, and glutathione S-transferase. The conjugation reactions normally target hydroxyl, carboxyl, amino, or thiol groups. Glucuronidation is the most common conjugation method. There are four classes of glucuronide metabolites: O-, N-, S-, and C-glucuronides. Sulphate conjugations occur less frequently.

Molecular building blocks to physiochemical reactions

When designing and optimizing a drug molecule within the multiphore conceptualization, it is important to remember that the drug molecule is constructed from molecular building blocks that add function and structure to the molecule.

It is important that the vulnerability of each of these building blocks to metabolic attack be appreciated during the drug design process. This section lists the major molecular building blocks and briefly outlines their susceptibility to metabolism.

Alkanes

Alkyl functional groups tend to be metabolically nonreactive and to be excreted unchanged. Therefore, alkanes can be used to build the framework of a molecule or as lipophilic functional groups. There are a few exceptions to the rule of alkanes being metabolically inert. Rarely, a linear alkyl group will be oxidized in a process that is catalyzed by a mixed-function oxidase enzyme. When this occurs, it does so either at the end of the hydrocarbon chain or adjacent to the final carbon (the "omega-minus-one carbon"). The metabolism of butylbarbital is a good example of this.

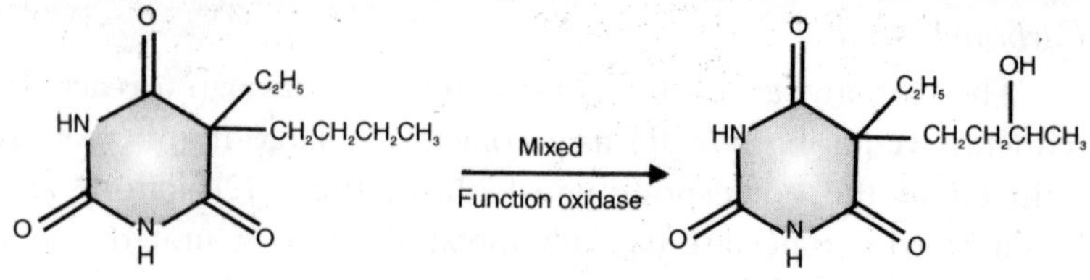

Fig. 13.8. Metabolism of butylbarbital by mixed-function oxidase.

Cycloalkanes

Cycloalkanes are conformationally restricted alkanes. Three rings are employed in drug design: cyclopropane, cyclopentane, cyclohexane (the latter two are used quite commonly). While cyclopropane may be reactive, due to ring strain, cyclopentane and cyclohexane are metabolically inert. These substituents are useful for increasing lipid solubility.

Alkenes

Alkenes are, in general, metabolically stable. The majority of alkene-containing drugs do not exhibit significant rapid metabolism at the double bond. There are some isolated examples of alkene-containing compounds that undergo epoxidation, catalyzed by mixed-function oxidase, or that add water across the double bond to give an alcohol.

Halogenated hydrocarbons

Halogenated hydrocarbons are not easily metabolized and show significant stability *in vivo*. The addition of halogens tends to increase the lipophilicity and to prolong the half-life of the drug.

Aromatic hydrocarbons

Aromatic rings are very susceptible to oxidation, in particular to aromatic hydroxylation. The oxidation of aromatic rings frequently proceeds via an epoxide intermediate, which may actually be stable enough to be isolated. The hydroxylation of an aromatic ring increases hydrophilicity, thus promoting renal excretion and slightly decreasing the half-life of the drug. Aromatic hydrocarbons are oxidized in a number of organs, but the liver is a preferred location.

Alcohols

The alcohol functional group tends to be metabolized by a variety of enzymes. Primary and secondary alcohols are oxidized to carboxylic acids or ketones. Tertiary alcohols are substantially more stable to oxidation. Rather than oxidase-mediated oxidation, alcohols can also be conjugated. If the alcohol is conjugated with glucuronic acid, a glucuronide forms; if it is conjugated with sulfuric acid, a sulfate is formed. Regardless, both of these conjugations increase hydrophilicity and decrease the half-life of the drug molecule.

Phenols

Phenols are oxidized via hydroxylation to yield a diphenolic molecule. This hydroxylation is either *ortho* or *para* to the primary alcohol. Phenols may also be conjugated with either glucuronic acid or sulfuric acid.

Ethers

In general, ethers are very stable and are excreted unchanged. Sometimes, an ether that involves a small alkyl group (occasionally a methyl, rarely an ethyl) will be dealkylated, with the small alkyl group being excreted as an aldehyde; the remainder of the drug molecule is left as an alcohol.

Aldehydes and ketones

Many metabolic routes are possible, including both oxidation and reduction. However, oxidations are more common. Aldehydes are very susceptible to oxidation, which is catalyzed by various enzymes including aldehyde oxidase and aldehyde dehydrogenase; this oxidation yields a carboxylic acid.

Ketones, on the other hand, tend to be stable to oxidation. Conversely, aldehydes are seldom metabolized by reduction. Ketones, however, frequently undergo reduction to a secondary alcohol; this is particularly true for α,β-unsaturated ketones.

Carboxylic acids

Metabolism of carboxylic acids is relatively straightforward. First and foremost, carboxylic acids undergo a diverse variety of conjugations. Carboxylic acids conjugate with glucuronic acid, glutamine, and glycine; the resulting conjugates are water soluble and more easily excreted. Alternatively, carboxylic acids may be oxidized, especially beta to the carboxyl group.

Esters

Not surprisingly, esters are not metabolically stable. They are readily converted to the corresponding free acid and alcohol via hydrolysis, a process that may be either base- or acid-catalyzed. Esters are rapidly hydrolyzed throughout many locations within the body. Orally administered ester-based drugs are quickly hydrolyzed within the stomach.

Amides

Although similar to esters in terms of being a functional derivative of a carboxylic acid, amides, unlike esters, are relatively metabolically stable. In general, amides are stable to acid- and base-catalyzed hydrolysis. This stability is related to the overlapping electron clouds within the amide functionality and the corresponding multiple resonance forms. Amidases are enzymes that can catalyze the hydrolysis of amides. Nevertheless, amides are much more stable than esters.

Carbonates, carbamates

The ester portion of both carbonates and carbamates is hydrolyzed to give the monosubstituted carbonic acid, which is unstable and decomposes with loss of carbon dioxide. Thus, carbonates are hydrolyzed to give alcohol and carbon dioxide; carbamates are hydrolyzed to yield an alcohol, an amine, and carbon dioxide.

Ureides

Compounds that contain urea functionalities are stable and are not commonly metabolized or hydrolyzed.

Amines

An important metabolic route for primary and secondary amines is conjugation with either glucuronic acid or sulfuric acid to yield the corresponding water-soluble glucuronides and sulfates. Another common reaction is dealkylation of secondary and tertiary amines. The alkyl groups are sequentially removed and then "lost" as either aldehydes or ketones. The amine is thus converted from a tertiary amine to a secondary amine and then to a primary amine. This form of metabolism occurs most favorably if the alkyl group is small, such as methyl, ethyl, or propyl. Finally, amines may be acetylated, with the resulting amide undergoing typical amide metabolism.

Thio ethers

Thio ethers are commonly oxidized to a sulfoxide or sulfone. The conversion of chlorpromazine to chlorpromazine sulfoxide is a good example of this.

Sulfonic acids

Drugs that contain sulfonic acids are quite acidic. They have high water solubility and are rapidly excreted, leading to a rather short half-life. They can also be irritating to the gastrointestinal tract during the pharmaceutical phase.

Sulfonamides

In drug design, sulfonamides are frequently attached to aromatic rings. Thus many drugs are constructed around a benzenesulfonamide nucleus. This nucleus is quite stable to acid-, base-, or enzyme-catalyzed hydrolysis.

Epoxides

Epoxides or oxiranes are quite reactive metabolically. These three- membered rings will readily open in the presence of most nucleophiles. They will react with most biopolymers, including proteins and nucleic acids. Compounds with epoxides have very short half-lives and tend to be toxic.

Aziridines

These are similar to epoxides. They are extremely reactive to most nucleophiles and open readily. Aziridines tend to behave as alkylating agents and thus have carcinogenic effects. In harmony with this, drugs containing aziridines have been used to treat cancer.

Tetrahydrofuran and furan

Some drugs containing tetrahydrofurans (oxalanes) tend to be readily air oxidized, and thus have a poor shelf life and are unstable even prior to being administered; other tetrahydrofuran-containing drugs are metabolically stable. Furans (oxoles) behave more like aromatic compounds than ethers; accordingly, they undergo aromatic hydroxylation.

Pyrrole and pyrrolidine

Pyrroles (azoles) undergo aromatic hydroxylation; pyrrolidines (azolidines) undergo conjugation with either glucuronic acid or sulfuric acid.

Thiophenes and tetrahydrothiophenes

Thiophenes (thioles) are subject to aromatic hydroxylation; tetrahydrothiophenes (thiolanes) undergo oxidation of the sulfur to give sulfoxides or sulfones.

Oxazole and isoxazole

These are five-membered ring heterocycles containing oxygen and a nitrogen. Two common forms are oxazole (1,3-oxazole) and isoxazole (1,2-oxazole), both of which undergo aromatic hydroxylation.

Pyrazole and imidazole

These are five-membered ring heterocycles containing two nitrogens, one basic and one neutral. Two common forms are pyrazole (1,2-diazole) and imidazole (1,3-diazole); both are prone to aromatic hydroxylation.

Pyridine and piperidine

These are six-membered heterocyclic rings containing one nitrogen. Piperidine (azine) is aromatic and undergoes aromatic hydroxylation. Piperidine functions as a secondary amine and undergoes hydroxylation with either glucuronic acid or sulfuric acid.

Indole

This is a bicyclic (one five-membered ring plus one six-membered ring) aromatic heterocycle. Since most indole-containing drugs are substituted with the remainder of the drug being positioned at the 3 position, the aromatic hydroxylation tends to occur at the 4–7 position.

Coumarin

The coumarin moiety is found in a number of important drugs. This is a bicyclic heterocycle containing two six-membered rings and two oxygens, one endocyclic, one exocyclic. Since the coumarin contains an intramolecular lactone ester, it undergoes hydrolysis to yield a carboxylic acid and a phenol.

Site of Delivery Considerations

The next major consideration during optimization for the pharmacokinetic/pharmaceutical phase concerns the design of drugs to overcome barriers during their distribution. Of these barriers, the blood–brain barrier is by far the most important to the drug designer.

Friend and drug designer

The blood–brain barrier (BBB) is both a friend and a foe to the drug designer. When designing a drug for a non-neurologic condition, the BBB is a friend in that it can be used to preclude entry of the drug into the brain, thereby decreasing side effects. However, when designing a drug for a neurological indication, the BBB is a true barrier that must be crossed. Moreover, if a drug has been designed to cross the BBB, it typically will also cross the maternal–fetal placental barrier – a consideration when designing drugs that may be used in women of childbearing age. The notion of a BBB first emerged in the early years of the twentieth century when it was observed that organic dyes injected into animals stained all tissues except the brain.

There appeared to be some invisible barrier that prevented certain molecules from entering into the brain. Over the past twenty to thirty years, the structural basis of the BBB has been more carefully delineated. Many different structural components contribute to the BBB. The small blood vessels, or capillaries, are the first structural level of the BBB. Drug molecules are distributed throughout the body by the bloodstream and the capillary is the point at which a drug leaves the bloodstream to bind to a receptor. Within the brain, capillaries are composed of cells, called *endothelial cells,* that are connected to each other by *tight junctions.* These junctions are a first-line impediment, slowing the journey of the drug molecule from within the capillary to a receptor site on a neuron. The next component of the BBB is the *astrocyte.* The astrocyte is a *glial cell* (helper cell) within the brain.

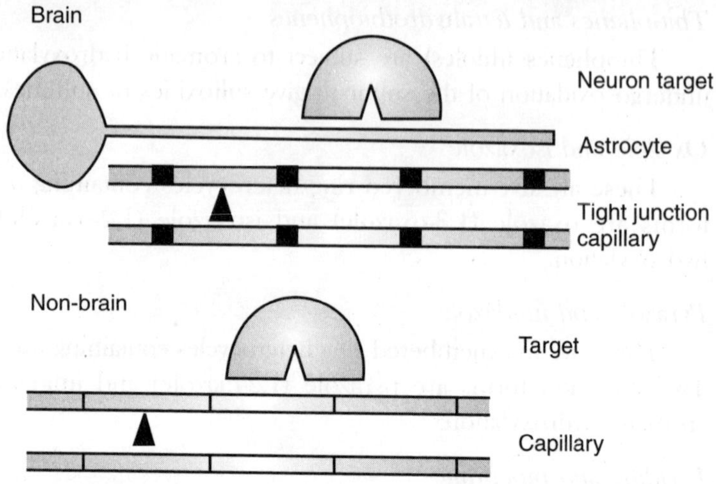

Fig. 13.9. The blood-brain barrier (BBB) is a major impediment to the delivery of drugs to the brain.

The astrocyte wraps itself around the capillary to provide yet another line of defense between the drug in the capillary and the neuronal receptor to which it is traveling. The tight junction–astrocyte barrier is unique to the brain and forms the BBB. There are two methods for crossing the BBB: passive diffusion or active transport. Passive diffusion is the route preferred by most neuroactive drugs. Lipid solubility is a desired chemical property for a molecule to diffuse across the BBB. For a molecule to cross the BBB by passive diffusion it should have a molecular weight less than 650 g/mol and should have a log*P* (logarithm of the octanol–water partition coefficient) value between 1.5 and 3.0; in accord with these requirements, the molecule should not be too polar, probably having no charged groups and fewer than three hydrogen-bonding donors or acceptors. The second route for traversal across the BBB is by active transport. This is an energy-dependent process whereby a protein carrier physically shuttles the drug molecule across the BBB.

There are a number of molecular substrates that the brain requires for its normal functioning; these substances are not biosynthesized within the brain and are not able to enter the brain by passive diffusion. Because of their importance to normal brain neurochemistry, evolution has resulted in the existence of protein carriers to transport them into the brain. D-glucose and L-phenylalanine are two such molecules, and there are a number of others. Therefore, if a drug is designed as a structural analog of D-glucose or L-phenylalanine, there is a possibility that it will be actively transported across the BBB.

ADME Considerations

Sometimes, ADME considerations can be used to the advantage of the drug designer. This is particularly true when applied to the concept of a *prodrug.* A prodrug is a drug molecule that is biologically inactive until it is activated by a metabolic process. The active compound is released as a metabolite. The purpose of a pro-drug can be to:

1. Increase or decrease the metabolic stability of a drug
2. Interfere with transport characteristics
3. Mask side effects or toxicity
4. Improve the flavor of a drug

An ester, for example, can be used to "mask" a carboxylate. Within the body, the ester is hydrolyzed, releasing the drug in its bioactive carboxylate form.

Managing stability

The regulation of drug stability can take two directions: a prodrug can increase the *in vivo* stability of an active compound and prolong its action, or it can automatically limit its duration and prevent potential toxicity. There are many examples of drug stabilization. Among local anesthetics, procaine is an ester and is therefore very easily hydrolyzed by esterases. By conversion of the ester into an amide in lidocaine, the duration of action is increased several fold. Lidocaine is also used intravenously as an antiarrhythmic agent. In that application, it must pass through the liver–the principal drug-metabolizing organ –in which it loses an N-ethyl group to become a convulsant and emetic.

To minimize these unwanted and toxic effects, tocainide–whose α-methyl group prevents degradation, and which lacks the vulnerable N-ethyl groups–was prepared. This compound is not a prodrug in the strict sense, but rather represents a molecular modification. Replacement of a "*vulnerable moiety*" such as a methyl group by a less readily oxidized chlorine was used to transform the short-acting tolbutamide, an oral antidiabetic, into the long-acting chlorpropamide, with a half-life sixfold greater than its parent. A *decrease* in stability is often a desirable modification. For example, succinylcholine (suxamethonium)–a neuromuscular blocking agent used in surgery–has a self-limiting activity, since the ester is hydrolyzed in about 10 minutes, preventing the potential for overdose, which could be fatal with more stable curarizing agents. An ester group can be introduced into a local anesthetic, such as tolycaine, to prevent the drug from reaching the CNS if it is injected intravascularly by accident or abuse. The ester group is fairly stable in the tissues but is very rapidly hydrolyzed in the serum to the polar carboxylic acid, which cannot penetrate the blood-brain barrier.

Disposable moiety

Interference with transport characteristics can serve many purposes. The introduction of a hydrophilic "*disposable moiety*" can restrict a drug to the gastrointestinal tract and prevent its absorption. Such a type of drug is represented by the intestinal disinfectant succinyl-sulfathiazole. On the other hand, lipophilic groups can ensure peroral activity, as in the case of the penicillin derivative pivampicillin, which enters the circulation and then slowly releases the antibiotic in its free acid form, producing high blood levels of the latter. Lipophilic groups that are not easily hydrolyzed are used extensively for *depot preparations*, which liberate the active drug molecule slowly, for a period of days or weeks.

Steroid hormone palmitates and pamoates, and antimalarial esters (e.g., cycloguanil pamoate), can deliver the active drugs over a prolonged time; cycloguanil, for example, is released over a period of several months. This can be a great convenience for the patient, especially in areas with remote medical facilities. Drug designers have attempted for many years to use selective drug-transport moieties, and have met with moderate success. The idea is to attach a drug, such as an antitumor agent, to a natural product that will accumulate selectively in a specific organ and act as a "Trojan horse" for the drug. The attachment of alkylating agents to estrogens has been tried in the treatment of ovarian cancer, and amino acids have also been used as drug carriers. A recent ingenious application of the carrier concept is the utilization of *antibodies* – which can, at least in principle, be tailored to any site – as drug carriers. In this regard, antitumor agents such as adriamycin and methotrexate have been linked covalently to leukemia antibodies and melanoma antibodies, with some initial success. The large-scale preparation of antibodies is, of course, a major difficulty in this approach; however, the new monoclonal antibodies hold great promise.

Minimizing toxicity

Masking of the side effects or toxicity of drugs was historically the first application of the prodrug concept. This concept goes back to the turn of the twentieth century, and in fact many prodrugs were not at the time really recognized as such. For instance, castor oil is a laxative because it is hydrolyzed intestinally to the active ricinoleic acid. However, the classical example is prontosil, which undergoes a reduction to sulfanilamide. The analgesic phenacetin acts in the form of its hydrolysis product, p-acetaminophenol. Another classical example of side-effect masking occurs in aspirin and its many analogs–the result of a considerable effort to eliminate the gastric bleeding caused by salicylic acid. Selective bioactivation (toxification) is illustrated in the case of the insecticide malathion. This acetylcholinesterase inhibitor is desulfurized selectively to the toxic malaoxon, but only by insect and not mammalian enzymes. Malathion is therefore relatively nontoxic to mammals (LD_{50} = 1500 mg/kg, rat; p.o.). Higher organisms rapidly detoxify malathion by hydrolyzing one of its ester groups to the inactive acid, a process not readily available to insects. This makes the compound doubly toxic to insects since they cannot eliminate the active metabolite.

Drug modification

Taste improvement is quite an important aspect of drug modification, especially in pediatric medicine. The extremely bitter taste of some antibiotics, such as clindamycin or chloramphenicol, can be masked successfully by preparing esters or pamoate salts of these drugs, which are very insoluble and therefore have no taste.

Novel Drug Delivery

Novel drug delivery systems can also have a profound effect on pharmacokinetics, even if they do not involve the use of prodrugs in the classical sense. Novel polymers have permitted the development of membranes with controlled diffusion rates. For example, pilocarpine used in the treatment of glaucoma, can be applied in a steady-release ocular insert that lasts for a week. The intrauterine release of progesterone as a contraceptive has also been achieved, with a single insert lasting a year. The great advantage of this is that the constant release rate of 65 μg/day means that much less drug is released than with the use of oral contraceptive tablets. The transdermal delivery of scopolamine as an antiemetic for motion sickness represents another successful application of microporous membrane technology. Here the drug is applied in a plastic strip similar to a "Band-Aid," usually behind the ear.

Low-density lipoproteins and liposomes (drug-filled lipid–cholesterol vesicles measuring a fraction of a micrometer) are also being used to protect drugs from enzymatic destruction during transport in the bloodstream. Osmotic minipumps—cylinders measuring about 25 × 5 mm – are widely used to deliver constant amounts of drug solutions to experimental animals. They require surgical implantation. The osmotic compartment swells in contact with tissue fluid and squeezes the drug reservoir, displacing the drug solution in a continuous flow. The rate of delivery is specified by the size of the opening in the container and the swelling rate of the osmotic "syringe." The great advantage of these systems is the uniform drug delivery they permit, as opposed to the enormous drug level fluctuations inherent with the traditional oral or injected parenteral modes of drug administration. Patient compliance and convenience of use are also ensured. Although these interesting developments in bioengineering are not, strictly speaking, in the realm of drug design or even medicinal chemistry, they can nevertheless contribute substantially to the success of drug therapy.

CLINICAL APPROACH

The ultimate long-term goal of medicinal chemistry is to design a "*useful drug*". A useful drug is more than a molecule that is safe and efficacious. It is more than a compound that produces papers in scholarly journals. A useful drug is a drug molecule that is not only safe and efficacious, but also one that can pass government regulations, pass through multiple levels of human clinical trials, be economically produced in large quantities, be successfully marketed, and can ultimately help people with disease. Successfully treating humans with disease is the "*bottom line*" in drug design. Perhaps the greatest hurdle along the pathway of a molecule becoming a useful drug is the need to sequentially pass *clinical trials*. However, before a drug can be evaluated in human clinical trials, it must first successfully negotiate *preclinical testing*. This frequently involves five or six types of test, and is completed in non-human animals:

1. Acute toxicity – acute dose that is lethal in 50% of animals; usually two species, usually two routes of administration
2. Subacute toxicity – physiology, histology, autopsy studies; two species, sometimes with dosings over a 6 month time period
3. Chronic toxicity – detailed organ evaluation; two species, sometimes studied for 1–2 years
4. Mutagenic potential – effects on genetic stability of bacteria (Ames test) of mammalian cells in culture
5. Carcinogenic potential – required if drug is to be administered for prolonged periods of time
6. Reproductive performance effects – effects on animal progeny, production of birth defects

Once a molecule successfully passes the preclinical testing, it is ready for human clinical trials. There are four phases of human clinical trial.

Table 13.2 Drugs producing adverse effects on the fetus

Drug	*Effect*
ACE inhibitors	Kidney damage
Amphetamines	Abnormal developmental patterns
Androgens	Masculinization of female
Busulfan	Congenital malformations
Carbamazepine	Neural tube defects affecting brain formation
Cocaine	Stroke in fetus
Cyclophosphamide	Congenital malformations
Cytarabine	Congenital malformations
Diethylstilbestrol	Vaginal adenocarcinoma in child
Ethanol	Risk of fetal alcohol syndrome
Etretinate	High risk of multiple congenital malformations
Iodine	Congenital goiter, hypothyroidism
Isotretinoin	High risk of face, ear, and other malformations
Methotrexate	Multiple congenital abnormalities
Methylthiouracil	Hypothyroidism in child
Metronidazole	May be mutagenic (animal studies show no evidence for mutagenic or teratogenic effects in humans)
Penicillamine	Congenital skin malformations
Phenytoin	Fetal hydantoin syndrome
Propylthiouracil	Congenital goiter
Streptomycin	Eighth nerve toxicity (deafness) in child
Tamoxifen	Increased risk of spontaneous abortion or fetal damage
Tetracycline	Discoloration and defects of teeth and altered bone growth
Thalidomide	Phocomelia (shortened bones of the limbs)
Trimethadione	Multiple congenital abnormalities
Valproic acid	Neural tube defects of the brain

Phase 1: the effects of the drug as a function of dose are measured in a small number (25–45) of healthy volunteers who do not have the disease under study; safety is primarily evaluated.

Phase 2: the drug is studied in a small number of people (20–150) who have the disease under study; both safety and efficacy are evaluated.

Phase 3: the drug is studied in a large number of people (hundreds to thousands) who have the disease under study, typically using a multi-center, double-blind, placebo-controlled, randomized clinical trial (RCT) protocol.

Phase 4: once the drug has been approved for market, vigilant post-marketing surveillance is done to ascertain the possible appearance of previously undetected toxicities or problems.

During the drug development phases, toxicity is one of the most important hurdles to the success of a drug molecule. Toxicity can affect the person who is taking the medication (causing skin rashes, liver problems, bone marrow failure, etc.) or, in the case of women, can affect their developing fetus if they are pregnant.

Table 13.3 Approved drugs withdrawn because of toxicity

Drug	*Year*	*Adverse reaction*
Astemizole	1998	Interactions (e.g., with grapefruit juice)
Benoxaprofen	1982	Liver damage
Centoxin	1993	Increased mortality
Cerivastatin	2001	Muscle breakdown
Cisapride	2000	Cardiac arrhythmias
Clioquinol	1975	Optic neuropathy (eye problem)
Dexfenfluramine	1997	Cardiac valve abnormalities
Fenfluamine	1997	Cardiac valve abnormalities
Flosequinan	1993	Increased mortality
Indoprofen	1984	Gastrointestinal bleeding/perforation
Metipranolol 0.6% eyedrops	1990	Anterior uveitis (eye problem)
Mibefradil	1998	Many drug interactions
Nomifensine	1986	Hemolytic anemia
Noscapine	1991	Gene toxicity
Remoxipride	1994	Aplastic anemia
Sertindole	1998	Cardiac arrhythmias
Suprofen	1987	Renal impairment
Temafloxacin	1992	Various serious adverse effects
Terodiline	1991	Cardiac arrhythmias
Tolcapone	1998	Hepatobiliary disorders
Triazolam	1991	Psychiatric disorders
Troglitazone	1997	Hepatic disorders
Zimeldine	1983	Hypersensitivity
Zomepirac	1983	Anaphylaxis

Successful completion of the four phases of human clinical trials enables a drug to be widely distributed for the treatment of human disease. However, a simple reality of drug discovery is that drugs are developed by industry. The lead compound may have been identified in an academic university-based laboratory, but the clinical trials are invariably completed by the industrial sector. Academic institutions, governments, or international organizations (e.g., the World Health Organization, WHO) do not develop drugs. Because of this, drug molecules tend to be developed only if they have a good prospect for being profitable. In order to be profitable, a drug molecule should be patented so that the vendor can enjoy exclusive rights to its marketing.

14

Alternative Biological Specimens

In recent years, the drug testing community has seen a significant increase in the use of "*alternative biological specimens*" such as oral fluid, hair, and sweat for the detection of drugs subject to use and abuse. These alternative biological specimens offer different detection times and in most instances, significantly different metabolic profiles when compared to traditional urine testing. In addition, the fact that there continues to be issues with urine drug testing related to substitution, dilution, and specimen adulteration is the very reason that there has been an increase in the popularity and use of these alternative biological specimens for the detection of drugs subject to abuse. The purpose of this chapter will be to examine the use of sweat as one of these alternative biological specimens.

HISTOLOGY OF SKIN

The skin is composed of two major layers: the *epidermis* and the *dermis*. The epidermis is the outermost layer of the skin and is composed of stratified epithelium. This layer varies between 75 and 150 μm in thickness over most of the body, except the palms of the hands and soles of the feet. The outer surface of the epidermis is called the stratum corneum. This layer acts as a barrier to restrain the passage of water and solutes in either direction across the skin. Beneath the epidermis lies the dermis, the second major layer, and it is comprised of dense fibroelastic connective tissue. The dermis supports extensive vascular and nerve networks in addition to specialized excretory and secretory glands.

Biochemistry of Sweat

Moisture is lost from the skin by two distinct pathways. The first of these pathways appears to be caused by the diffusion of moisture through the dermal and epithelial layers. Moisture lost through this mechanism is referred to as insensible sweat and results from the passive diffusion of volatiles, including water through the skin. The rate of fluid loss through insensible sweat appears to depend on body location, ambient temperature, body temperature, and the relative humidity of the environment. The second pathway for moisture lost is called sensible sweat and is most commonly referred to as sweat. This moisture is secreted from the eccrine glands and the apocrine glands. The innermost cells of the eccrine gland serve as secretory elements that then empty into either the lumen or the duct of the glands.

The duct terminates at the surface of the skin. There are a variety of triggers that stimulate the eccrine glands, which include but are not limited to exercise, stress (both mental and emotional), and thermal stress. The maximum rates of sensible sweat produced by the human body can be as high as 2 L/h in average subjects and as high as 4 L/h in trained athletes. The amount secreted can be influenced by a variety of factors such as physical, thermal, and emotional stress. The uneven distribution of sweat glands and the variability of these factors makes it difficult to systematically obtain sweat specimens. Two other types of secretory glands common in human skin are the apocrine and sebaceous gland. Apocrine glands are located in the axillae, pubic, and mammary areas. They are not thought to have significant thermoregulatory use, and their role and mechanism in humans is unclear.

The sebaceous gland secretes a substance called sebum, that is primarily comprised of lipids. While these glands are located over most of the body, they do not appear to be involved in thermoregulation, and their function does not seem to be directly related to the production of sweat. The mechanism by which drugs are incorporated into sweat is as yet unclear. Sweat originates from the blood/plasma, and its composition is largely determined by reabsorption and exchange mechanisms. As the pH of sweat may differ considerably from individual to individual, so may the composition. There is a strong correlation between the pKa of a drug and the amount found in the sweat.

SAMPLING OF SWEAT

A variety of methods have been employed for the collection of sweat. These approaches have included thermal stress followed by the collection of sweat using dry gauze covered with waterproof plastic and the use of salt-impregnated pads. Other sweat collection devices have included the use of cotton swabs, polyvinyl shirts, perspiration stains from clothing, drug wipes, and more elaborate devices such as a transcutaneous chemical collection device that incorporated a band-aid-like device with a water/gel matrix to absorb compounds and prevent back diffusion of these compounds into the skin. Other devices have incorporated pilocarpine to stimulate localized sweat production, and at least one device incorporated electrodes that dispensed pilocarpine for iontophoresis. However, these devices had a variety of restrictions and limitations.

Those that utilized a plastic covering prevent the skin from breathing and significantly altered the physiology of the underlying epidermis. Other approaches have limited use as they are primarily used to collect neat sweat or used as a surface wipe. Devices that utilize pilocarpine work well to stimulate the production of neat sweat but are limited in the amount of sweat they collect. The limitations observed with these sweat collection devices and approaches led to the development of a non-occlusive sweat collection device that permitted the evaporation of the trapped water content of the sweat. Eliminating the build up of this moisture minimizes skin irritation caused by moisture trapped against the skin. Because it would be possible to collect sweat components other than water over time, the potential to investigate drug metabolism and dose-dependent studies increased significantly. In the late 1980s, Sudormed developed a device known as the Sudormed Sweat Patch Specimen Container. This device was cleared by the US Food and Drug Administration (FDA) as a collection device in 1990 and is comprised of three significant components. The first of these components is a polyurethane/adhesive layer consisting of 3M's Tegaderm 1625 transparent wound dressing. This dressing has been used since 1980 as a medical dressing and is widely used in hospitals and clinics to secure and protect catheters and intravenous lines.

The adhesive used by 3M (Minneapolis, MN) is described as a hypoallergenic, water-resistant adhesive and has a tamper-evident characteristic. The adhesive used by 3M infiltrates the exfoliated stratum corneum cells. When the adhesive is removed from the skin, these cells adhere to the adhesive and prevent the adhesive from resticking. Consequently, this provides a tamper-evident mechanism to diminish the removal and reapplication of the sweat patch. A unique identification number is printed on the 3M Tegaderm, which aids in the identification of the sweat patch and is a significant deterrent toward substitution. The polyurethane/adhesive layer is approximately 6 cm wide, 7 cm long, and 0.025 mm in thickness. The second component of the sweat patch is the release liner. This is a very thin medical grade tissue paper that allows the release of the collection pad from the adhesive after patch wear.

The release liner is approximately 3 cm wide, 5 cm long, and 0.003 mm thick. Without this release liner, the collection pad will adhere to the adhesive. After wearing the patch, the collection pad is retained and the release liner along with the Tegaderm polyurethane/adhesive layer is discarded. The third component of the sweat patch is the collection pad. It is composed of medical grade cellulose supplied by Ahlstrom Filtration. The collection pad is approximately 3 cm wide and 5 cm long. It is approximately 0.7 mm thick and is the main component for the collection of the non-volatile components of sweat. The collection pad absorbs a minimum of 300 μl of insensible perspiration at ambient temperature.

Studies involving controlled dosing have demonstrated that under these conditions, the sweat patch must be worn for a minimum of 24 h to collect adequate drugs for analysis. This wear time may be shortened if other factors such as exercise are present that increase sweat production. The sweat patch can be worn on the upper arm, lower rib cage area, and the upper back. Because of the aggressive nature of the adhesive, care should be taken to ensure that the application area is not subject to vast amounts of flexing. The upper arm is the most common area for sweat patch application as it easily presents itself, and the bicep can be flexed prior to the sweat patch application to reduce irritation caused by tension between the skin and the adhesive. Prior to application of the sweat patch, the skin should be cleaned using two isopropanol wipes.

Care must be taken to allow the alcohol to completely evaporate, otherwise a skin irritation could develop due to isopropanol trapped beneath the sweat patch. In 1992, the worldwide marketing and distribution rights for the sweat patch for the detection of drugs and alcohol were purchased by PharmChem, Inc. The name PharmChek was chosen for this application of the Sudormed Sweat Patch Specimen Container. Later that same year, Sudormed submitted a pre-market notification 510(k) to the FDA for the use of the sweat patch for the detection of cocaine (COC) and metabolites in sweat. This pre-market notification focused on the detection of COC and metabolites using gas chromatography-mass spectrometry (GC/MS). The FDA subsequently notified Sudormed that this COC submission could not be approved without an approved method for screening for COC in sweat. The FDA requested that a separate 510(k) submission for a screening assay for the detection of COC be submitted along with data to support the use of the sweat patch as a collection device.

In addition, the FDA requested that Sudormed provide similar separate 510(k) submissions for screening and confirmation for each of the other drugs [opiates, phencyclidine (PCP), amphetamines, and marijuana] for which it intended to have the sweat patch approved as a collection device. To comply with the FDA's request, Sudormed met with a number of manufacturers of screening reagents to assess their ability to detect the presence of drugs in sweat. As a result of these discussions, SolarCare Technologies Corporation (STC) in Bethlehem, PA, agreed to modify its enzyme-linked immunoassay microplate screening assay (ELISA) to detect the presence of drugs in sweat. Sudormed and SolarCare Technologies subsequently submitted co-dependent 510(k) documents for the detection of COC, opiates, PCP, amphetamines, and marijuana by the end of 1993. By the end of 1996, all five submissions had been cleared by the FDA. STC underwent a merger in the late 1990s and is now know as OraSure Technologies.

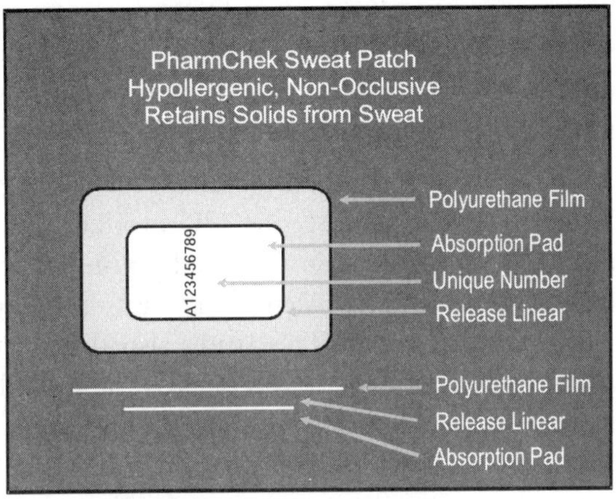

Fig. 14.1. Diagram of a sweat patch.

FINDING DRUGS IN SWEAT

There have been many studies designed to detect the presence of alcohol and other drugs in sweat. Pawan et al. studied alcohol levels using polyvinyl shirts to collect sweat. Vree et al. administered controlled oral doses of amphetamines and measured the amphetamine levels in sweat. This study indicated that the elimination of amphetamines in sweat was largely independent of the sweat pH. Ishiyama et al. determined that methamphetamine was excreted by sweat glands and that the analysis of sweat would be a valuable tool in forensic practices. Henderson et al. measured methadone and metabolites in both the sweat and urine of patients that were in methadone treatment programs, primarily to determine the optimal doses of methadone.

The data suggested that the drug levels in sweat could not be used to determine optimal doses of methadone but indicated that sweat may be a significant route of elimination for drugs. Cook et al. studied the detection of ^{3}H PCP in sweat by means of filter papers placed on the arm below the axilla. They determined that significant levels of radioactivity were present in the filter papers up to 54 h following drug administration. Parnas et al. studied the excretion in sweat of phenytoin, phenobarbitone, and carbamazepine using an occlusive bandage consisting of one to three layers of filter paper covered by sampling padding. This study indicated that while all three drugs were found to be present in sweat, the phenytoin sweat concentrations was found to correspond to the free fraction in plasma and was independent of sweat flow. Phenobarbitone sweat concentrations were found to increase with increasing sweat flow. This data suggested that drug level monitoring, under changing climatic conditions, may be of clinical significance. There have been numerous studies that have entailed controlled dosing studies for COC, at least one study involving a controlled dose for heroin, at least one study involving a controlled does for codeine, and at least one study involving the administration of PCP.

Cone et al. examined the effect the route of administration and dose on sweat COC levels. A 25 mg intravenous dose of COC produced sweat COC levels that ranged from 11.0 to 44.7 ng/patch. A 32 mg dose of COC administered intranasally produced sweat COC levels that ranged from 31.2 to 39.8 ng/patch. A 42 mg smoked dose produced sweat COC levels that ranged from 4.7 to 73.6 ng/patch. This study also demonstrated that COC appears in sweat within 1-2 h of administration and that levels peaked within 24 h. In addition, it was noted that COC was detectable in trace amounts following the administration of 1 mg of the drug. In all cases, COC was the major analyte excreted in sweat, although smaller amounts of ecgonine methyl ester (EME) and benzoylecgonine (BE) were present. Cone et al. also demonstrated that a 20 mg intravenous dose of heroin produced sweat heroin levels that ranged from 6.9 to 53.3 ng/patch.

Heroin was the major analyte excreted in sweat in this study although smaller amounts of 6-acetyl morphine were present. Morphine was not detected in any of these samples. Kacinko et al. administered three subcutaneous doses of COC (75 mg/70 kg) over a 1-week period and observed levels that ranged from 3.8 to 84.7 ng/patch. In a similar study, three subcutaneous doses of COC (150 mg/70 kg) were administered over a 1-week period. In this study, COC levels were as high as 375 ng/patch. Uemura et al. administered 210 mg/70 kg of COC intravenously to volunteers and observed sweat COC levels as high as 375 ng/patch. Of particular interest in this study was the observation that sweat patches placed on the back of the individuals showed sweat COC levels that were up to eight times higher that those patches placed on the shoulders. Burnes et al. administered either 50 or 126 mg of COC hydrochloride intranasally to volunteers. A 1-week interval separated dosing periods, and the order of the doses were counterbalanced.

Sweat patch levels for the 50 mg dose ranged from 0 to 105 ng/mL (0 to 263 ng/patch) while individuals receiving the 126 mg dose demonstrated sweat patch COC levels that ranged from 0 to 140 ng/mL (0 to 315 ng/patch). Schwilke et al. administered three doses of codeine sulfate within a 7-day period. Those individuals who were administered the low does of 60 mg/70 kg demonstrated peak sweat patch codeine levels that ranged from 0 to 225 ng/patch. Those individuals that received the high dose of 120 mg/70 kg of codeine sulfate demonstrated sweat patch codeine levels that ranged from 0 to 96 ng/patch. Cook et al. administered 0.5 mg of a deuterated PCP (D_5) to individuals intravenously using a bolus approach. The analysis was conducted using RIA coupled with HPLC confirmation and indicated a concentration of PCP D_5 in sweat of 5.9 ng/mL. In this study, the sweat was not collected using a sweat patch but was collected as neat dripped off the upper body of one of the subjects who was exercising vigorously on an exer-cycle.

ANALYZING SAMPLES

Upon arival at the testing facility, the sweat patch must be reconstituted to extract the drugs off the collection pad and suspend them in an appropriate matrix using the process described by Spiehler et al. The sweat patch was placed in a 6-mL polypropylene specimen vial with screw cap.

The sweat patch reconstitution is accomplished by adding 2.5 mL of a 25:75 0.25 M sodium acetate buffer: methanol solution to each sweat patch specimen. Each vial is capped and shaken on a horizontal shaker for 30 min at 2400 rpm. Following this extraction, the tube may be centrifuged to force the collection pad to the bottom, or a device similar to a serum separator may be used to force the collection pad to the bottom of the tube.

Other investigators have used a similar approach but have used between 4 and 6 mL of the sodium acetate buffer: methanol solution to reconstitute the sweat patch to allow for a greater volume of eluate for subsequent analysis. Previously, the results for sweat patch analysis have been reported in ng/mL units using 2.5 mL as described above to reconstitute the sweat patch. The current approach is to present the sweat patch drug concentrations in a ng/patch format. The current testing levels used by the industry for sweat testing are listed below. With the exception of PCP, all testing levels were established using receiver operating characteristic approaches as described by Spiehler et al.

Table 14.1 Testing levels in sweat

Screening levels		*Confirmation levels*	
Amphetamines	25 (25)	Amphetamine	25 (25)
		Methamphetamine	25 (25)
		MDMA	(25)
		MDA	(25)
		MDEA	(25)
Cocaine	25 (25)	Cocaine	25 (25)
		Benzoylecgonine	25 (25)
Cannabinoids	3.75 (4)	Δ9 THC	1.25 (1)
Opiates	25 (25)	Morphine	25 (25)
		Heroin	25
		6-Acetyl Morphine	25 (25)
		Codeine	25 (25)
Phencyclidine	18.75 (20)	Phencyclidine	18.75 (20)

Testing levels in parenthesis are levels proposed by the US Department of Health and Human Services, Substance Abuse and Mental Health Services Administration. Currently, sweat patch drug results are expressed as ng/mL of reconstituted acetate buffer/methanol. The proposed SAMHSA guidelines express the drug levels as ng/patch. For example, using this approach, the cutoff for amphetamines would be 25 ng/patch (10 ng/mL × 2.5 mL of reconstitution matrix).

Immunoassays

The OraSure Drugs-Of-Abuse Assays are micro-plate competitive immunoassays for the determination of drugs-of-abuse or their metabolites in PharmChek sweat patch eluate. Sample or standard is added to microtiter wells along with enzyme-labeled hapten conjugate. There is a competition to bind the antibody fixed onto the well. The wells are washed six times with deionized water to remove excess enzyme conjugate and drug once the competition for the antibody coated on the surface of the microplate well is complete. A substrate (3,3″ ,5,5″-tetramethylbenzidine) is added and incubated for an additional 30 min. Following this incubation, the reaction is stopped using 100 µL of 2 N sulfuric acid. The microplates are then read at two wavelengths 450 and 630 nm using a microplate reader. The readings at 630 nm are used to negate any wavelength distortions due to scratches or fingerprints on the microplates.

The results of the 630 nm reading are subtracted from the results of the 450 nm reading. These absorbance reading are inversely proportional to the amount of drug or drug metabolite(s) present in the sample or calibrator/control. Because EIA assays are based on the principle of an antibody recognizing a drug or class of drugs, metabolites or different members of a drug class cross-react with antibody to varying degrees. Therefore, EIA assays are used as qualitative procedures. A positive test result indicates the presence of the drug/drug class at or above the stated cutoff concentration. Specimens that produce an absorbance reading less than or equal to the cutoff value established by the calibration procedure are designated as an initial test "positive." Specimens that produce absorbance readings that are greater than or equal to the cutoff calibrator value are designated as "negative." Initial results are sent to the Data Review Department for evaluation. Specimens testing positive are sent to GC/MS for confirmation analysis. Specimens testing negative are not subjected to further analysis.

Additional Testing

After a sample has been screened positive by EIA for a drug or drug group, GC/MS is performed on another aliquot of the specimen. The purpose of the additional testing is threefold: First, for those cases in which the EIA assays tests for a general class of drugs, GC/MS identifies the specific drug(s) that are present. Second, GC/MS quantifies the drug to compare each specimen to an established cutoff concentration, below which the specimen will be reported as negative. Third, GC/MS confirms all positive EIA results with a technology that is chemically different from EIA. In many cases, the drug is derivatized to enhance its chromatographic properties. Following extraction and derivatization, if applicable, the specimen is injected into a GC equipped with a fused silica capillary column and coupled to an electron impact (EI) mass spectrometer. Because of the lower confirmation levels for cannabinoids, *negative chemical ionization* (NCI) or coupled mass spectrometry (MS/MS) is required.

Because the reconstitution matrix is an acetate buffer and methanol solution, confirmation using LC/MS/MS may be a more attractive approach than conventional GC/MS or GC/MS/MS. Each analysis incorporates an internal standard that is added to the specimen. The internal standard is selected on the basis of chemical similarity to the drug(s) to be tested. In many cases, the internal standard is a deuterated derivative of the drug(s) to be analyzed. The internal standard is selected so as to extract, derivatize, and chromatograph in a manner similar to the analyte. It should also yield spectral masses that are close to, but different from, those of the analyte. As they flow through the capillary column, constituent drugs are separated from interference and other drugs on the basis of their volatility and varying affinities to the column.

As the effluent enters the mass spectrometer, it is subjected to an electron beam, which fragments the molecules, creating a "*fingerprint*" pattern unique to the drug(s) tested. The response of three ion fragments for each analyte and two ion fragments for each internal standard is monitored and the area measured using *selected ion monitoring* (SIM) mode. The compound retention time, the presence of the three-analyte ions, and their abundance ratios enable the conclusive identification of drugs in the unknown samples as compared to the standards. Quantitation is achieved by comparing the ratio of the area of the base peak of the analyte to the area of the base peak of the internal standard in a specimen to the analogous ratio derived from the calibration standard. The quotient of that calculation is multiplied by the known concentration of the calibration standard to determine the quantitative value of the unknown.

Extraction of Amphetamines

Amphetamine and methamphetamine are eluted off the sweat patch using a methanol/acetate buffer. Amphetamine and methamphetamine are extracted, with deuterated analogs as internal standards, from alkalinized patch extracts. They are back-extracted into acid and alkalinized. Finally, they are reextracted into *n*-butyl chloride that is treated with 4-carbethoxyhexafluorobutyryl chloride (CB) to form the acyl derivatives. They are identified and quantitated by capillary GC EI three-ion SIM MS. No compounds have been identified which interfere with this assay although several other primary and secondary amines may be identified by this procedure. Methamphetamine is metabolized into amphetamine; however, little amphetamine is detected in sweat. Methamphetamine can be detected in sweat using the patch in as short a time as 4 h. Levels in the patch are dose related but vary considerably between individuals.

Detection of Cannabinoids

Δ-9-Tetrahydrocannabinol (THC) is eluted off the sweat patch using a 75/25 methanol/acetate buffer combination. The mixture is diluted with an aqueous buffer, and the THC is extracted using a solid phase column. The extracted THC is derivatized using trifluoroacetic anhydride and analyzed by GC/MS using NCI. Quantitation is achieved using a deuterated internal standard.

Detection of Cocaine

COC and its metabolites, BE, and EME are eluted off the sweat patch using a methanol/acetate buffer mixture (75:25). The drugs are extracted from the mixture using a solid-phase column. The BE and EME are derivatized using HFIP and PFPA and analyzed by GC/MS using SIM. COC is analyzed in this same chromatographic assay in its underivatized form. Quantitation is achieved using deuterated internal standards. COC and its two major metabolites, BE and EME, can be detected in sweat as early as 2 h after a single dose. The levels in the patch will reach a maximum between 2 and 3 days. Continued use of COC will result in increasing levels in the patch.

Detection of Opiates

Codeine, morphine, 6-acetylmorphine (6-AM) and heroin are eluted off the sweat patch using a 75/25 methanol/acetate buffer mixture. The eluate is made alkaline, and the drugs are extracted into an organic solvent. The drugs are back-extracted into acid that is alkalinized and then reextracted into organic solvent that is evaporated to dryness. The residue is treated with BSTFA and analyzed by GC/MS using SIM. Heroin is analyzed in its underivatized form. Quantitation is achieved using deuterated internal standards.

Table 14.2 PharmChem personal communication

Drug	*Extraction*	*Derivative*	*Method*	*Ions*	*LOD/LOQ*
Amphetamine	Liq/liq	4-CB	GC/MS EI	248,266,294	5/5
Methamphetamine	Liq/liq	4-CB	GC/MS EI	262,280,308	5/5
Cannabinoids	Solid phase	TFAA	GC/MS NCI	410.3, 413.3	0.5/0.5
Cocaine	Solid phase	HFIP/PFPA	GC/MS EI	182,272,303	5/5
Benzoylecgonine	Solid phase	HFIP/PFPA	GC/MS EI	318,334,439	5/5
Codeine	Solid phase	N/A	GC/MS EI	371,372,343	7.5/7.5
Morphine	Solid phase	BSTFA	GC/MS EI	429,414,430	10/12.5
6-MAM	Solid phase	BSTFA	GC/MS EI	399,340,287	7.5/7.5
Heroin	Solid phase	BSTFA	GC/MS EI	327,268,310	7.5/10
Phencyclidine	Liq/liq	None	GC/MS EI	200,242,243	2.2/2.5

Detection of Phencyclidine

PCP is eluted off the sweat patch using a 75/25 methanol/acetate buffer combination. The eluate is made alkaline, and PCP is extracted into an organic solvent. The drug is back-extracted into acid that is then alkalinized. Finally, PCP is extracted into a small volume of chloroform: isoamyl alcohol (9:1) and is analyzed by GC/MS using SIM. Quantitation is achieved using deuterated internal standard. At present, there is only one laboratory providing sweat patch testing services. PharmChem Inc. located in Fort Worth, TX, owns the world-wide marketing rights to the PharmChek Sweat Patch. Analytical parameters for sweat patch testing are summarized in Table 6.2 and were provided by PharmChem Inc.

CORELATIVE APPROACH

Numerous studies have confirmed that the predominant drug detected in sweat is the parent. Following administration, drugs may be detected in sweat in as little as 2 h. These controlled dose studies concluded that a sweat patch must be worn for a minimum of 24 h to collect sufficient drug for analysis. In addition, it appears that drugs in sweat lag behind the detection of the same drug in urine by 1-2 days. The sweat patch is a collection device and serves as a reservoir for drugs and other compounds secreted in sweat. Consequently, it cannot detect multiple uses of drugs that may have occurred while the sweat patch was worn and currently cannot detect what route of administration was used.

Little research has been conducted regarding the lowest dose required to obtain a positive sweat patch result. Cone et al. has reported that COC was detectable in the sweat patch in trace amounts following the administration of 1 mg of COC. Concern has been raised regarding the susceptibility of the sweat patch to contamination from the incomplete cleaning of the skin with the isopropanol wipes. A subsequent study concerning the incomplete cleaning of the skin considered the efficiency of the two isopropanol wipes at removing 1000 ng of methamphetamine and COC applied to the skin such that no drug was subsequently transferred to the sweat patch during normal wear. The authors also looked at the effects of a 5% soap solution followed by water and a single isopropanol wipe at removing up to 1000 ng of methamphetamine and COC from the skin. They concluded that cleansing with soap and isopropanol was more effective at removing the residual drug from the skin than the two isopropanol wipes alone and that there was a relationship between the dose applied to the skin and the residual drug detected on the patch.

Although the parent drug was sometimes detected following these treatments, neither the drug nor the metabolite was detected in the same patch. Consequently, under the existing procedures and the proposed SAMHSA guidelines, none of these sweat patches would have been reported as positive for either COC or methamphetamine, regardless of the cleaning procedure used. Concern has been raised regarding the susceptibility of the sweat patch to vapor phase contamination. This concern arose out of several criminal justice court cases in which the defendants claimed that they had not used COC and that their positive sweat patch test results must be due to the COC vapor resulting from other individuals in the household smoking crack COC.

Kidwell et al. designed a series of laboratory experiments to test the susceptibility of the sweat patch to environmental contamination using COC, methamphetamine, and heroin. These investigators concluded that the sweat patch was resilient to external vapor phase contamination if the sweat patch and the external surface of the patch were kept dry. However, if the pad was moistened with basic buffer, water, or artificial sweat, both COC and methamphetamine could be detected in the absorbent pad. The authors acknowledged that this project was primarily undertaken for mechanistic studies rather than reflecting a realistic real-life contamination scenario as the amounts of drug vapor used were large.

A subsequent study was conducted by Crouch et al. to determine whether the sweat patch absorbent pad was susceptible to external contamination (in vitro) when known amount of drug were volatilized in the presence of the sweat patch. The absorbent pad was left dry, moistened with artificial sweat, harvested drug-free sweat, or water, covered with the polyurethane Tegaderm material, and subjected to known amounts of volatilized drug. The authors concluded that the polyurethane membrane of the sweat patch serves as an effective barrier to vaporized COC and methamphetamine. Under a variety of physiologically relevant conditions, the sweat patch absorbent pad remained drug free.

It would require gross exposure of the patch (equivalent to volatizing 200 g of COC and 100 g of methamphetamine) in a 20′ × 20′ × 8′-sized room and remaining in this room for at least 1 h to duplicate these experimental conditions such that either COC or methamphetamine could be detected in the sweat patch. These conditions are certainly not representative of normal environmental conditions. In addition, while these extreme conditions did demonstrate the presence of either COC or methamphetamine on the absorbent pad, no BE or amphetamine were detected. Therefore, under the existing procedures and the proposed SAMHSA guidelines, none of these patches would have been reported as positive.

CONCLUDING REMARKS

Sweat is an easy specimen to collect, with several methodologies available for use. Most drugs-of-abuse may be extracted from the sweat using current testing methodologies. Screening techniques for urine are generally not appropriate for sweat due to low-drug concentrations detected in sweat and the incorrect compound targeted. One study demonstrated in a prison setting that wearing the sweat patch proved to be a deterrent to drug use and provides a means to monitor individuals in a correction or probation/parole setting for an extended period of time. Disadvantages include the need to wear the patch for a minimum of 24 h to achieve measurable drug concentrations. The sweat patch device appears to be resistant to substitution and contamination under realistic "*real-life*" conditions. Although the concentration of drugs measured in sweat appear to be dose dependent, future research should focus on determining the factors contributing to the variability in concentrations. Other drugs-of-abuse such as hydrocodone and oxycodone should also be studied.

15

Detecting Drugs in Eye Fluid

The human eye is recessed in the pyramid-shaped bony orbit and is connected to the brain by the optic nerve. The eyeball provides protection for the retina, the photosensitive portion of the eye. Both the retina and the optic nerve are enclosed by dense fibrous tissue, the sclera, and the dura mater. There are three chambers of the eye: the anterior chamber, the posterior chamber, and the vitreous. The dura mater surrounds the optic nerve and merges with the sclera to occupy the posterior chamber that constitutes five-sixths of the globe. The remaining one-sixth of the globe, the anterior portion, contains the cornea that refracts incident light.

The vitreous is located between the lens and the retina and fills the center of the eye. It constitutes 80% of the eye and has a volume of approximately 4 mL. There are two fluids contained within the eye. The anterior chamber contains a clear, watery fluid, the aqueous humor. The aqueous humor is produced in the posterior chamber and has a volume of approximately 250 μL. The vitreous is comprised of a transparent, delicate connective tissue gel called the gel vitreous or a transparent liquid called the liquid vitreous. The gel vitreous is a collagen gel that is water insoluble and liquefies with age such that the adult eye contains only liquid vitreous. For the purposes of this chapter, the gel vitreous and the liquid vitreous will be considered as one specimen and will be referred to as the vitreous humor.

CHEMICAL NATURE OF VITREOUS HUMOR

The vitreous humor weighs approximately 4 g. It consists of 99% water and has a specific gravity of 1.0050-1.0089. It has a refractive index of 1.3341, which is slightly lower than the aqueous humor. Its viscosity is approximately two times that of water but with an osmotic pressure close to that of aqueous humor. The pH of the vitreous humor is 7.5. The osmolality of the vitreous humor ranges from 288 to 323 mOsm/kg, slightly higher than the osmolality of serum (75-295 mOsm/kg). Collagen is the major structural protein of the vitreous humor. Although similar to cartilage collagen, there are some distinct differences. Another major component of the vitreous humor is *hyaluronic acid* (HA). HA is a glycosamineglycan, a polysaccharide composed of repeating disaccharide units; each unit contains a hexosamine linked to uronic acid. The vitreous humor is composed of interpenetrating networks of HA molecules and collagen fibrils. In addition to collagen and HA, there are six specific non-collagenous proteins and two types of glycoproteins in the human vitreous humor. Besides the larger molecules, there are a number of low-molecular weight substances that are found in the vitreous humor.

Vitreous humor concentrations of sodium and chloride will approximate the serum concentrations of these ions in healthy adults, especially in the early postmortem period. Potassium concentrations in the vitreous humor increase rapidly after death as potassium leaves the cells into nearby fluids. Vitreous humor calcium concentrations are comparable to serum calcium concentrations. The vitreous glucose concentration is similar to the serum concentration. Glucose is rapidly broken down after death in non-diabetics but remains elevated in uncontrolled diabetics. Two nitrogenous compounds, urea and creatinine, are also present in concentrations similar to serum and are stable during the early postmortem period. In postmortem cases, these substances are routinely measured as indicators for the serum concentrations of these substances at death.

RETINAL CIRCULATION

There is a dual blood supply to the human eye. One is the retinal circulation that is mediated by the central artery and vein of the retina. The other source is the uveal circulation that is a system of ciliary arteries that run in the middle coat of the eye, the uvea. The retinal capillaries are virtually impermeable to molecules with a molecular weight greater than 1900 Da. Movement across the blood-retinal barrier occurs primarily by diffusion. There also appears to be an active transport mechanism across the blood-retinal barrier. The uveal circulation supplies the ciliary body and the iris. The ciliary body produces the vitreous humor. The ciliary capillaries have openings that permit macromolecules to escape. Conversely, the iris capillaries have much lower permeability. Equilibrium between blood and vitreous is slower than between the blood and other extracellular fluids. This suggests the presence of a barrier and is called the blood–vitreous barrier.

This barrier consists of vascular endothelium and its basement membrane, stroma, and two layers of ciliary epithelium. The movement of molecules in and out of the vitreous occurs by a number of mechanisms - diffusion, hydrostatic pressure, osmotic pressure, convection, and active transport. Water movement is significant, as approximately 50% of the water is replaced every 10-15 min. High-molecular weight substances and colloidal particles travel by convection. Low-molecular weight substances move in and out of the vitreous primarily by diffusion; however, there is evidence that bulk flow also contributes to their movement. Common drugs are small molecular weight molecules and therefore, move in and out of the vitreous by diffusion. Only free drug is able to leave the blood and enter the vitreous. Therefore, drugs that are not highly protein bound would be expected to have significant concentrations in the vitreous humor. Conversely, drugs that are highly protein bound would not appear to any significant extent in the vitreous humor.

SAMPLING

The vitreous humor should be collected using a syringe and a 20-gauge needle. The needle is placed against the eye at the lateral aspect just above the junction between the upper and lower eyelids. The needle is inserted into the eye approximately 2 cm and the vitreous humor is gradually withdrawn. It is recommended that the vitreous humor from both eyes be collected. No preservatives need to be added to the specimen; however, it should be stored in the refrigerator until analyzed.

DETECTING DRUG IN EYE FLUID

The analysis of drugs in vitreous humor is similar to the analysis in other postmortem fluids. Most of the published methods that were developed for blood and urine have been employed successfully with vitreous humor. These methods have used either solid-phase or liquid-liquid extraction depending on the characteristics of the analytes. Detection systems, such as gas chromatography, gas chromatography/mass spectrometry, liquid chromatography, and liquid chromatography/mass spectrometry have all been used to identify and quantify drugs in vitreous humor. Because vitreous humor is approximately 99% water, some methods of analysis have been used following little or no sample preparation. For example, Chronister et al. analyzed benzoylecgonine (BE) in vitreous humor by cloned enzyme donor immunoassay (CEDIA) without any specimen preparation.

The manufacturer's urine calibrators were used, but a lower cutoff was programmed into the autoanalyzer. Felscher et al. used fluorescence polarization immunoassay for vitreous humor drug analysis without pretreatment. Logan and Stafford tested vitreous humor for cocaine and BE by liquid chromatography by diluting the vitreous humor 1:1 with water and filtering through a 3 cm pre-column.

CORELATIVE APPROACH

Limited data was identified in the scientific literature pertaining to the distribution of amphetamine and methamphetamine into vitreous humor. There were several single case reports of methylenedioxymethamphetamine (MDMA), where MDMA and/or methylenedioxyamphetamine (MDA) were measured in the vitreous humor. Crifasi and Long reported MDMA and MDA concentrations in blood and vitreous humor in a traffic fatality. MDMA concentrations in the blood and vitreous humor were 2.14 and 1.11 mg/L, respectively; MDA concentrations were less than 0.25 mg/L in both specimens. Decaestecker et al. measured MDMA concentrations in an overdose from 4-methyl-thioamphetamine.

The subclavian blood, femoral blood, and vitreous humor MDMA concentrations were 0.013, 0.010, and 0.067 mg/L, respectively. DeLetter et al. published an extensive fluid and tissue distribution study of MDMA and MDA in an overdose from MDMA. Blood MDMA concentrations ranged from 3.1 to 7.6 mg/L depending on the site of the blood specimen. The vitreous humor MDMA concentration was 3.4 mg/L. Blood MDA concentrations ranged from 0.09 to 0.29 mg/L; the vitreous humor MDA concentration was 0.06 mg/L. In another case report from the same laboratory, blood MDMA and MDA concentrations ranged from 0.27 to 0.81 mg/L and 0.009 to 0.043 mg/L, respectively. Vitreous humor MDMA and MDA concentrations were 0.36 and 0.015 mg/L, respectively. These studies indicate that MDMA and MDA were detectable in the vitreous humor when identified in the blood, but the data is too limited to draw any other conclusions.

Detection of Cannabinoids

As many postmortem laboratories do not test for delta-9-tetrahydrocannabinol (THC) or its metabolites, the data related to the detection of these compounds is limited. Lin and Lin measured 11- nor-delta-9-tetrahydrocannabinol-9-carboxylic acid (THC-COOH) in blood and vitreous humor specimens from 50 driver fatalities. Alkaline hydrolysis was performed on all specimens. Vitreous humor concentrations of THC-COOH in all cases were less than 0.01 mg/L. Heart blood concentrations of THC-COOH ranged from 0.01 to 0.33 mg/L and exceeded the vitreous humor concentration in all cases. This is expected because the polarity of THC-COOH makes it difficult to cross the blood-vitreous barrier. THC was not tested in these cases. From this limited data, it appears that vitreous humor is not a useful specimen to screen for marijuana use.

Detection of Cocaine and Metabolites

There have been a number of studies published on the distribution of cocaine and metabolites in vitreous humor. Several of these studies describe single cases where blood and vitreous humor were tested for cocaine and/or BE. Sturner and Garriott reported a single case of a cocaine overdose where the blood cocaine concentration was 8.5 mg/L and the vitreous humor cocaine concentration was 3.8 mg/L. BE concentrations were not reported. In another overdose case, Poklis et al. found a blood cocaine concentration of 1.8 mg/L and a vitreous humor cocaine concentration of 2.4 mg/L. The cocaine was administered intravenously. Furnari published a tissue distribution study in a cocaine intoxication death in a body packer.

The blood and vitreous humor cocaine concentrations were 4.0 and 7.1 mg/L, respectively; the BE concentrations were 17.0 and 5.8 mg/L, respectively, in the blood and the vitreous humor. In addition to these single case reports, a number of researchers have collected blood and vitreous humor concentrations in multiple cases. Logan and Stafford used liquid chromatography to measure cocaine and BE in 28 paired blood and vitreous humor specimens. Cocaine concentrations greater than 0.01 mg/L were detected in 17 of the 28 blood specimens and 22 of the 28 vitreous humor specimens. All six specimens where the vitreous humor cocaine concentration was less than 0.01 mg/L were associated with a blood cocaine concentration less than 0.01 mg/L. The mean vitreous humor to blood cocaine concentration ratio was 1.61 with a range between 0.1 and 2.6. A comparison of the blood and vitreous humor cocaine concentrations showed considerable spread, with a correlation of $R = 0.70$.

A high blood cocaine concentration was generally associated with a high vitreous humor cocaine concentration. BE was quantitated in only five of the 28 specimens, so a meaningful comparison to vitreous humor BE concentrations could not be performed. BE was tested in 24 of the 28 vitreous humor specimens at a limit of quantitation of 0.05 mg/L. The vitreous humor BE concentration exceeded the vitreous humor cocaine concentration in only three of these 24 specimens. Mackey-Bojack et al. quantitated cocaine, BE, and cocaethylene in blood and vitreous humor in 62 medical examiner cases. The average blood cocaine concentration in these cases was 0.489 mg/L with a range of 0-6.28 mg/L. This was not statistically different from the average vitreous humor cocaine concentration of 0.613 mg/L (range 0-4.50 mg/L). Differences between the average blood and vitreous humor cocaethylene concentrations were also found not to be statistically different. The mean blood cocaethylene concentration was 0.022 mg/L and the mean vitreous humor cocaethylene concentration was 0.027 mg/L. However, there was a statistically significant difference between the blood and vitreous humor BE concentrations; the average blood BE concentration was 1.941 mg/L and the average vitreous humor BE concentration was 0.989 mg/L.

The blood BE concentrations exceeded the vitreous humor cocaine concentrations in 46 of the 53 cases. There were linear relationships between blood and vitreous humor concentrations for cocaine and BE but not for cocaethylene. Nevertheless, the authors of the study warned that prediction of a blood concentration from a vitreous humor concentration could not be reliably performed as the magnitude and direction of the concentration differences between the two specimens was too variable. Chronister et al. analyzed 392 vitreous humor specimens for BE by a commercially available immunoassay. Twenty-three specimens were positive at a cutoff of 0.1 mg/L. Twenty-two of these confirmed positive for BE by gas chromatography/mass spectrometry in the blood. The blood BE concentrations exceeded the vitreous humor concentrations in 19 of the 22 cases. A number of conclusions can be made from the above discussion. Vitreous humor is a useful specimen to identify cocaine use. Either cocaine or BE can be used as the target compound to screen for cocaine use.

Vitreous humor cocaine concentrations are generally in the same range as blood cocaine concentrations. However, vitreous humor BE concentrations are generally lower than blood BE concentrations. This is expected, as BE, being more hydrophilic than cocaine, would encounter greater difficulty in crossing the blood-vitreous barrier. It is generally unreliable to predict a blood cocaine or BE concentration from a vitreous humor cocaine or BE concentration.

Detection of Opioids

There have been a number of published studies that include data on the distribution of opioids into vitreous humor. The specific opioids for which data are available include morphine/heroin, oxycodone, methadone, propoxyphene, fentanyl, and sufentanil.

Detection of morphine/heroin

Ziminski et al. investigated 49 drug-related medical examiner cases, of which 13 were positive for morphine. Morphine was detected in vitreous humor from 12 of the 13 cases, ranging in concentration from 0.03 to 0.14 mg/L. The case in which morphine was not detected in vitreous humor was classified as an "*acute opiate intoxication,*" and the authors suspected that death occurred prior to morphine equilibration with vitreous humor. There was also a single case in which morphine was detected in vitreous humor but not in blood. Of the 11 cases in which morphine was detected in both vitreous humor and blood, the vitreous humor to blood morphine concentration ratios ranged from 0.03 to 6.0, with a mean of 0.18 (median = 0.11). A study of 10 cases of heroin overdose quantitated free and total morphine in vitreous humor by RIA. Following treatment with β-glucuronidase, higher concentrations of morphine in vitreous humor were observed, demonstrating that morphine conjugates distribute into the vitreous humor.

The ratio of free to total morphine ranged from 0.14 to 0.77. The authors also analyzed corresponding blood, urine and bile specimens and found that vitreous humor concentrations did not correlate well with the measured concentrations in these matrices. The vitreous humor to blood morphine concentration ratios for these 10 cases were all less than one, ranging from 0.08 to

0.83. Gerostamoulos and Drummer reported a series of 40 heroin-related deaths. The authors analyzed femoral blood, bile, brain, cerebrospinal fluid, kidney, liver, plasma, urine, and vitreous humor for free morphine (MOR), morphine-3-glucuronide (M3G), and morphine-6-glucuronide (M6G). The mean concentrations (mg/L) of morphine species in vitreous humor were M3G, 0.08; M6G, 0.02; free MOR, 0.09; and total MOR, 0.16. In addition, the authors calculated the percentage of M3G and Free MOR as a proportion of total morphine.

Of all the specimens analyzed, vitreous humor contained the lowest proportion of M3G (32%) and the highest proportion of free morphine (70%). When comparing blood and vitreous humor-free morphine concentrations, a correlation coefficient of 0.261 was calculated, with a 95% confidence interval of -0.219-0.639, indicating that vitreous humor morphine concentrations may be of limited value for toxicological interpretation. Pragst et al. investigated the distribution of morphine in blood, vitreous humor, and cerebrospinal fluid in 154 opiate-related fatalities. In general, they found free and conjugated morphine concentrations to be lower in vitreous humor than in blood. Most frequently, vitreous humor to blood concentration ratios were in the range of 0.4 to 0.6 for free morphine and less than 0.3 for conjugated morphine.

The authors also suggested that survival time significantly influenced vitreous humor to blood morphine concentration ratios. In cases for which a long (>5 h) survival time was known, they observed lower concentrations of morphine in vitreous humor and blood and greater vitreous humor to blood concentration ratios compared to the cases with a shorter survival time. However, the authors cautioned that the interpretative value of this finding is complicated by the effects of repeated use, which may result in the accumulation of morphine in the vitreous humor. A smaller study completed by Bogusz quantitated free morphine and M6G in four cases of heroin overdose. The author observed poor correlation between the free morphine to M6G ratios in blood and vitreous humor and no relationship between the estimated survival time and the free morphine to M6G ratio in either specimen type. However, the author indicated that any observations made with this study must be confirmed by a study including a greater number of cases.

A second study by Bogusz et al. analyzed vitreous humor, cerebrospinal fluid, and blood for morphine, M3G, and M6G in specimens collected from 21 heroin-related deaths; vitreous humor was available for 12 cases. It was observed that the M3G concentration was greater than the M6G concentration in all vitreous humor specimens analyzed. Vitreous humor M3G concentrations ranged from 36 to 250 ng/mL while M6G concentrations were between 3 and 46 ng/mL. It was also observed that vitreous humor concentrations of morphine, M3G, and M6G were lower than blood and cerebrospinal fluid concentrations of these analytes. Lin et al. analyzed vitreous humor from 223 opiate-positive cases to determine whether vitreous humor codeine and morphine concentrations could be useful for differentiating death due to codeine overdose from death due to heroin or morphine use.

Heroin use was confirmed in 41 cases by the presence of 6-acetylmorphine (6-AM) in vitreous humor. All of these heroin-related cases had a vitreous humor-free codeine to free morphine ratio significantly less than 1. In contrast, 20 cases that were attributed to codeine overdose had a free codeine to free morphine ratio greater than or equal to 1. Morphine and codeine concentrations in blood were not reported. The authors concluded that the codeine-morphine distribution pattern in vitreous humor resembles that observed in blood and urine. When the codeine concentration in vitreous humor is sufficiently high, it can serve as a valid indicator for the differentiation of codeine- and morphine-induced fatalities. However, the authors noted that the generally low concentration of codeine in vitreous humor might limit the usefulness of these findings. A recent study by Wyman and Bultman analyzed specimens from 25 heroin-related deaths. 6-AM was detected in all vitreous humor specimens, and it was present at a greater concentration than blood or CSF in 21 of the 25 vitreous humor specimens. In addition, vitreous humor was the only specimen that was positive for 6-AM in nine cases. The authors concluded that the detection of 6-AM in postmortem cases could be greatly improved by the analysis of vitreous humor.

Detection of oxycodone

There have only been a few reports of oxycodone distribution in vitreous humor. Drummer et al. published a study of nine oxycodone-related deaths. Vitreous humor was analyzed for one case, with a vitreous humor to blood oxycodone concentration ratio of 1.2 (B = 1.5 mg/L, VH = 1.8 mg/L). Anderson et al. have also completed a study of oxycodone-related deaths. Vitreous humor oxycodone concentrations were reported for seven cases. The mean vitreous humor to heart blood oxycodone concentration ratio for these cases was 1.1 (median = 1.0), with a range of 0.2-2.0. A ratio greater than or equal to 1 was observed in five of the seven cases. Vitreous humor oxycodone concentrations ranged from 0.18-0.82 mg/L. In a study of 14 cases presented by Winterling et al., the mean vitreous humor to blood oxycodone concentration ratio was 1.7 (median = 1.7), with a range of 0.24-3.9. This ratio was ≥1.0 in eleven of the 14 cases. The vitreous humor oxycodone concentrations ranged from 29 to 900 ng/mL.

Detection of methadone

Methadone distribution into vitreous humor has been described in several studies. Sturner and Garriott reported two cases of methadone intoxication. The blood methadone concentrations were 0.7 and 0.6 mg/L, whereas the vitreous humor methadone concentration was 0.1 mg/L in both cases. A study completed by Jennings included the analysis of vitreous humor and blood specimens from 47 methadone-positive medical examiner cases. Methadone was the primary analyte detected in vitreous humor; EDDP was detected in only two of the 47 vitreous humor specimens. Vitreous humor methadone concentrations ranged from 32 to 840 ng/mL. Peripheral blood methadone concentrations ranged from 82 to 1800 ng/mL. The average vitreous humor to peripheral blood methadone concentration ratio was 0.29 (median = 0.25), with a range of 0.08 to 0.98. In general, the vitreous humor to blood methadone concentration ratio was less than 0.50, with only four of the 47 cases having a ratio exceeding 0.5.

Detection of propoxyphene

Limited data exists relating to the distribution of propoxyphene into vitreous humor. Sturner and Garriott published data from 10 cases of propoxyphene intoxication. Propoxyphene was detected in vitreous humor from all cases in concentrations ranging from 0.5 to 5.0 mg/L. In these cases, the vitreous humor to blood propoxyphene concentration ratio ranged from 0.05 to 0.34. The authors suspected that low vitreous humor to blood propoxyphene concentration ratios were detected in some cases because death occurred prior to equilibrium between vitreous humor and blood.

Detection of fentanyl

Anderson and Muto published data from a series of 25 fentanyl-positive postmortem cases. Vitreous humor fentanyl concentrations were available for three of these cases and ranged from 8 to 20 ng/mL. Vitreous humor to blood fentanyl concentration ratios were greater than unity in each of these cases and ranged from 1.25 to 2.08.

Detection of sufentanil

Ferslew et al. published a single case report involving a subject who committed suicide by the intravenous administration of midazolam and sufentanil. In this case, the vitreous humor sufentanil concentration was 1.2 ng/mL and the blood concentration was 1.1 ng/mL, resulting in a vitreous humor to blood concentration ratio close to unity.

Distribution of opioid

Opioid distribution into vitreous humor is variable and dependent on the protein binding characteristics and lipophilicity of the specific opioid detected. Opioids including morphine, 6-AM, codeine, oxycodone, methadone, propoxyphene, and fentanyl have been detected in vitreous humor and may serve as indicators of use. The limited number of case studies available indicate that methadone, morphine, and propoxyphene are generally detected in vitreous humor at concentrations considerably less than blood concentrations of these analytes, while vitreous humor oxycodone and fentanyl concentrations are closer to blood concentrations.

In addition, vitreous humor has been shown to be a useful marker for heroin use when urine is unavailable for 6-AM confirmation. As vitreous humor is devoid of esterase activity, 6-AM is relatively stable in this matrix compared to blood. The interpretative value of opioid concentrations in vitreous humor is limited and are most valuable when they are available in addition to blood concentrations.

CONCLUDING REMARKS

There are a number of advantages to the analysis of drugs in vitreous humor. The specimen is easily collected during the postmortem examination; even if a complete autopsy is not performed vitreous humor can be obtained. The specimen is clear and serous and consists essentially of water. As a result, the specimen is easy to work with analytically. Any method developed for urine or more complex postmortem specimens should be amenable to vitreous humor analysis. Commercially available immunoassays developed for urine can be used on vitreous humor. For chromatographic procedures, less specimen preparation is required. As vitreous humor is located in an anatomically isolated position, it is more protected from putrefaction, charring, and trauma than more centrally located fluids and tissues.

During the putrefactive process, decomposition products such as tyramine and phenethylamine may interfere with both the extraction and analysis of many drugs in blood and tissue specimens. This is less of a factor with vitreous humor drug analysis. However, when the body decomposes, the vitreous becomes desiccated, making it more difficult to collect the specimen. Moreover, if death involves trauma to central organs, the only specimen that may be available is cavity blood, with its potential contamination from tissues and stomach contents. In these cases, vitreous humor may be a useful alternative specimen. There are also some disadvantages in using vitreous humor for drug analysis. The main drawback is specimen volume. At best, only about 5 mL of vitreous humor can be collected on a case. If multiple drug analyses are required, it may be necessary to use less sample volume for each assay, limiting assay sensitivity. Some assays may not be performed at all because of limited sample volume. In addition, the need to perform electrolyte and glucose analysis may over ride drug testing decisions. Another disadvantage to the use of vitreous humor for drug analysis is a limited database from which interpretive assessments of the analytical results can be made. At the present time, the utility of vitreous humor in drug analysis is as an adjunct to blood or tissue drug analysis.

16

Detection of Drugs in Osteogenic Tissues

Bone and bone marrow have received relatively little attention compared to other alternative specimens. This is most likely due to the restriction of obtaining this specimen in post-mortem cases. Nevertheless, a number of case reports and studies have been conducted that clearly show that drugs are present in both bone and bone marrow. The presence of drugs in reported cases have aided the investigation into these deaths that was not possible given the often skeletonized state of the remains. This chapter reviews the current state of knowledge both in terms of the drugs detected in these specimens and how drugs are recovered from these most unusual matrices.

HISTOMORPHOLOGY OF OSTEOGENIC ORGANS

Bone marrow is a vascular tissue that is present in the central cavities of bones. Marrow is most easily obtained from the major bones, ribs, and vertebrae. There are two types of marrow: red and yellow. Red marrow supports clusters of hemapoietic cells, white blood cells, macrophages, and has a rich blood supply. Yellow marrow supports numerous blood vessels and fat cells. Bone is a highly vascularized tissue consisting of porous mineralized structure consisting of hydroxyapatite. The structure and composition varies according to the location. Cortical bones have a low turn-over rate and represents about 80% of the overall skeletal mass and provide the strength of the skeleton.

Trabecular or cancellous bone is less dense, is spongy, and has a higher turn-over rate and largely consists of epiphyseal and metaphyseal parts of long bones and within smaller bones. The degree of contact of drugs to the bone structures depends on the anatomical location of the bone and the local blood supply.

The long bones (i.e., femur) receive the most blood supply while the short, flat, or irregular bones receive more superficial supply through the periosteum. Bone is not a uniform structure. For example, bone consists of layers or bundles of bone. Bone contains channels that contain small blood vessels and nerves (haversian canals).

METHODOLOGY

Marrow when collected as a fluid can be diluted with water or a buffer, the mixture macerated, and drugs extracted with a solvent in a very similar manner to other fluid specimens. Bone marrow has a high fat content that may cause some difficulties, but this may be overcome by treating the dried solvent extract with hexane/ethanol (7:2) (e.g., 5 mL) and a small volume of water (0.2 mL). The hexane layer is discarded and the drugs isolated from the ethanol fraction. It is difficult to separate bone marrow from the bone itself in substantially decomposed or skeletonized remains when only dry material is present in the core of the bone. Indeed it may not be necessary to separate dried material in the core from the bone if evidence of exposure to drugs only is desired.

Table 16.1 Extraction techniques in bone and bone marrow

Tissue	*Preparation and extraction technique*
Bone including teeth	Prolonged soaking of annular rings or fragments in methanol
	Crushing bone and solvent extraction
Bone marrow	Solvent extraction of diluted marrow
	Treatment of marrow or dried marrow with methanol

Treatment of the crushed material with a solvent can be sufficient to isolate drugs particularly if contact time with the solvent is at least several hours. The extraction of drugs from bone is much more difficult given the hard matrix. Although any bone can potentially be used, the more useful bones appear to be the femur or other long bones, where sectioning of the bones can occur to increase the surface area. Other than the long bones, the vertebrae and iliac crest are also samples collected at autopsy. Bones should first be thoroughly cleaned of any attached tissue and any surface contamination before attempting to extract drugs from the internal structures.

Then annular rings of femur can be soaked in methanol (or another solvent) over an extended period of time (overnight). This usually extracts much of the drug contained within the porous matrix. Alternatively, bone can be crushed and treated with solvent. This will probably increase the extractability of drugs but can introduce some occupational health and safety dangers for the analyst when crushing bone. As quantitative measurements are not likely to be meaningful, it is not necessary to overdo the preparation of bone. Extracts of marrow or bone can be subject to the usual screening techniques, such as immunoassay or chromatographic techniques, providing appropriate comparisons are made with known drug-free control cases.

FINDING DRUG

A range of drugs have been detected in both marrow and bone. Drugs-of-abuse such as a number of amphetamines, cocaine, morphine and a number of benzodiazepines have all been detected in both marrow and bony structures. Benzoylecgonine has also been detected in bone; however 6-acetylmorphine, the heroin metabolite has not been detected in bone of heroin users. Numerous antidepressants and other opioids have also been detected suggesting that there is no apparent restriction in the ability of these tissues to take up drugs from the circulation. Studies that have examined both blood and bone collected from the same decedent have not always shown good concordance. That is, some drugs in blood were not found in bone and some drugs in bone were not found in blood taken at autopsy. This is not surprising as drugs absorbed into bone from previous exposures will almost certainly remain in bone longer than in blood.

The time frame for retention of drugs in bone is unknown but is likely to be quite long since the turnover of bone is relatively slow. A number of studies have shown relatively good correlation of bone marrow concentration with circulating blood in animal experiments using controlled doses. Indeed, marrow concentrations were detected within minutes of exposure of the animals to drug. This confirms the good perfusion of bony structures with the circulating blood and should enable drugs to be deposited in bone relatively quickly after exposure. However, there is little known about the extent of uptake of most drugs. Controlled animal experiments and human cases where paired bone and blood specimens have been available show some drugs that are present in much higher concentrations than blood. For example, desipramine, flurazepam, and pentobarbital are present in concentrations some 30 times > that of blood.

DRUGS IN BONE AFTER DEATH

A number of case reports of remains found years after death have yielded drugs when the bone or bone containing dried marrow was subject to extraction techniques. For example, methamphetamine and amphetamine were detected after 5 years in dry marrow, and triazolam was detected in a suspected homicide victim 4 years after death. The relative resistance of drugs in bones to be degraded or leached out by prolonged exposure is also confirmed by the detection of methamphetamine and amphetamine in marrow of bones deliberately stored under water for 1 year. It is likely, however, that the benzodiazepines have limited stability in bone structures over prolonged periods of time (months). Raikos et al. reported 54% loss of morphine (155 ng/g compared with 340 ng/g) in bone buried for 1 year in a suspected heroin intoxication case.

DETECTION OF DRUGS IN TEETH

As teeth are another form of bone, it is not surprising that drugs have been found in this tissue. Solvent extraction of pulverized teeth of known drug users has resulted in the detection of 6-acetylmorphine and morphine, as well as cocaine and benzoylecgonine. It is likely that other drugs consumed by the owner will also be deposited. However, the rate and degree of entry into the various structures of teeth (dentine and enamel) is not known and likely to be slower than for long bones due to the lesser degree of vascularization.

CONCLUDING REMARKS

Bones of various type and size including teeth contain evidence of past drug use. Specimens are relatively easy to collect at autopsy and occasionally are the only specimens available. Preparation by soaking is easy although crushing or pulverization requires additional tools. Conventional isolation and chromatographic techniques demonstrate the necessary sensitivity to detect the presence of drugs. Drug concentrations in bone may not correlate with postmortem blood concentrations. Indeed, some drugs such as olanzapine, may not be detected in bone. Although it is not possible at the present time to estimate the time or duration of exposure, drug presence in bone structures may at least provide evidence of prior contact with these substances.

17

Drugs in Central Nervous System

A brief overview of the anatomy of the brain is provided, focusing primarily on the regional distribution of the binding sites of drugs-of-abuse in the human brain whenever published data is available. Generally, the brain can be divided into the following parts: telencephalon, diencephalon, mesencephalon (or midbrain), pons and cerebellum, and medulla oblongata. The telencephalon is composed of the two large cerebral hemispheres containing the basal ganglia. The diencephalon is situated between the cerebral hemispheres and is divided into the thalamus and the hypothalamus. The remaining parts of the brain are grouped together to form the brainstem: The mesencephalon and the medulla are divided by the pons, which is separated from the cerebellum by a cavity called the fourth ventricle.

The border between the medulla oblongata and the spinal cord is defined by the pyramidal decussation (in the height of the foramen magnum). These anatomical subdivisions are interconnected, creating the necessity to define various substructures that describe spatial relationships (nuclei, tracts, etc.). A detailed illustration would be beyond the scope of this chapter and can be found elsewhere. The *central nervous system* (CNS) is comprised of gray matter (mainly composed of nerve cell bodies), white matter (mainly composed of white axon fibers), and glial cells. About 67-84% of the total weight of the brain is water. The remaining 16-33% is mainly proteins and lipoids - in nearly equal amounts. A structural protein unique to the CNS is neurokeratin, and the most common amino acid found in the brain is glutamic acid. The concentration of lipoids is higher in white matter than in gray matter.

The lipoid fraction contains phosphatides (cephalines, lecithines, and sphingomyelines), cerebrosides (glycolipids), cholesterol, and gangliosides. The most common fatty acids found in the brain are unsaturated C20- and C22-fatty acids, palmitic acid, stearic acid, oleic acid, arachidonic acid, docosanoic acid, and C24-fatty acids. Phosphatides and cerebrosides tend to form viscous, stringy solutions in water, which can create problems during the extraction of brain specimens. Because most drugs-of-abuse establish their effects through specific receptors in the brain, their regional distribution is of high interest. The location of binding sites for drugs-of-abuse can be studied by quantitative autoradiography in vitro. After proven successful in animal studies, this method was further applied to human brain tissue. Biegeon et al. performed a binding study with nanomolar and micromolar (representative for behaviorally active doses) concentrations of cocaine in brain sections of three drug-free subjects. For low concentrations of tritiated cocaine, the highest density of binding sites was found in the basal ganglia (caudate and putamen, also referred to as the striatum). The thalamus showed moderate density, and density was low in the cortex and hippocampus (parts of the telencephalon).

High concentrations of tritiated cocaine showed more homogeneous binding throughout the brain. A study to determine the regional distribution of the three different opiate receptor types (mu, delta, and kappa) in human brain tissue was performed by Maurer et al. Opiate-binding sites were generally predominant in grey matter and almost absent in white matter. A very high density of μ-opiate receptors - through which morphine exerts most of its pharmacological actions - was found in the brainstem (area tegmentalis ventralis, grisum centrale mesencephali, inferior colliculus and nucleus inter-peduncularis); moderate density was found in the molecular layer of the cerebellar cortex (also a part of the brainstem) and the neocortex (laminae I-V) and hippocampus (gyrus dentatus) of the telencephalon. More recently, quantitative autoradiography detected μ-opiate receptors in the telencephalon, with high density in the laminae I-III of the neocortex, and the nucleus caudatus; moderate density in the laminae III-IV of the neocortex, the nucleus basalis of Meynert, and the corpus amygdaloideum; and low density in the laminae V-VI of the neocortex and the claustrum.

Moderate density was also found in the thalamus (diencephalon) and the cerebellum (brainstem). Low density was found in the hypothalamus (diencephalon). Gabilondo et al. demonstrated that in the postmortem brain of heroin addicts, no apparent alterations in the densities and affinities of μ-opiate receptors in various brain regions could be observed when compared to the control group. Brain α_2-adrenoceptor densities, however, appeared to be down-regulated during opiate dependence in humans. The highest density of specific cannabinoid receptors was found in the telencephalon - in the basal ganglia (substantia nigra pars rediculata and globus pallidus) and in the hippocampus - and in the cerebellum (brainstem). The very low density in the brainstem areas controlling cardiovascular and respiratory functions may explain why there are no reports of fatal cannabis intoxication in humans.

PHARMACOKINETICS OF DRUG

Upon absorption into the body, the inherent properties of drugs-of-abuse (such as molecular size, lipid solubility, degree of ionization, and plasma protein binding) determine their capacity to cross capillary walls and penetrate into cells. Only the unbound fraction circulating in the blood has the ability to penetrate the blood-brain barrier, and the permeability of this barrier determines the amount of drug reaching the site of action in the brain. Oldendorf et al. studied the uptake of different opiates in the brains of rats, and predicted rates of penetration through the blood-brain barrier. In the extremely short time frame of 15 s between bolus injection into the carotid artery and decapitation of the rats, the permeability of the blood-brain barrier for morphine was less than measurable. The uptake of codeine, heroin, and methadone was 24, 68, and 42%, respectively.

Bjoerkman et al. characterized the cerebral uptake of morphine in pigs by measuring changes in the arteriovenous plasma concentration gradient over the brain.

Over a much longer observation period (90 min) than in the previously mentioned study, it was shown that morphine uptake in the brain reached a maximum approximately 3 min after the start of the infusion before changing to a slow, steady release. These results are in agreement with a study performed by Mullis et al., who determined the half-life of morphine in rat brain tissue to be biphasic, with an initial half-life of approximately 2 h followed by a slower half-life of about 5 h between 4 and 48 h after a single subcutaneous injection. It could be expected that the redistribution of morphine back into blood would also be retarded by low permeability of the blood-brain barrier. Indeed, unconjugated morphine persisted in the rat brains in nanomolar concentrations for at least 24 h after a single analgesic dose.

Wu et al. demonstrated that the brain uptake of morphine-6-glucuronide was 32 times lower than that of morphine. This reduced blood-brain barrier permeability is consistent with the much lower lipid solubility of morphine-6-glucuronide relative to morphine. Som et al. demonstrated the rapid, intense uptake of cocaine in rat brains after a pharmacological dose was injected intravenously. Mule et al. confirmed the rapid entry of cocaine into the brain matrix and demonstrated that this drug also leaves the brain rapidly.

The penetration of benzoylecgonine was considerably slower; that would be expected due to its high polarity, lower lipid solubility, and the partition coefficient of 0.15. Furthermore, the blood-brain barrier seems to have a limiting effect on the access and accumulation of tetrahydrocannabinol (THC) in the brain. After penetration of the blood–brain barrier, biotransformation can take place within the brain. Examples include heroin, which largely survives the 10-15 s required to reach and enter the brain after intravenous injection and is then rapidly hydrolyzed into 6-acetylmorphine and morphine; another example would be the hydrolysis of cocaine into benzoylecgonine. Although drugs-of-abuse are distributed throughout the brain, for several there appears to be an uneven distribution. Therefore, the brain cannot be regarded as a single pharmacokinetic compartment.

Morphine concentrations in the hypothalamus and in the hippocampus of rats, for example, were significantly higher than those found in the midbrain-thalamus and the striatum. The highest concentration of cocaine in rat brains was found in the cortex, followed by the striatum and the cerebellum. Animal studies, however, are not always representative of human conditions, and in contrast to the previously mentioned animal study, distribution of cocaine in post-mortem human brain appears to be uniform. These findings and the question as to whether there is a correlation between the concentration of drugs-of-abuse in different brain regions.

SAMPLING TECHNIQUE

Appropriate sample preparation is one of the most important pre-requisites for the successful identification and quantification of drugs-of-abuse in brain specimens. It is a multi-step process consisting of sampling, sample pretreatment, sample extraction, chromatographic separation, and detection. A high level of accuracy at each step is essential to ensure the correct interpretation of results.

Collecting Specimen

Sampling of the brain is crucial and demands particular care by the professional conducting the autopsy. Specimen amount, as well as the area of the brain from which the specimen is taken, must be carefully documented. To minimize specimen degradation, specimens should be analyzed as soon as possible and kept frozen; unstable drugs-of-abuse (e.g., cocaine) must be properly preserved.

Treatment Before Sampling

The goal of pre-treatment is to make drugs-of-abuse accessible for the extraction process that is to follow while simultaneously removing interfering compounds. Tissue specimens such as brain specimens require careful homogenization (e.g., using an electric-powered blender or equivalent device) to obtain a sample that is suitable for extraction. This process can be followed by protein precipitation, but potential irreversible loss of significant compounds due to adsorption and occlusion must be considered. If protein precipitation cannot be avoided, possible adsorption losses must be compensated for by adding stable isotope internal standards in the expected concentrations of the targeted drugs-of-abuse.

Moreover, digestion of the homogenates with lipases or proteases gives rise to increased background due to the co-extraction of an increased number of artifacts. Centrifugation prior to extraction should be performed in a cooled centrifuge so as to prevent any loss of analytes from the warming of the sample. Generally, a major problem in the work-up procedure of tissue specimens is the impossibility of determining the actual extraction efficiency and hence the absolute content of drugs in the tissue specimen. Nevertheless, Drummer et al. pointed out that when suitable precautions are taken, there is little evidence to suggest that extraction efficiencies of drugs from solid tissues are likely to be much worse than with fluid specimens. Therefore, internal standards – with chemical and physical properties as similar as possible to the analytes – should be added to the sample at the earliest conceivable stage, in any event before buffering and extraction of the sample, to compensate for possible loss of target compounds throughout the entire sample preparation procedure.

Today, stable isotope internal standards are generally recommended and should be added in similar concentrations to the expected drugs-of-abuse, but they require assays based on mass spectrometry (MS). Whenever available, certified reference materials should be used as controls. However, in many forensic cases, the matrix is unique (e.g., a decomposed brain specimen); in these cases, the control can never reach the ideal of being identical to the unknown specimen, and quantitative results must be interpreted cautiously.

Extraction Procedure

Sophisticated extraction procedures are needed for the isolation and concentration of targeted drugs-of-abuse from the complex matrix of brain specimens. At the same time, as many interferences as possible originating from the specimen should be excluded. In the past, *liquid-liquid extraction* (LLE) and, more recently, *solid-phase extraction* (SPE) have been applied to reach those goals. Today, SPE procedures are often favored because automated devices for sample extraction are available, which minimize systematic errors and human-caused variances and improve reproducibility.

Method of liquid-liquid extraction

The advantages of LLE are its wide dynamic range and that it is based on strictly defined thermodynamic relationships, which simplifies method development. LLE can be performed at pH values optimized for the targeted drugs-of-abuse, and basic or acidic back-extraction can be applied to remove interferences. Because LLE is frequently combined with protein precipitation, extraction yields are often highly variable due to adsorption and occlusion. Moreover, during extraction, stable emulsions occur and multiple interferences are co-extracted from the brain matrix. To prevent emulsions and eliminate the need for centrifugation, the aqueous phase has been immobilized onto inert support materials, which improves both extraction efficiency and reproducibility.

However, the applicability of this technique for brain specimens is limited, because homogenization of the potential infectious specimens with the support material is difficult. The need for an inert carrier to support the sample during extraction also seems to be the major drawback for the application of both *accelerated solvent extraction* (ASE) and *supercritical fluid extraction* (SFE) to brain specimens. In the past, several procedures for LLE of animal and human brain specimens have been published for opiates, for methadone, for cocaine and metabolites, and for amphetamines.

Method of solid-phase extraction

In SPE, analytes are isolated from the aqueous matrix onto a solid sorbent followed by selective washing - to remove interferences from the biological matrix - and elution with appropriate solvents. The high extraction efficiency allows for small specimen amounts, which is crucial in forensic cases because often only a limited sample is available. Moreover, SPE offers the advantage of reduced solvent consumption, decreased evaporation volumes, as well as reduced environmental waste and personnel exposure to harmful substances. Furthermore, there is no problem with emulsions, and each step of the extraction process can be automated. This increases both reproducibility and laboratory turnover. The reliability of SPE has generally improved since manufacturers began to implement quality control systems for the sorbents.

Although method development of SPE is more complex than that for LLE, SPE procedures covering a broad range of drugs-of-abuse have been developed primarily for body fluids such as urine or blood. Scheurer et al. reviewed the few existing SPE procedures for tissue specimens and concluded that the problems unique to brain tissue were not insurmountable and that existing procedures published for animal brain would be equally applicable to human tissue. Browne et al. reported that lipids present in brain homogenates cause C18- and C8-extraction columns to become blocked as a result of their interaction with the lipophilic packing materials and showed that polar C2-columns worked better. Another approach to this problem would be the digestion of the brain specimens with triacylglycerol lipase prior to extraction.

The main drawback of SPE for postmortem brain specimens is the inability of tissue homogenates to pass easily through tightly packed bonded cartridges. To cope with the increased viscosity of brain specimens, high-flow SPE columns were developed, but the high speed at which the samples pass through the packing material reduces the contact time between drug and sorbent, resulting in a lower extraction efficiency compared to procedures using more common tightly packed sorbents. Silica-based bonded sorbents and polystyrene-divinylbenzene copolymers have been used in the extraction of human brain specimens.

Procedures have been published for opiates, for cocaine and metabolites, for amphetamines and analogs, and for cannabinoids. For the extraction of brain specimens, polystyrene resins offer a major advantage compared to other sorbents. The porous structure enables the exclusion of micelles containing monoglycerides, diglycerides, and phospholipids. These micelles are produced by dispersion during homogenization of the specimen and, because they are prevented from adsorption, lipoid interferences are reduced in the resulting extract.

Thin Layer Chromatography

To enable accurate interpretation of brain drug concentrations in determining cause of death, these substances must be identified and the quantitative results must be sufficient with regard to sensitivity and precision. This is the reason why methods such as immunoassay, fluorometric assay, and thin-layer chromatography, although applied in the past, are no longer recommended. Differential radioimmunoas say, for instance, allows for the separate determination of a targeted drug-of-abuse and its metabolites, yet other cross-reacting substances cannot be ruled out. Fluorometric assays are limited by quenching interferences. Thin-layer chromatography lacks sensitivity and does not produce precise quantitative results. Methods such as *high-performance liquid chromatography* (HPLC) with electro-chemical, ultraviolet, and fluorescence detection, as well as *gas chromatography* (GC) with electron capture, flame ionization, and nitrogen phosphorus detection, have been successfully applied for the detection of drugs-of-abuse in brain specimens.

To obtain reliable quantitative results with these procedures, internal standards with chemical and physical properties as similar as possible to the analytes are needed. Because internal standards that meet all these demands are not easily available, quantitative results must be interpreted with caution. For the majority of drugs-of-abuse, stable isotope analogs are available and can be used as "ideal" internal standards, provided MS is utilized. If added at the earliest possible stage of the work-up procedure in a similar concentration to the targeted drugs-of-abuse, stable isotope internal standards correct any loss of the target compound during the whole process of sample preparation and quantification. In the heterogeneous brain matrix, MS, in combination with chromatographic techniques, offers the highest selectivity, sensitivity, and most precise quantitative results. Therefore, combinations of GC and MS or LC-MS have developed into the standard techniques for the detection of drugs-of-abuse today and should be applied whenever possible. For GC-MS, appropriate derivatization may be necessary to achieve sufficient volatility for drugs-of-abuse. At the same time, care must be taken to prevent injection port temperatures that are too high, leading to the production of artifacts (e.g., anhydroecgonine methyl ester). These problems do not exist when LC-MS is applied; instead, matrix effects - which interfere with the ionization process - must be taken into consideration.

FINDING DRUGS IN BRAIN

Morphine, 6-acetylmorphine, cocaine, and cocaine metabolites have been detected in human brain specimens, and issues concerning the interpretation of these results have been discussed in the literature. In contrast, little data on the concentration of methadone, amphetamine, methamphetamine, 3,4-methylenedioxymethamphetamine (MDMA), 3,4-methylenedioxy-amphetamine (MDA), 3,4-methylenedioxyethylamphetamine (MDEA), cannabinoids, and phencyclidine (PCP) in human brain specimens are available, and interpretational issues are unaddressed. For all other drugs-of-abuse, there is no data available regarding human brain specimens.

The majority of the publications were written in the 1980s, and numerous extraction and detection procedures were applied. Moreover, different amounts of specimen were used and only a few publications gave the precise region of the brain assayed and the time between death and sampling. Data, therefore, are not directly comparable, and interpretation or conclusions should be drawn very cautiously.

CORELATIVE APPROACH

Detection of Opiates

In the human body, heroin is rapidly deacetylated to 6-acetylmorphine and then to morphine. In vitro studies have demonstrated that human liver homogenates possessed the greatest hydrolytic activity and human brain tissue the least. Heroin has not been detected in brain specimens though 6-acetylmorphine levels in the brain were found to be substantially higher than in blood, liver, lung, and kidney specimens. Sticht et al. detected 6-acetylmorphine in all brain specimens of heroin intoxication cases where the proportion of unconjugated morphine to total morphine in the blood exceeded 45%. The presence of 6-acetylmorphine in the brain denotes heroin consumption and suggests that consumption took place within 4.5 h of the time of death. Morphine is further metabolized by conjugation, which results in the formation of morphine-glucuronides (morphine-6-glucuronide is physiologically active), and by N-demethylation to normorphine.

Based on lower lipid solubility, morphine-glucuronides show reduced blood–brain barrier permeability relative to morphine, which is supported by the findings of Reed et al., where morphine concentrations increased approximately 260% in blood specimens following hydrolysis, whereas brain hydrolysis did not increase the yield more than 25%. Published data includes concentrations of total morphine (morphine and morphine glucuronides) as well as concentrations of unconjugated morphine in brain specimens, involving a wide concentration range, thus making interpretation difficult. The question of distribution of morphine within the brain has been addressed in different studies, but it is difficult to compare these results because different analytical procedures were utilized. A study by Spiehler et al. analyzed of the following regions of the CNS: cerebellum, cerebral cortex, mid-brain, pons, medulla, spinal cord, and *cerebrospinal fluid* (CSF).

Results were presented for 33 cases, where death was mainly attributed to intravenous administration of heroin. Although total morphine concentrations in the brain were found to vary from region to region in each case, no consistent overall pattern was found, and there was also no correlation of total morphine distribution between brain regions. Moreover, no correlation between total morphine concentrations in brain fluid (CSF) and brain tissue could be established. In cases of slow death (death occurring 2–48 h after administration of heroin), total morphine concentrations in the CSF, spinal cord, and medulla tended to be elevated. The authors concluded that there was a lack of correlation between distribution of morphine in the brain and proposed sites of action. In another study by Spiehler et al., 52 brain specimens (cerebral cortex) from cases of acute intoxication (defined in this study as death occurring less than 3 h after heroin or morphine injection) showed a mean total morphine concentration of 0.47 μg/g in the brain with a wide range of 0.03-3.4 μg/g.

In 92 brain specimens where death occurred more than 3 h after administration or the time interval was unknown, total morphine concentrations showed a mean of 0.28 μg/g and a range of 0-2.5 μg/g. The study focused on the ratio of unconjugated morphine to total morphine in the blood, and therefore, no conclusions were drawn from the total morphine concentrations in the brain specimens. To establish possible patterns or relationships, these data were analyzed in a follow-up study in 1989 using artificial intelligence computer software. All programs applied in this study found total morphine useful in diagnosing morphine intoxication if the concentration in the brain was greater than 0.08 μg/g or greater than the concentration of unconjugated morphine in the blood. Short, moderate, and long-time intervals between last dose and death were differentiated.

In addition to other factors, rapid deaths were characterized by total morphine concentrations in the brain greater than 0.16-0.22 μg/g. The author concluded that, for cases with a known history, the artificial intelligence programs were successful 70-90% of the time in classifying the case according to response and time. Pare et al. investigated whether regional morphine concentrations in the brain correlate with opiate receptor density. The following brain sections from 21 suspected heroin-associated fatalities were analyzed for total morphine: brain stem (medulla, pons, and midbrain), cerebellum, cerebral cortex (basal ganglion), hypothalamus, and the thalamus. In all of the cases where death was attributed to narcotic intoxication, the total morphine concentration exceeded 0.2 μg/g in one or more of the brain sections.

In more than 50% of the specimens analyzed, the highest total morphine concentration was found in the brain stem or the thalamus – both regions have high concentrations of opiate receptors. The authors concluded that the total morphine concentration in the thalamus could be used to predict concentration in the blood. Based on investigations of 62 cases of intoxication with opiates where information about the survival time was available, Vycudilik found that comparison of the unconjugated morphine concentration in the medulla and the cerebellum provided information on the interval between administration and death. Calculating the ratio between the unconjugated morphine concentration in the medulla and the cerebellum resulted in three categories: A ratio less than one indicated short survival time (<1 h between administration of opiates and death); a ratio value of approximately one indicated a moderate survival time (1-2 h between administration of opiates and death); and a ratio greater than one (approximately two or more) indicated a long survival time (more than 6 h between administration of opiates and death). A similar correlation was observed for 6-acetylmorphine. The analysis of the unconjugated morphine concentration in the medulla provided information about consumption of opiates even when morphine was no longer detectable in the blood.

Detection of Cocaine

In the human body cocaine is rapidly metabolized by esterases and spontaneously hydrolyzed at physiological pH to form benzoylecgonine, ecgonine methyl ester and, through N-demethylation, norcocaine. Norcocaine possesses pharmacological activity comparable to cocaine. In cases where cocaine is used in combination with ethanol, cocaethylene is also found as an active metabolite, but its role in the fatal outcome of such cases is not totally understood. Nevertheless, cocaethylene has been detected in brain, and because of its longer half-life, cocaethylene concentrations can exceed the concentration of cocaine. Numerous publications were found reporting concentrations of cocaine (and cocaine metabolites) in brain specimens, describing a wide concentration range and thus making interpretation difficult. The question of distribution of cocaine within the brain, as well as its major metabolite, benzoylecgonine, was addressed in several studies.

Spiehler et al. reviewed 37 cocaine intoxication cases and 46 cases in which cocaine was incidental to the cause of death. In two of the cases, cocaine and benzoylecgonine were evenly distributed throughout the brain (in one case, the cocaine was injected and in the other, snorted), with similar concentrations found in dopamine-rich (substantia nigra) and dopamine-poor brain regions (frontal and occipital cortex, cerebellum, medulla, and spinal cord). Cases of fatal intoxication showed a mean cocaine concentration of 13.3 μg/g in the brain with a wide range of 0.17-31.0 μg/g; in cases where cocaine was incidental to the cause of death, the mean cocaine concentration was 0.12 μg/g in the brain with a range of 0-0.7 μg/g. Cocaine readily crosses the blood-brain barrier; benzoylecgonine, however, is restricted. When significant amounts are detected in the brain, it would appear to be derived from cocaine. Therefore, the ratio between cocaine and benzoylecgonine in brain specimens for cases of fatal intoxication (mean 14.7) was different than that found in incidental cases (mean 0.87). The corresponding ranges overlapped, 0.12-100 for fatal intoxications and 0.04-6 for incidental cases.

The ratio between brain and blood concentration for cocaine and benzoylecgonine provided information as to when the drug was used. In fatalities related to intoxication, the mean ratio between brain and blood concentration for cocaine was 9.60 (range: 0.65-155, median: 3.8) compared to 2.5 for incidental cases (range: 0.6-9.2). For benzoylecgonine, the mean ratio between brain and blood concentrations for cases of fatal intoxication was 0.36 (range: 0.04-1.0, median: 0.38) compared to 1.4 for incidental cases (range: 0.05-6.5). Again, there was an overlap in the corresponding ranges, and the high ratio of 1.4 for benzoylecgonine in incidental cases was attributed to accumulation of the metabolite in the brain tissue after chronic cocaine abuse. In another study of 10 cases, the ratio between brain and blood concentrations for cocaine was lower in cases where cocaine intoxication was determined as the cause of death than for cases where death was attributed to excited delirium (ratios >4). These observations differed from the trend noted by Spiehler even though all ratios for cases of cocaine intoxication fell within the previously published range.

Different regions of the brain were examined in two cocaine intoxication cases and in one additional case in which a gunshot wound was the cause of death. In each case, five to eight different brain regions were investigated (basal ganglia, cerebellum, cerebral white, cerebral gray, hypothalamus, hippocampus, motor cortex, frontal cortex, lenticular nuclei, thalamus, occipital cortex, pons, temporal cortex, and medulla). Contrary to previous findings, the cocaine and benzoylecgonine distribution throughout the brain was dissimilar. For example, the region of the brain containing the highest concentration of cocaine differed in all three cases that were included in the study, namely the hippocampus, the thalamus, and the basal ganglia. In a larger study, the concentration of cocaine and its major metabolites was determined in 15 different brain regions of 14 chronic cocaine users, including subjects who had recently ingested a high dose of cocaine as well as subjects who did not use the drug immediately before death. Because cocaine has a high affinity to dopamine transporters, dopamine-rich areas of the brain were included. The study also included dopamine-transporter-poor areas of the brain.

The following regions of the brain were investigated: frontal cortex, temporal cortex, occipital cortex, parietal cortex, cingulated cortex, cerebellar cortex, caudate, putamen, parolefactory cortex, internal globus pallidus, external globus pallidus, Ammon's horn of hippocampus, hypothalamus, medial-dorsal thalamus, and medial pulvinar thalamus. Although the concentrations of cocaine and its metabolites in the brain varied markedly among the chronic users of the drug, they showed little regional heterogeneity. It could not be clarified whether an initial selective binding of cocaine to the striatal dopamine transporter occurred in vivo followed by rapid redistribution to other brain areas or whether an initial non-selective distribution of cocaine to all areas of the brain was the reason for the relatively homogenous regional distribution pattern. Moreover, some redistribution between death and freezing of the brain specimens could not be excluded.

Detection of Amphetamines

Little data exist reporting concentrations of amphetamine and 3,4-MDEA in brain specimens. Interpretation of results has not been addressed. The regional distribution within the brain has been investigated for methamphetamine as well as 3,4-MDMA and 3,4-methylenedioyxamphetamine (MDA). Concentrations of methamphetamine in brain specimens of autopsy cases vary with a range of 0.02–20.9 μg/g. In a larger study, the concentration of methamphetamine and its metabolite, amphetamine, was determined in 15 brain regions of 14 chronic methamphetamine users. Dopamine-rich areas as well as dopamine-poor areas of the brain were investigated including the parietal cortex, frontal cortex, occipital cortex, temporal cortex, cingulated cortex, white matter, cerebellar cortex, Ammon's horn of hippocampus, putamen, caudate, internal globus pallidus, external globus pallidus, hypothalamus, medial-dorsal thalamus, and medial pulvinar thalamus.

Although the concentrations of methamphetamine in the brain varied markedly among chronic users of the drug (0.24–56.6 μg/g), methamphetamine was distributed homogeneously within the brain. The authors concluded that any marked preferential uptake or retention of methamphetamine or amphetamine in dopamine-rich (striatum) versus dopamine-poor brain areas of chronic methamphetamine users seemed unlikely. However, any redistribution occurring after death and before freezing of the brain specimens could not be excluded. To study the regional distribution of MDMA and its metabolite, MDA, in a fatal intoxication case, different areas of the brain (temporal cortex, parietal cortex, frontal cortex, occipital cortex, cerebellum, and brainstem) were investigated. Regional differences were found in the various brain regions with the highest concentrations of MDMA as well as MDA being in the frontal and parietal cortex.

Detection of Cannabinoids

One publication reporting the concentration of cannabinoids in the brain was found in the literature. The concentration in the cortex of delta-9-THC was 0.0 14 μg/g, for 1 1-hydroxy-delta-9-tetrahydrocannabinol (11-OH-THC) it was 0.0 12 μg/g, and for 11-nor-delta-9-tetrahydrocannabinol carboxylic acid (THC-COOH) it was 0.024 μg/g. The authors of this case report noted that concentrations in the brain might be less influenced by the release of cannabinoids from fat compartments than concentrations in blood.

Detection of Phencyclidine

A single publication describing the concentration of PCP in 11 brain specimens was found in the literature. The concentration of PCP in the brain ranged from 0.03 to 0.81 μg/g. Because PCP has a very large volume of distribution, the correlation coefficient between blood and brain concentration of 0.26 found in this study indicated incomplete distribution of PCP throughout the body in some of the cases.

CONCLUDING REMARKS

In forensic toxicology, conclusions about the effects of a drug-of-abuse on the deceased have to be drawn from the results of postmortem specimens. These conclusions are based either on reference data generated by investigators in the testing laboratories or the literature (animal studies, clinical studies, systematic postmortem studies, case reports, etc.). This task requires trained experts with experience because there are numerous practical problems that have to be taken into consideration. The drug concentrations found in postmortem cases span a wide range, reflecting first-time users as well as subjects with tolerance to the drug. Moreover, inherent physiological differences between individuals must be considered, and specific case details - which could explain the individual circumstances under which death occurred - could be either absent or misleading.

In drug abuse cases, multiple drug use is frequently observed, resulting in complex interactions that make the drug effects difficult to predict. Therefore, all measurable drugs should be considered significant when determining the cause of death. The majority of published extraction procedures and data collections for drugs-of-abuse utilized serum, plasma, or blood specimens. In postmortem forensic toxicology, however, the isolated investigation of a blood specimen is often insufficient in determining whether a drug-of-abuse caused the death of an individual. This may be demonstrated for cocaine-related deaths, where blood concentrations of cocaine are indistinguishable from concentrations in recreational users. Blood concentrations of morphine in heroin-related deaths also overlap with those found in non-drug-related deaths. Furthermore, blood concentrations of methadone in fatalities are frequently below the plasma concentrations targeted in clinical practice.

Moreover, the phenomenon of postmortem drug redistribution into the blood from reservoirs with a high concentration, which results in site-dependent differences and time dependent changes, is well documented in the literature for methadone, cocaine, benzoylecgonine, and cocaethylene. To provide the appropriate foundation on which an expert opinion can be rendered in fatal cases of drug abuse, additional tissue analyses are often required. A variety of tissue specimens have been used in postmortem forensic toxicology. Brain specimens have some advantages over all other tissues in determining cause of death. In respect to the decomposition of postmortem specimens in forensic cases, brain specimens show greater stability and retarded putrefaction compared to other tissues or blood. In addition, following death, bacteria that possess enzymes, which add to the decomposition of labile molecules, transmigrate from the gastrointestinal tract and the lungs throughout the body. Because the brain is a separate compartment, the effects of these bacteria upon the brain are delayed. Also, metabolic activity is lower in the brain than in other tissues or blood. Moriya et al. demonstrated that cocaine was stable in decomposed homogenates of human brain at 20-25 °C as well as at 37°C over an observed period of 24 h.

In specimens of rabbit brains, cocaine degraded much more slowly than in blood or liver over a period of 5 days. These factors increase the likelihood of the detection of drugs-of-abuse in brain specimens compared with other tissues, especially when the drugs possess low stability (e.g., cocaine). Drugs-of-abuse establish their effects, including fatal side effects such as respiratory depression, through the CNS. It can be assumed that concentrations of drugs-of-abuse measured in postmortem brain specimens should be close or equal to perimortem concentrations of the drug at its site of action. The superiority of brain specimens over other tissues has been demonstrated in numerous publications for opiates and for cocaine and its metabolites. In the case of opiates, the likelihood of detecting 6-acetylmorphine as evidence of heroin consumption was higher in human brain compared with other tissues and blood. Moreover, the presence of 6-acetylmorphine in the brain was used as an indicator of a short survival time.

A total morphine concentration in the brain that is greater than the concentration of unconjugated morphine in the blood was defined as a possible indicator of acute intoxication. Rapid deaths were characterized by total morphine concentrations in the brain that were greater than 0.16-0.22 μg/g. However, a correlation between morphine distribution in different brain regions and proposed sites of action could not be established though frequently the highest concentrations of total morphine were found in the brainstem and the thalamus - areas with high concentrations of opiate receptors. Comparison of unconjugated morphine concentrations in the medulla and the cerebellum provided information on the interval between administration of opiates and death although this time-dependent concentration gradient between two parts of the brain is currently unexplained. Similar to opiates, when establishing cocaine intoxication, brain specimens appear to be superior to other tissue specimens or blood.

The mean cocaine concentration in the brain after fatal intoxication was clearly different from cases where cocaine was incidental to the cause of death; nevertheless, the observed ranges overlapped. The same phenomenon was observed for the ratio between cocaine and benzoylecgonine in brain specimens and, to a lesser degree, for the ratio between brain and blood concentrations of cocaine and benzoylecgonine. These ratios provided information on the interval between administration of cocaine and death, the latter with a proposed peak brain/blood cocaine ratio of approximately 10 between 1 and 2 h after cocaine administration. A high concentration of benzoylecgonine in the brain due to accumulation of this metabolite could be used as an indicator of chronic cocaine use. Throughout the brain of chronic cocaine users, distribution of cocaine and benzoylecgonine appeared to be uniform and did not correlate with proposed sites of action.

Similar results (homogeneous drug distribution) could be found for chronic methamphetamine users; in contrast, MDMA and MDA showed regional differences in brain distribution (highest concentrations in frontal and parietal cortex). Although reliable procedures for the extraction of complex matrices such as brain specimens are available today and have been standardized through the use of automated devices (especially for SPE), relatively few papers on drugs-of-abuse in brain specimens have been published recently. Modern analytical instruments, in particular the combination of chromatographic techniques with MS, offer sufficient selectivity, sensitivity, and produce precise quantitative results.

When applying GC-MS or LC-MS, stable isotopes as internal standards added at the earliest stage of sample preparation increase reproducibility. Therefore, comparable results are produced, which could be used in the future to create a data collection of reliable reference values to enhance accurate interpretation of results. Moreover, these procedures can be successfully applied to small amounts of brain specimen, also allowing for the detection of drugs-of-abuse in the small structures of the brain. In combination with recently developed immunohistochemical techniques to determine the topographic distribution of the receptors as well as the drugs, a range of opportunities for the systematic investigation of drugs-of-abuse in postmortem human brain specimens are now available.

Index

P

Q

R

S